The Facts On File

# DICTIONARY
## of
# MATHEMATICS

The Facts On File

# DICTIONARY
## of
# MATHEMATICS

### Fourth Edition

**Edited by
John Daintith
Richard Rennie**

Checkmark Books®
*An imprint of Facts On File, Inc.*

**The Facts On File Dictionary of Mathematics**
Fourth Edition

Copyright © 2005, 1999 by Market House Books Ltd

Checkmark Books
An imprint of Facts On File, Inc.
132 West 31st Street
New York NY 10001

For Library of Congress Cataloging-in-Publication Data,
please contact Facts On File, Inc.
   ISBN 0-8160-5652-8

Checkmark Books are available at special discounts when purchased in bulk quantities for businesses, associations, institutions, or sales promotions. Please call our Special Sales Department in New York at (212) 967-8800 or (800) 322-8755.

You can find Facts On File on the World Wide Web at
http://www.factsonfile.com

Compiled and typeset by Market House Books Ltd, Aylesbury, UK

Printed in the United States of America

MP PKG 10 9 8 7 6 5 4 3 2 1

This book is printed on acid-free paper.

# PREFACE

This dictionary is one of a series designed for use in schools. It is intended for students of mathematics, but we hope that it will also be helpful to other science students and to anyone interested in science. Facts On File also publishes dictionaries in a variety of disciplines, including biology, chemistry, forensic science, marine science, physics, space and astronomy, and weather and climate.

*The Facts On File Dictionary of Mathematics* was first published in 1980 and the third edition was published in 1999. This fourth edition of the dictionary has been extensively revised and extended. The dictionary now contains over 2,000 headwords covering the terminology of modern mathematics. A totally new feature of this edition is the inclusion of over 800 pronunciations for terms that are not in everyday use. A number of appendixes have been included at the end of the book containing useful information, including symbols and notation, symbols for physical quantities, areas and volumes, expansions, derivatives, integrals, trigonometric formulae, a table of powers and roots, and a Greek alphabet. There is also a list of Web sites and a bibliography. A guide to using the dictionary has also been added to this latest version of the book.

We would like to thank all the people who have cooperated in producing this book. A list of contributors is given on the acknowledgments page. We are also grateful to the many people who have given additional help and advice.

# ACKNOWLEDGMENTS

*Contributors*

Norman Cunliffe B.Sc.
Eric Deeson M.Sc., F.C.P., F.R.A.S.
Claire Farmer B.Sc.
Jane Farrill Southern B.Sc., M.Sc.
Carol Gibson B.Sc.
Valerie Illingworth B.Sc., M.Phil.
Alan Isaacs B.Sc., Ph.D.
Sarah Mitchell B.A.
Roger Adams B.Sc.
Roger Picken B.Sc.
Janet Triggs B.Sc.

*Pronunciations*

William Gould B.A.

# CONTENTS

# GUIDE TO USING THE DICTIONARY

The main features of dictionary entries are as follows.

## Headwords
The main term being defined is in bold type:

> **absolute** Denoting a number or measurement that does not depend on a standard reference value.

## Plurals
Irregular plurals are given in brackets after the headword.

> **abscissa** (*pl.* **abscissas** or **abscissae**) The horizontal or *x*-coordinate in a two-dimensional rectangular Cartesian coordinate system.

## Variants
Sometimes a word has a synonym or alternative spelling. This is placed in brackets after the headword, and is also in bold type:

> **angular frequency (pulsatance)** Symbol: ω The number of complete rotations per unit time.

Here, 'pulsatance' is another word for angular frequency. Generally, the entry for the synonym consists of a simple cross-reference:

> **pulsatance** *See* angular frequency.

## Abbreviations
Abbreviations for terms are treated in the same way as variants:

> **cosecant** (**cosec; csc**) A trigonometric function of an angle equal to the reciprocal of its sine ....

The entry for the synonym consists of a simple cross-reference:

> **cosec** *See* cosecant.

## Multiple definitions
Some terms have two or more distinct senses. These are numbered in bold type

> **base 1.** In geometry, the lower side of a triangle, or other plane figure, or the lower face of a pyramid or other solid.
> **2.** In a number system, the number of different symbols used, including zero.

## Cross-references

These are references within an entry to other entries that may give additional useful information. Cross-references are indicated in two ways. When the word appears in the definition, it is printed in small capitals:

> **Abelian group** /ă-**beel**-ee-ăn / (commutative group) A type of GROUP in which the elements can also be related to each other in pairs by a commutative operation.

In this case the cross-reference is to the entry for 'group'.

Alternatively, a cross-reference may be indicated by 'See', 'See also', or 'Compare', usually at the end of an entry:

> **angle of depression** The angle between the horizontal and a line from an observer to an object situated below the eye level of the observer. *See also* angle.

## Hidden entries

Sometimes it is convenient to define one term within the entry for another term:

> **arc** A part of a continuous curve. If the circumference of a circle is divided into two unequal parts, the smaller is known as the *minor arc* and...

Here, 'minor arc' is a hidden entry under arc, and is indicated by italic type. The entry for 'minor arc' consists of a simple cross-reference:

> **minor arc** *See* arc.

## Pronunciations

Where appropriate pronunciations are indicated immediately after the headword, enclosed in forward slashes:

> **abacus** /**ab**-ă-kŭs/ A calculating device consisting of rows of beads strung on wire and mounted in a frame.

Note that simple words in everyday language are not given pronunciations. Also headwords that are two-word phrases do not have pronunciations if the component words are pronounced elsewhere in the dictionary.

## Pronunciation Key

Bold type indicates a stressed syllable. In pronunciations, a consonant is sometimes doubled to prevent accidental mispronunciation of a syllable resembling a familiar word; for example, /**ass**-id/ (acid), rather than /**as**-id/ and /ul-tră- **sonn**-iks/ (ultrasonics), rather than /ul-tră-**son**-iks/. An apostrophe is used: (a) between two consonants forming a syllable, as in /**den**-t'l/ (dental), and (b) between two letters when the syllable might otherwise be mispronounced through resembling a familiar word, as in /**th'e**-ră-pee/ (therapy) and /tal'k/ (talc). The symbols used are:

/a/ *as in* back /bak/, active /**ak**-tiv/
/ă/ *as in* abduct /ăb-**dukt**/, gamma /**gam**-ă/
/ah/ *as in* palm /pahm/, father /**fah**-*th*er/
/air/ *as in* care /kair/, aerospace /**air**-ŏ-spays/
/ar/ *as in* tar /tar/, starfish /**star**-fish/, heart /hart/
/aw/ *as in* jaw /jaw/, gall /gawl/, taut /tawt/
/ay/ *as in* mania /**may**-niă/ ,grey /gray/
/b/ *as in* bed /bed/
/ch/ *as in* chin /chin/
/d/ *as in* day /day/
/e/ *as in* red /red/
/ĕ/ *as in* bowel /**bow**-ĕl/
/ee/ *as in* see /see/, haem /heem/, caffeine /**kaf**-een/, baby /**bay**-bee/
/eer/ *as in* fear /feer/, serum /**seer**-ŭm/
/er/ *as in* dermal /**der**-măl/, labour /**lay**-ber/
/ew/ *as in* dew /dew/, nucleus /**new**-klee-ŭs/
/ewr/ *as in* epidural /ep-i-**dewr**-ăl/
/f/ *as in* fat /fat/, phobia /**foh**-biă/, rough /ruf/
/g/ *as in* gag /gag/
/h/ *as in* hip /hip/
/i/ *as in* fit /fit/, reduction /ri-**duk**-shăn/
/j/ *as in* jaw /jaw/, gene /jeen/, ridge /rij/
/k/ *as in* kidney /**kid**-nee/, chlorine /**klor**-een/, crisis /**krў**-sis/
/ks/ *as in* toxic /**toks**-ik/
/kw/ *as in* quadrate /**kwod**-rayt/
/l/ *as in* liver /**liv**-er/, seal /seel/
/m/ *as in* milk /milk/
/n/ *as in* nit /nit/

/ng/ *as in* sing /sing/
/nk/ *as in* rank /rank/, bronchus /**bronk**-ŭs/
/o/ *as in* pot /pot/
/ô/ *as in* dog /dôg/
/ŏ/ *as in* buttock /**but**-ŏk/
/oh/ *as in* home /hohm/, post /pohst/
/oi/ *as in* boil /boil/
/oo/ *as in* food /food/, croup /kroop/, fluke /flook/
/oor/ *as in* pruritus /proor-ў-tis/
/or/ *as in* organ /**or**-găn/, wart /wort/
/ow/ *as in* powder /**pow**-der/, pouch /powch/
/p/ *as in* pill /pil/
/r/ *as in* rib /rib/
/s/ *as in* skin /skin/, cell /sel/
/sh/ *as in* shock /shok/, action /**ak**-shŏn/
/t/ *as in* tone /tohn/
/th/ *as in* thin /thin/, stealth /stelth/
/*th*/ as in then /*th*en/, bathe /bay*th*/
/u/ *as in* pulp /pulp/, blood /blud/
/ŭ/ *as in* typhus /**tў**-fŭs/
/û/ *as in* pull /pûl/, hook /hûk/
/v/ *as in* vein /vayn/
/w/ *as in* wind /wind/
/y/ *as in* yeast /yeest/
/ў/ *as in* bite /bўt/, high /hў/, hyperfine /**hў**-per-fўn/
/yoo/ *as in* unit /**yoo**-nit/, formula /**form**-yoo-lă/
/yoor/ *as in* pure /pyoor/, ureter /yoor-**ee**-ter/
/ўr/ *as in* fire /fўr/

Abacus: the number 3258 is shown on the right-hand side.

**abacus** /ab-ă-kŭs/ A calculating device consisting of rows of beads strung on wire and mounted in a frame. An abacus with nine beads in each row can be used for counting in ordinary arithmetic. The lowest wire counts the digits, 1, 2, ... 9, the next tens, 10, 20, ... 90, the next hundreds, 100, 200, ... 900, and so on. The number 342, for example, would be counted out by, starting with all the beads on the right, pushing two beads to the left hand side of the bottom row, four to the left of the second row, and three to the left of the third row. Abaci, of various types, are still used for calculating in some countries; experts with them can perform calculations very rapidly.

**Abelian group** /ă-**beel**-ee-ăn / (**commutative group**) A type of GROUP in which the elements can also be related to each other in pairs by a commutative operation. For example, if the operation is multiplication and the elements are rational numbers, then the set is an Abelian group because for any two elements $a$ and $b$, $a \times b = b \times a$, and all three numbers, $a$, $b$, and $a \times b$ are elements in the set. All cyclic groups are Abelian groups. *See also* cyclic group.

**abscissa** /ab-**siss**-ă/ (*pl.* **abscissas** or **abscissae**) The horizontal or $x$-coordinate in a two-dimensional rectangular Cartesian coordinate system. *See* Cartesian coordinates.

**absolute** Denoting a number or measurement that does not depend on a standard reference value. For example, absolute density is measured in kilograms per cubic meter but relative density is the ratio of density to that of a standard density (i.e. the density of a reference substance under standard conditions). *Compare* relative.

**absolute convergence** The convergence of the sum of the absolute values of terms in a series of positive and negative terms. For example, the series:
$$1 - (1/2)^2 + (1/3)^3 - (1/4)^4 + \ldots$$
is absolutely convergent because
$$1 + (1/2)^2 + (1/3)^3 + (1/4)^4 + \ldots$$
is also convergent. A series that is convergent but has a divergent series of absolute values is *conditionally convergent*. For example,
$$1 - 1/2 + 1/3 - 1/4 + \ldots$$
is conditionally convergent because
$$1 + 1/2 + 1/3 + 1/4 + \ldots$$

is divergent. *See also* convergent series.

**absolute error** The difference between the measured value of a quantity and its true value. *Compare* relative error. *See also* error.

**absolute maximum** *See* maximum point.

**absolute minimum** *See* minimum point.

**absolute value** The MODULUS of a real number or of a complex number. For example, the absolute value of –2.3, written |–2.3|, is 2.3. The absolute value of a complex number is also the modulus; for example the absolute value of 2 + 3i is $\sqrt{(2^2 + 3^2)}$.

**abstract algebra** *See* algebra; algebraic structure.

**abstract number** A number regarded simply as a number, without reference to any material objects or specific examples. For example, the number 'three' when it does not refer to three objects, quantities, etc., but simply to the abstract concept of 'three'.

**acceleration** Symbol: $a$ The rate of change of speed or velocity with respect to time. The SI unit is the meter per second per second (m s$^{-2}$). A body moving in a straight line with increasing speed has a positive acceleration. A body moving in a curved path with uniform (constant) speed also has an acceleration, since the velocity (a vector depending on direction) is changing. In the case of motion in a circle the acceleration is $v^2/r$ directed toward the center of the circle (radius $r$).

For constant acceleration:
$$a = (v_2 - v_1)/t$$
$v_1$ is the speed or velocity when timing starts; $v_2$ is the speed or velocity after time $t$. (This is one of the equations of motion.) This equation gives the mean acceleration over the time interval $t$. If the acceleration is not constant
$$a = dv/dt, \text{ or } d^2x/dt^2$$
*See also* Newton's laws of motion.

**acceleration due to gravity** *See* acceleration of free fall.

**acceleration of free fall** (**acceleration due to gravity**) Symbol: $g$ The constant acceleration of a mass falling freely (without friction) in the Earth's gravitational field. The acceleration is toward the surface of the Earth. $g$ is a measure of gravitational field strength – the force on unit mass. The force on a mass $m$ is its weight $W$, where $W = mg$.

The value of $g$ varies with distance from the Earth's surface. Near the surface it is just under 10 meters per second per second (9.806 65 m s$^{-2}$ is the standard value). It varies with latitude, in part because the Earth is not perfectly spherical (it is flattened near the poles).

**acceptance region** When considering a hypothesis, the sample space is divided into two regions – the *acceptance region* and the *rejection region* (or *critical region*). The acceptance region is the one in which the sample must lie if the hypothesis is to be accepted.

**access time** The time needed for the reading out of, or writing into, the memory of a computer, i.e. the time it takes for the memory to transfer data from or to the CPU (*see* central processor).

**accumulation point** (**cluster point**) For a given set $S$, a point that can be approached arbitrarily closely by members of that set. Another way of saying this is that an accumulation point is the limit of a sequence of points in the set. An accumulation point of a set need not necessarily be a member of the set itself, although it can be. For example, any rational number is an accumulation point of the set of rationals. But 0 is an accumulation point of the set $\{1, \frac{1}{2}, \frac{1}{4}, \frac{1}{8}, \ldots\}$ although it is not itself a member of the set.

**accuracy** The number of significant figures in a number representing a measurement or value of a quantity. If a length is written as 2.314 meters, then it is normally assumed that all of the four figures are meaningful, and that the length has been measured to the nearest millimeter. It is incorrect to write a number to a precision of, for example, four significant figures when the accuracy of the value is only three significant figures, unless the error in the estimate is indicated. For example, 2.310 ± 0.005 meters is equivalent to 2.31 meters.

**acre** A unit of area equal to 4840 square yards. It is equivalent to 0.404 68 hectare.

**action** An out-dated term for *force. See* reaction.

**action at a distance** An effect in which one body affects another through space with no apparent contact or transfer between them. *See* field.

**actuary** An expert in statistics who calculates insurance risks and relates them to the premiums to be charged.

**acute** Denoting an angle that is less than a right angle; i.e. an angle less than 90° (or π/2 radian). *Compare* obtuse; reflex.

**addend** /**ad**-end, ă-**dend**/ One of the numbers added together in a sum. *See also* augend.

**adder** The circuitry in a computer that adds digital signals (i.e. the ADDEND, AUGEND and carry digit) to produce a sum and a carry digit.

**addition** Symbol: + The operation of finding the SUM of two or more quantities. In arithmetic, the addition of numbers is commutative ($4 + 5 = 5 + 4$), associative ($2 + (3 + 4) = (2 + 3) + 4$), and the identity element is zero ($5 + 0 = 5$). The inverse operation to addition is subtraction. In *vector addition*, the direction of the two vectors affects the sum. Two vectors are added by placing them head-to-tail to form two sides of a triangle. The length and direction of the third side is the VECTOR SUM. *Matrix addition* can only be carried out between matrices with the same number of rows and columns, and the sum has the same dimensions. The elements in corresponding positions in each MATRIX are added arithmetically.

**addition formulae** Equations that express trigonometric functions of the sum or difference of two angles in terms of separate functions of the angles.

$$\sin(x + y) = \sin x \cos y + \cos x \sin y$$
$$\sin(x - y) = \sin x \cos y - \cos x \sin y$$
$$\cos(x + y) = \cos x \cos y - \sin x \sin y$$
$$\cos(x - y) = \cos x \cos y + \sin x \sin y$$
$$\tan(x + y) = (\tan x + \tan y)/(1 - \tan x \tan y)$$
$$\tan(x - y) = (\tan x - \tan y)/(1 + \tan x \tan y)$$

They are used to simplify trigonometric expressions, for example, in solving an equation. From the addition formulae the following formulae can be derived:
The *double-angle formulae*:
$$\sin(2x) = 2 \sin x \cos x$$
$$\cos(2x) = \cos^2 x - \sin^2 x$$
$$\tan(2x) = 2\tan x/(1 - \tan^2 x)$$
The *half-angle formulae*:
$$\sin(x/2) = \pm\sqrt{[(1 - \cos x)/2]}$$
$$\cos(x/2) = \pm\sqrt{[(1 + \cos x)/2]}$$
$$\tan(x/2) = \sin x/(1 + \cos x) = (1 - \cos x)/\sin x$$
The *product formulae*:
$$\sin x \cos y = \tfrac{1}{2}[\sin(x + y) + \sin(x - y)]$$
$$\cos x \sin y = \tfrac{1}{2}[\sin(x + y) - \sin(x - y)]$$
$$\cos x \cos y = \tfrac{1}{2}[\cos(x + y) + \cos(x - y)]$$
$$\sin x \sin y = \tfrac{1}{2}[\cos(x - y) - \cos(x + y)]$$

**addition of fractions** *See* fractions.

**addition of matrices** *See* matrix.

**address** *See* store.

**ad infinitum** /ad in-fă-**nÿ**-tŭn/ To infinity; an infinite number of times. Often abbreviated to *ad inf*.

**adjacent** 1. Denoting one of the sides forming a given angle in a triangle. In a right-angled triangle it is the side between the given angle and the right angle. In trigonometry, the ratios of this adjacent side to the other side lengths are used to define the cosine and tangent functions of the angle.

**2.** Denoting two sides of a polygon that share a common vertex.

**3.** Denoting two angles that share a common vertex and a common side.

**4.** Denoting two faces of a polyhedron that share a common edge.

**adjoint** (of a matrix) *See* cofactor.

**admissible hypothesis** Any hypothesis that could possibly be true; in other words, an hypothesis that has not been ruled out.

**aether** /ee-th'er/ *See* ether.

**affine geometry** /ă-**fÿn**/ The study of properties left invariant under the group of affine transformations. *See* affine transformation; geometry.

**affine transformation** A transformation of the form
$$x' = a_1x + b_1y + c_1,$$
$$y' = a_2x + b_2y + c_2$$
where $a_1b_2 - a_2b_1 \neq 0$. An affine transformation maps parallel lines into parallel lines, finite points into finite points, leaves the line at infinity fixed, and preserves the ratio of the distances separating three points that lie on a straight line. An affine transformation can always be factored into the product of the following important special cases:

1. *translations:* $x' = x + a$, $y' = y + b$
2. *rotations:* $x' = x\cos\theta + y\sin\theta$, $y' = -x\sin\theta + y\cos\theta$
3. *stretchings* or *shrinkings:* $x' = tx$, $y' = ty$
4. *reflections in the x-axis or y-axis:* $x' = x$, $y' = -y$ or $x' = -x$, $y' = y$
5. *elongations* or *compressions:* $x' = x$, $y' = ty$ or $x' = tx$, $y' = y$

**aleph** /**ah**-lef, **ay**-/ The first letter of the Hebrew alphabet, used to denote transfinite cardinal numbers. $\aleph_0$, the smallest transfinite cardinal number, is the number of elements in the set of integers. $\aleph_1$, is the number of subsets of any set with $\aleph_0$, members. In general $\aleph_{n+1}$ is defined in the same way as the number of subsets of a set with $\aleph_n$ members.

**algebra** The branch of mathematics in which symbols are used to represent numbers or variables in arithmetical operations. For example, the relationship:
$$3 \times (4 + 2) = (3 \times 4) + (3 \times 2)$$
belongs to arithmetic. It applies only to this particular set of numbers. On the other hand the equation:
$$x(y + z) = xy + xz$$
is an expression in algebra. It is true for any three numbers denoted by $x$, $y$, and $z$. The above equation is a statement of the distributive law of arithmetic; similar statements can be written for the associative and commutative laws.

Much of elementary algebra consists of methods of manipulating equations to put them in a more convenient form. For example, the equation:
$$x + 3y = 15$$
can be changed by subtracting $3y$ from both sides of the equation, giving:
$$x + 3y - 3y = 15 - 3y$$
$$x = 15 - 3y$$
The effect is that of moving a term ($+3y$) from one side of the equation to the other and changing the sign. Similarly a multiplication on one side of the equation becomes a division when the term is moved to the other side; for example:
$$xy = 5$$
becomes:
$$x = 5/y$$
'Ordinary' algebra is a generalization of arithmetic. Other forms of *higher algebra* also exist, concerned with mathematical entities other than numbers. For example, matrix algebra is concerned with the relations between matrices; vector algebra with vectors; Boolean algebra is applicable to logical propositions and to sets; etc. An algebra consists of a number of mathematical entities (e.g. matrices or sets) and operations (e.g. addition or set inclusion) with formal rules for the relationships between the mathematical entities. Such a system is called an *algebraic structure*.

**algebra, Boolean** *See* Boolean algebra.

**algebraic structure** A structure imposed on elements of a set by certain operations that act on or combine the elements. The combination of the set and the operations satisfy particular axioms that define the structure. Examples of algebraic structures are FIELDS, GROUPS, and RINGS. The study of algebraic structure is sometimes called *abstract algebra*.

**algorithm** /al-gŏ-ri*th*-'m/ A mechanical procedure for performing a given calculation or solving a problem in a series of steps. One example is the common method of long division in steps. Another is the Euclidean algorithm for finding the highest common factor of two positive integers.

**allometry** /ă-**lom**-ĕ-tree/ A relation between two variables that can be expressed by the equation

$$y = ax^b$$

where $x$ and $y$ are the variables, $a$ is a constant and $b$ is a growth coefficient. it is used to describe the systematic growth of an organism, in which case $y$ is the mass of a particular part of the organism and $x$ is its total mass.

**alphanumeric** Describing any of the characters (or their codes) that stand for the letters of the alphabet or numerals, especially in computer science. Punctuation marks and mathematical symbols are not regarded as alphanumeric characters.

**alternate angles** A pair of equal angles formed by two parallel lines and a third line crossing both. For example, the two acute angles in the letter Z are alternate angles. *Compare* corresponding angles.

**alternate-segment theorem** A result in geometry stating that the angle between a tangent to a CIRCLE and a chord drawn from the point of contact of the tangent is equal to any angle subtended by the chord in the alternate segment of the circle, where the alternative segment is on the side of the chord opposite (alternate to) the angle.

**alternating series** A series in which the terms are alternately positive and negative, for example:

$$S_n = -1 + 1/2 - 1/3 + 1/4 \ldots + (-1)^n/n$$

Such a series is convergent if the absolute value of each term is less than the preceding one. The example above is a convergent series.

An alternating series can be constructed from the sum of two series, one with positive terms and one with negative terms. In this case, if both are convergent separately then the alternating series is also convergent, even if the absolute value of each term is not always smaller than the one before it. For instance, the series:

$$S_1 = 1/2 + 1/4 + 1/8 + \ldots + 1/2^n$$

and

$$S_2 = -1/2 - 1/3 - 1/4 - 1/5 - \ldots(-1)/(n + 1)$$

are both convergent, and so their sum:

$$S_n = S_1 + S_2$$
$$= 1/2 - 1/2 + 1/4 - 1/3 + 1/8 - 1/4 + \ldots$$

is also convergent.

**alternation** *See* disjunction.

**altitude** The perpendicular distance from the base of a figure (e.g. a triangle, pyramid, or cone) to the vertex opposite.

**ambiguous** Having more than one possible meaning, value, or solution. For example, an ambiguous case occurs in finding

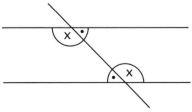

Alternate angles: alternate angles formed by a line cutting two parallel lines.

the sides and angles of a triangle when two sides and an angle other then the included angle are known. One solution is an acute-angled triangle and the other is an obtuse-angled triangle.

**ampere** /**am**-pair/ Symbol: A The SI base unit of electric current, defined as the constant current that, maintained in two straight parallel infinite conductors of negligible circular cross-section placed one meter apart in vacuum, would produce a force between the conductors of $2 \times 10^{-7}$ newton per meter. The ampere is named for the French physicist André Marie Ampère (1775–1836).

**amplitude** The maximum value of a varying quantity from its mean or base value. In the case of a simple harmonic motion – a wave or vibration – it is half the maximum peak-to-peak value.

**analog computer** A type of COMPUTER in which numerical information (generally called data) is represented in the form of a quantity, usually a voltage, that can vary continuously. This varying quantity is an analog of the actual data, i.e. it varies in the same manner as the data, but is easier to manipulate in the mathematical operations performed by the analog computer. The data is obtained from some process, experiment, etc.; it could be the changing temperature or pressure in a system or the varying speed of flow of a liquid. There may be several sets of data, each represented by a varying voltage.

The data is converted into its voltage analog or analogs and calculations and other sorts of mathematical operations, especially the solution of differential equations, can then be performed on the voltage(s) (and hence on the data they represent). This is done by the user selecting a group of electronic devices in the computer to which the voltage(s) are to be applied. These devices rapidly add voltages, and multiply them, integrate them, etc., as required. The resulting voltage is proportional to the result of the operation. It can be fed to a recording device to produce a graph or some other form of permanent record. Alternatively it can be used to control the process that produces the data entering the computer.

Analog computers operate in real time and are used, for example, in the automatic control of certain industrial processes and in a variety of scientific experiments. They can perform much more complicated mathematics than digital computers but are less accurate and are less flexible in the kind of things they can do. *See also* hybrid computer.

**analog/digital converter** A device that converts analog signals (the output from an ANALOG COMPUTER) into digital signals, so that they can be dealt with by a digital computer. *See* computer.

**analog electronics** A branch of electronics in which inputs and outputs can have a range of voltages rather than fixed values. A frequently used circuit in analog electronics is the *operational amplifier*, so called because it can perform mathematical operations such as differentiation and integration.

**analogy** A general similarity between two problems or methods. Analogy is used to indicate the results of one problem from the known results of the other.

**analysis** The branch of mathematics concerned with the limit process and the concept of convergence. It includes the theory of differentiation, integration, infinite series, and analytic functions. Traditionally, it includes the study of functions of real and complex variables arising from differential and integral calculus.

**analytic** A function of a real or complex variable is *analytic* (or *holomorphic*) at a point if there is a neighborhood $N$ of this point such that the function is *differentiable* at every point of $N$. An alternative (and equivalent) definition is that a function is analytic at a point if it can be represented in a neighborhood of this point by its Taylor series about it. A function is said to be analytic in a region if it is analytic at every point of that region.

**analytical geometry** (coordinate geometry) The use of coordinate systems and algebraic methods in geometry. In a plane Cartesian coordinate system a point is represented by a set of numbers and a curve is an equation for a set of points. The geometric properties of curves and figures can thus be investigated by algebra. Analytical geometry also enables geometrical interpretations to be given to equations.

**anchor ring** *See* torus.

**and** *See* conjunction.

**AND gate** *See* logic gate.

**angle** (plane angle) The spatial relationship between two straight lines. If two lines are parallel, the angle between them is zero. Angles are measured in degrees or, alternatively, in radians. A complete revolution is 360 degrees ($360°$ or $2\pi$ radians). A straight line forms an angle of $180°$ ($\pi$ radians) and a right angle is $90°$ ($\pi/2$ radians).

The angle between a line and a plane is the angle between the line and its orthogonal projection on the plane.

The angle between two planes is the angle between lines drawn perpendicular to the common edge from a point – one line in each plane. The angle between two intersecting curves is the angle between their tangents at the point of intersection.

*See also* solid angle.

**angle of depression** The angle between the horizontal and a line from an observer to an object situated below the eye level of the observer. *See also* angle.

**angle of elevation** The angle between the horizontal and a line from an observer to an object situated above the eye level of the observer. *See also* angle of depression.

**ångstrom** /**ang**-ström/ Symbol: Å  A unit of length defined as $10^{-10}$ meter. The ångstrom is sometimes used for expressing wavelengths of light or ultraviolet radiation or for the sizes of molecules. The unit is named for the Swedish physicist and astronomer Anders Jonas Ångstrom (1814–74).

**angular acceleration** Symbol: $\alpha$  The rotational acceleration of an object about an axis; i.e. the rate of change of angular velocity with time:
$$\alpha = d\omega/dt$$
or
$$\alpha = d^2\theta/dt^2$$
where $\omega$ is angular velocity and $\theta$ is angular displacement. Angular acceleration is analogous to linear acceleration. *See* rotational motion.

**angular displacement** Symbol: $\theta$  The rotational displacement of an object about an axis. If the object (or a point on it) moves from point $P_1$ to point $P_2$ in a plane perpendicular to the axis, $\theta$ is the angle $P_1OP_2$, where O is the point at which the perpendicular plane meets the axis. *See also* rotational motion.

**angular frequency** (pulsatance) Symbol: $\omega$  The number of complete rotations per unit time. Angular frequency is often used to describe vibrations. Thus, a simple harmonic motion of frequency $f$ can be represented by a point moving in a circular path at constant speed. The foot of a perpendicular from the point to a diameter of the circle moves with simple harmonic motion. The angular frequency of this motion is equal to $2\pi f$, where $f$ is the frequency of the simple harmonic motion. The unit of angular frequency, like frequency, is the hertz.

**angular momentum** Symbol: $L$  The product of the moment of inertia of a body and its angular velocity. Angular momentum is analogous to linear momentum, moment of inertia being the equivalent of mass for ROTATIONAL MOTION.

**angular velocity** Symbol: $\omega$  The rate of change of angular displacement with time: $\omega = d\theta/dt$. *See also* rotational motion.

**anharmonic oscillator** /an-har-**mon**-ik/ A system whose vibration, while still periodic, cannot be described in terms of simple harmonic motions (i.e. sinusoidal

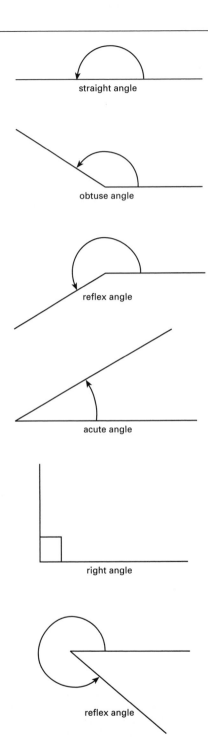

straight angle

obtuse angle

reflex angle

acute angle

right angle

reflex angle

Angle: types of angle

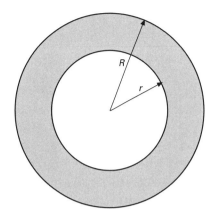

Annulus (shown shaded)

motions). In such cases, the period of oscillation is not independent of the amplitude.

**annuity** A pension in which an insurance company pays the annuitant fixed regular sums of money in return for sums of money paid to it either in installments or as a lump sum. An *annuity certain* is paid for a fixed number of years as opposed to an annuity that is payable only while the annuitant is alive.

**annulus** /an-yŭ-lŭs/ (*pl.* **annuli** or **annuluses**) The region between two concentric circles. The area of an annulus is $\pi(R^2 - r^2)$, where $R$ is the radius of the larger circle and $r$ is the radius of the smaller.

**antecedent** In logic, the first part of a conditional statement; a proposition or statement that is said to imply another. For example, in the statement 'if it is raining then the streets are wet', 'it is raining' is the antecedent. *Compare* consequent. *See also* implication.

**anticlockwise** (**counterclockwise**) Rotating in the opposite sense to the hands of a clock. *See* clockwise.

**antiderivative** /an-tee-dĕ-**riv**-ă-tiv/ A function $g(x)$ that is related to a real function $f(x)$ by the fact that the derivative of $g(x)$ with respect to x, denoted by $g'(x)$, is equal to $f(x)$ for all values of $x$ in the domain of f. The function $g(x)$ is said to be the antiderivative of $f(x)$. The indefinite integral $\int f(x)dx$ does not specify all the antiderivatives of $f(x)$ since an arbitrary constant $c$ can be added to any antiderivative. Thus, if both $g_1(x)$ and $g_2(x)$ are antiderivatives of a continuous function $f(x)$ then $g_1(x)$ and $g_2(x)$ can only differ by a constant.

**antilogarithm** /an-tee-**lôg**-ă-ri*th*-<m/ (**antilog**) The inverse function of a LOGARITHM. In common logarithms, the antilogarithm of $x$ is $10^x$. In natural logarithms, the antilogarithm of $x$ is $e^x$.

**antinode** /**an**-tee-nohd/ A point of maximum vibration in a stationary wave pattern. *Compare* node. *See also* stationary wave.

**antinomy** /an-**tin**-ŏ-mee/ *See* paradox.

**antiparallel** /an-tee-**pa**-ră-lel/ Parallel but acting in opposite directions, said of vectors.

**antisymmetric matrix** /an-tee-să-**met**-rik/ (**skew-symmetric matrix**) A square matrix $A$ that satisfies the relation $A^T = -A$, where $A^T$ is the TRANSPOSE of $A$. The definition of an antisymmetric matrix means that all entries $a_{ij}$ of the matrix have to satisfy $a_{ij} = -a_{ji}$ for all the $i$ and $j$ in the matrix. This, in turn, means that all diagonal entries in the matrix must have a value of zero, i.e. $a_{ii} = 0$ for all $i$ in the matrix.

**apex** The point at the top of a solid, such as a pyramid, or of a plane figure, such as a triangle.

**Apollonius' circle** /ap-ŏ-**loh**-nee-ŭs/ A circle defined as the locus of all points P that satisfy the relation AP/BP = $c$, where A and B are points in a plane and $c$ is a constant. In the case of $c = 1$ a straight line is obtained. This case can be considered to be a particular case of a circle or can be explicitly left out as a special case.

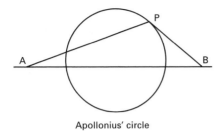

Apollonius' circle

The circle is named for the Greek mathematician Apollonius of Perga (*c.* 261 BC–*c.* 190 BC).

**Apollonius' theorem** /ap-ŏ-**loh**-nee-ŭs/ The equation that relates the length of a median in a triangle to the lengths of its sides. If *a* is the length of one side and *b* is the length of another, and the third side is divided into two equal lengths *c* by a median of length *m*, then:

$$a^2 + b^2 = 2m^2 + 2c^2$$

**apothem** /**ap**-ŏ-th'em/ (**short radius**) A line segment from the center of a regular polygon perpendicular to the center of a side.

**applications program** A computer program designed to be used for a specific purpose (such as stock control or word processing).

**applied mathematics** The study of the mathematical techniques that are used to solve problems. Strictly speaking it is the application of mathematics to any 'real' system. For instance, pure geometry is the study of entities – lines, points, angles, etc. – based on certain axioms. The use of Euclidean geometry in surveying, architecture, navigation, or science is applied geometry. The term 'applied mathematics' is used especially for mechanics – the study of forces and motion. *Compare* pure mathematics.

**approximate** Describing a value of some quantity that is not exact but close enough to the correct value for some specific purpose, as within certain boundaries of error. It is also used as a verb meaning 'to find the value of a quantity within certain bounds of accuracy, but not exactly'. For example, one can approximate an irrational number, such as π, by finding its decimal expansion to a certain number of places.

**approximate integration** Any of various techniques for finding an approximate value for a definite integral. There are many integrals that cannot be evaluated exactly and approximation techniques are used to estimate values for such integrals. Both analytical and numerical approximate integration techniques exist. In some analytical approximate integration techniques the value of the integral is expressed as an ASYMPTOTIC SERIES. Two examples of numerical approximate integration are SIMPSON'S RULE and the TRAPEZIUM RULE. *See also* numerical integration.

**approximately equal to** Symbol ≅ A symbol used to relate two quantities in which one of the quantities is a good approximation to the other quantity but is not exactly equal to it. An example of the use of this symbol is π ≅ 22/7.

**approximation** /ă-proks-ă-**may**-shŏn/ A calculation of a quantity that gives values

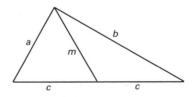

Apollonius' theorem: $a^2 + b^2 = 2m^2 + 2c^2$

Arc: major and minor arcs of a circle.

that are not, in general, exact but are close to the exact values. There are many integrals and differential equations in mathematics and its physical applications that cannot be solved exactly and require the use of approximation techniques. *See* Newton's method; numerical integration; Simpson's rule; trapezium rule.

**apse** /aps/ Any point on the orbit at which the motion of the orbiting body is at right angles to the central radius vector of the orbit. The distance from an apse to the centre of motion (the apsidal distance) equals the maximum or minimum value of the RADIUS VECTOR.

**arc** A part of a continuous curve. If the circumference of a circle is divided into two unequal parts, the smaller is known as the *minor arc* and the larger is known as the *major arc*.

**arc cosecant** (arc cosec; arc csc) An inverse cosecant. *See* inverse trigonometric functions.

**arc cosech** An inverse hyperbolic cosecant. *See* inverse hyperbolic functions.

**arc cosh** An inverse hyperbolic cosine. *See* inverse hyperbolic functions.

**arc cosine** (arc cos) An inverse cosine. *See* inverse trigonometric functions.

**arc cotangent** (arc cot) An inverse cotangent. *See* inverse trigonometric functions.

**arc coth** An inverse hyperbolic cotangent. *See* inverse hyperbolic functions.

**Archimedean solid** /ar-kă-**mee**-dee-ăn, -m…brevel-**dee**-ăn/ *See* semi-regular polyhedron. Archimedean solids are named for the Greek mathematician Archimedes (287 BC–212 BC).

**Archimedean spiral** A particular type of SPIRAL that is described in POLAR COORDINATES by the equation r = $a\theta$, where $a$ is a positive constant. It can be considered to represent the locus of a point moving along a radius vector with a uniform velocity while the radius vector itself is moving about a pole with a constant angular velocity. A spiral of this type asymptotically approaches a circle of radius $a$.

**Archimedes' principle** /ar-kă-**mee**-deez/ The upward force of an object totally or partly submerged in a fluid is equal to the weight of fluid displaced by the object. The upward force, often called the *upthrust*, results from the fact that the pressure in a fluid (liquid or gas) increases with depth. If the object displaces a volume V of fluid of density ρ, then:

$$\text{upthrust } u = V\rho g$$

where $g$ is the acceleration of free fall. If the upthrust on the object equals the object's weight, the object will float.

**arc secant** (arc sec) An inverse secant. *See* inverse trigonometric functions.

**arc sech** An inverse hyperbolic secant. *See* inverse hyperbolic functions.

**arc sine** (arc sin) An inverse sine. *See* inverse trigonometric functions.

**arc sinh** An inverse hyperbolic sine. *See* inverse hyperbolic functions.

**arc tangent** (arc tan) An inverse tangent. *See* inverse trigonometric functions.

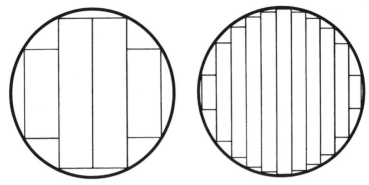

A curved area can be found by dividing it into rectangles and adding the areas of the rectangles. The more rectangles, the better the approximation.

**arc tanh** An inverse hyperbolic tangent. *See* inverse hyperbolic functions.

**are** /air/ A metric unit of area equal to 100 square meters. It is equivalent to 119.60 sq yd. *See also* hectare.

**area** Symbol: *A* The extent of a plane figure or a surface, measured in units of length squared. The SI unit of area is the square meter ($m^2$). The area of a rectangle is the product of its length and breadth. The area of a triangle is the product of the altitude and half the base. Closed figures bounded by straight lines have areas that can be determined by subdividing them into triangles. Areas for other figures can be found by using integral calculus.

**Argand diagram** /ar-gănd/ *See* complex number.

**argument** (amplitude) 1. In a complex number written in the form $r(\cos\theta + i\sin\theta)$, the angle $\theta$ is the argument. It is therefore the angle that the vector representing the complex number makes with the horizontal axis in an Argand diagram. *See also* complex number; modulus.
2. In LOGIC, a sequence of propositions or statements, starting with a set of premises (initial assumptions) and ending with a conclusion.

**arithmetic** The study of the skills necessary to manipulate numbers in order to solve problems containing numerical information. It also involves an understanding of the structure of the number system and the facility to change numbers from one form to another; for example, the changing of fractions to decimals, and vice versa.

**arithmetic and logic unit** (ALU) *See* central processor.

**arithmetic mean** *See* mean.

**arithmetic sequence** (arithmetic progression) A SEQUENCE in which the difference between each term and the one after it is constant, for example, {9, 11, 13, 15, ...}. The difference between successive terms is called the *common difference*. The general formula for the *n*th term of an arithmetic sequence is:
$$n_n = a + (n - 1)d$$
where *a* is the first term of the sequence and *d* is the common difference. *Compare* geometric sequence. *See also* arithmetic series.

**arithmetic series** A SERIES in which the difference between each term and the one after it is constant, for example, 3 + 7 + 11 + 15 + .... The general formula for an arithmetic series is:
$$S_n = a + (a + d) + (a + 2d) + \dots$$
$$+ [a + (n - 1)d]$$
$$= \Sigma[a + (n - 1)d]$$
In the example, the first term, *a*, is 3, the *common difference*, *d*, is 4, and so the

$n$th term, $a + (n - 1)d$, is $3 + (n - 1)4$. The sum to $n$ terms of an arithmetic series is $n[2a + (n - 1)d]/2$ or $n(a + l)/2$ where $l$ is the last ($n$th) term. *Compare* geometric series.

**arm** One of the lines forming a given angle.

**array** An ordered arrangement of numbers or other items of information, such as those in a list or table. In computing, an array has its own name, or *identifier*, and each member of the array is identified by a subscript used with the identifier. An array can be examined by a program and a particular item of information extracted by using this identifier and subscript.

**artificial intelligence** The branch of computer science that is concerned with programs for carrying out tasks which require intelligence when they are done by humans. Many of these tasks involve a lot more computation than is immediately apparent because much of the computation is unconscious in humans, making it hard to simulate. Programs now exist that play chess and other games at a high level, take decisions based on available evidence, prove theorems in certain branches of mathematics, recognize connected speech using a limited vocabulary, and use television cameras to recognize objects. Although these examples sound impressive, the programs have limited ability, no creativity, and each can only carry out a limited range of tasks. There is still a lot more research to be done before the ultimate goal of artificial intelligence is achieved, which is to understand intelligence well enough to make computers more intelligent than people. In fact there is considerable controversy about the whole subject, with many people postulating that the human thought process is different in kind to the computational operation of computer processes.

**assembler** *See* program.

**assembly language** *See* program.

**associative** Denoting an operation that is independent of grouping. An operation $\bullet$ is associative if
$$a \bullet (b \bullet c) = (a \bullet b) \bullet c$$
for all values of $a$, $b$, and $c$. In ordinary arithmetic, addition and multiplication are associative operations. This is sometimes referred to as the *associative law of addition* and the *associative law of multiplication*. Subtraction and division are not associative. *See also* commutative; distributive.

**astroid** /**ass**-troid/ A star-shaped curved defined in terms of the parameter $\theta$ by: $x = a \cos^3\theta$, $y = a \sin^3\theta$, where $a$ is a constant.

**astronomical unit** (au, AU) The mean distance between the Sun and the Earth, used as a unit of distance in astronomy for measurements within the solar system. It is defined as 149 597 870 km.

**asymmetrical** /ay-să-**met**-ră-kăl/ Denoting any figure that cannot be divided into two parts that are mirror images of each other. The letter R, for example, is asymmetrical, as is any solid object that has a left-handed or right-handed characteristic. *Compare* symmetrical.

**asymptote** /**ass**-im-toht/ A straight line towards which a curve approaches but never meets. A hyperbola, for example, has two asymptotes. In two-dimensional Cartesian coordinates, the curve with the equation $y = 1/x$ has the lines $x = 0$ and $y = 0$ as asymptotes, since $y$ becomes infinitely small, but never reaches zero, as $x$ increases, and vice versa.

**asymptotic series** /ass-im-**tot**-ik/ A series of the form $a_0 + a_1/x + a_2/x^2 + \dots a_n/x^n$ representing a function $f(x)$ is an asymptotic series if $|f(x) - S_n(x)|$ tends to zero as $|x| \to \infty$ for a fixed $n$, where $S_n(x)$ is the sum of the first $n$ terms of the series. Asymptotic series can be defined for either real or complex variables. They are usually divergent although some are convergent. Asymptotic series are used extensively in mathematical analysis and its physical applications. For example, many integrals that cannot

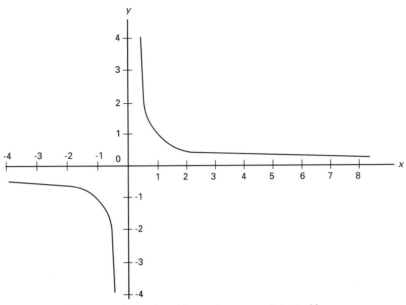

Asymptote: the *x*-axis and the *y*-axis are asymptotes to this curve.

be calculated precisely can be calculated approximately in terms of asymptotic series.

**atmosphere** A unit of pressure equal to 760 mmHg. It is equivalent to 101 325 newtons per square meters $(101\ 325\ N\ m^{-2})$.

**atmospheric pressure** *See* pressure of the atmosphere.

**atto-** Symbol: a A prefix denoting $10^{-18}$. For example, 1 attometer (am) = $10^{-18}$ meter (m).

**attractor** The point or set of points in phase space to which a system moves with time. The attractor of a system may be a single point (in which case the system reaches a fixed state that is independent of time), or it may be a closed curve known as a *limit cycle*. This is the type of behavior

 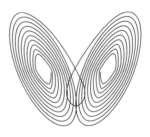

Attractor: an example of a strange attractor. The variable shown on the left displays chaotic behavior. The plot in phase space on the right shows a non-intersecting curve.

found in oscillating systems. In some systems, the attractor is a curve that is not closed and does not repeat itself. This, known as a *strange attractor*, is characteristic of chaotic systems. *See* chaos theory.

**AU** (au) *See* astronomical unit.

**augend** /**aw**-jend, aw-**jend**/ A number to which another, the ADDEND, is added in a sum.

**automorphism** /aw-tŏ-**mor**-fiz-ăm/ In very general terms, a transformation on elements belonging to a set that has some structure, in which the structural relations between the elements are unchanged by the transformations. The concept of automorphism applies to structure in algebra and geometry. An example of an automorphism is a transformation among the points of space that takes every figure into a similar figure. The set of automorphisms on a set of elements is a GROUP called the *automorphism group*. In physics the set of transformations between frames of reference forms the *physical automorphism group*.

**auxiliary equation** An equation that is used in the solution of an inhomogeneous second-order linear DIFFERENTIAL EQUATION of the form $ad^2y/dx^2 + bdy/dx + cy = f(x)$ where $a$, $b$, and $c$ are constants and $f(x)$ is a function of $x$. The solution is found in terms of the solution to the simpler homogeneous equation $ad^2y/dx^2 + bdy/dx + cy = 0$. A result in the theory of differential equations shows that the general solution of the first equation can be written in the form $y = g(x) + y_1(x)$, where $g(x)$ is a function, known as the *complementary function*, which is the general solution of the second equation, and $y_1(x)$ is a particular solution of the first equation, known as the *particular integral*. The complementary function can be found by taking $y = \exp(mx)$ to be a solution of the second equation. Taking this solution gives rise to the *auxiliary equation* $am^2 + bm + c = 0$. The form the complementary function takes depends on the nature of the roots of the auxiliary equation. If the equation has two different real roots, $m_1$ and $m_2$, then the complementary function is given by $y = A \exp(m_1x) + B \exp(m_2x)$, where $A$ and $B$ are constants. If the auxiliary equation has one root $m$ that occurs twice then the complementary function is given by $y = (A + Bx) \exp(mx)$.

**average** *See* mean.

**axial plane** /**aks**-ee-ăl/ A fixed reference plane in a three-dimensional coordinate system. For example, in rectangular Cartesian coordinates, the planes defined by $x = 0$, $y = 0$, and $z = 0$ are axial planes. The $x$-coordinate of a point is its perpendicular distance from the plane $x = 0$, and the $y$- and $z$-coordinates are the perpendicular distances from the $y = 0$ and $z = 0$ planes respectively. *See also* coordinates.

**axial vector** (pseudovector) A quantity that acts like a vector but changes sign if the coordinate system is changed from a right-handed system to a left-handed system. An example of an axial vector is a vector formed by the (cross) vector product of two vectors. *See also* polar vector.

**axiom** /**aks**-ee-ŏm/ (postulate) In a mathematical or logical system, an initial proposition or statement that is accepted as true without proof and from which further statements, or theorems, can be derived. In a mathematical proof, the axioms are often well-known formulae.

**axiom of choice** *See* choice; axiom of.

**axis** (*pl.* axes) 1. A line about which a figure is symmetrical.
2. One of the fixed reference lines used in a graph or a coordinate system. *See* coordinates.
3. A line about which a curve or body rotates or revolves.
4. The line of intersection of two or more coaxial planes.

**azimuth** /**az**-ă-mŭth/ The angle θ measured in a horizontal plane from the $x$-axis in spherical polar coordinates. It is the same as the longitude of a point.

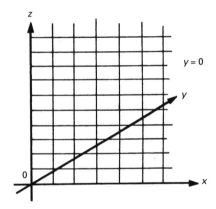

In three-dimensional rectangular Cartesion coordinates, the $x$- and $y$-axes lie in the axial plane $z = 0$, the $y$- and $z$-axes in the axial plane $x = 0$, and the $x$- and $z$-axes in the axial plane $y = 0$.

Axial planes

**backing store** *See* disk; magnetic tape; store.

**ballistic pendulum** A device for measuring the momentum (or velocity) of a projectile (e.g. a bullet). It consists of a heavy pendulum, which is struck by the projectile. The momentum can be calculated by measuring the displacement of the pendulum and using the law of constant momentum. If the mass of the projectile is known its velocity can be found.

**ballistics** The study of the motion of objects that are propelled by an external force (i.e. the motion of projectiles).

**bar** A unit of pressure defined as $10^5$ pascals. The *millibar* (mb) is more common; it is used for measuring atmospheric pressure in meteorology.

**bar chart** A GRAPH consisting of bars whose lengths are proportional to quantities in a set of data. It can be used when one axis cannot have a numerical scale; e.g. to show how many pink, red, yellow, and white flowers grow from a packet of mixed seeds. *See also* histogram.

**barn** Symbol: b A unit of area defined as $10^{-28}$ square meter. The barn is sometimes used to express the effective cross-sections of atoms or nuclei in the scattering or absorption of particles.

**barrel** A US unit of capacity used to measure solids or liquids. It is equal to 7056 cubic inches (0.115 6 m³). In the petroleum industry, 1 barrel = 42 US gallons, 35 imperial gallons, or 159.11 liters.

**barrel printer** *See* line printer.

**barycenter** /**ba**-ră-sen-ter/ *See* center of mass. Used in particular for the center of mass of a system of separate objects considered as a whole.

**barycentric coordinates** /ba-ră-**sen**-trik/ Coordinates that relate the center of mass of separate objects to the center of mass of the system of several objects as a whole. Consider three objects with masses $m_1$, $m_2$,

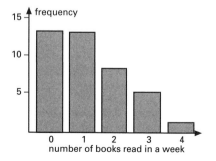

This bar chart shows the results when a group of 40 school students were asked how many books they each had read in the previous week. 13 had read none, 13 had read one, 8 had read two, 5 had read three, 1 had read four, and no-one had read five or more.

Bar chart

and $m_3$ with $m_1 + m_2 + m_3 = 1$, and their centers of mass at the points $p_1 = (x_1, y_1, z_1)$, $p_2 = (x_2, y_2, z_2)$, $p_3 = (x_3, y_3, z_3)$. Then the center of mass of the three objects together is the point

$$p = m_1 p_1 + m_2 p_2 + m_3 p_3 =$$
$$(m_1 x_1 + m_2 x_2 + m_3 x_3, \ m_1 y_1 + m_2 y_2 + m_3 y_3,$$
$$m_1 z_1 + m_2 z_2 + m_3 z_3)$$

$(m_1, m_2, m_3)$ are said to be the barycentric coordinates of the point $p$ with respect to the points $p_1$, $p_2$, and $p_3$.

**base** 1. In geometry, the lower side of a triangle, or other plane figure, or the lower face of a pyramid or other solid. The altitude is measured from the base and at right angles to it.
2. In a number system, the number of different symbols used, including zero. For example, in the decimal number system, the base is ten. Ten units, ten tens, etc., are grouped together and represented by the figure 1 in the next position. In binary numbers, the base is two and the symbols used are 0 and 1.
3. In logarithms, the number that is raised to the power equal to the value of the logarithm. In common logarithms the base is 10; for example, the logarithm to the base 10 of 100 is 2:

$$\log_{10} 100 = 2$$
$$100 = 10^2$$

**base unit** A unit that is defined in terms of reproducible physical phenomena or prototypes, rather than of other units. The second, for example, is a base unit (of time) in the SI, being defined in terms of the frequency of radiation associated with a particular atomic transition. Conventionally, seven units are chosen as base units in the SI. *See also* SI units.

**BASIC** *See* program.

**basis vectors** In two dimensions, two nonparallel VECTORS, scalar multiples of which are added to form any other vector in the same plane. For example, the unit vectors $i$ and $j$ in the directions of the $x$- and $y$-axes of a Cartesian coordinate system are basis vectors. The position vector $OP$ of the point P(2,3) is equal to $2i + 3j$.

Similarly, in three dimensions a vector can be written as the sum of multiples of three basis vectors.

**batch processing** A method of operation, used especially in large computer systems, in which a number of programs are collected together and input to the computer as a single unit. The programs forming a batch can either be submitted at a central site or at a *remote job entry* site; there can be any number of remote job entry sites, which can be situated at considerable distance from the computer. The programs are then executed as time becomes available in the system. *Compare* time sharing.

**Bayes' theorem** /bayz/ A formula expressing the probability of an intersection of two or more sets as a product of the individual probabilities for each. It is used to calculate the probability that a particular event $B_i$ has occurred when it is known that at least one of the set $\{B_1, B_2, \ldots B_n\}$ has occurred and that another event $A$ has also occurred. This conditional probability is written as $P(B_i|A)$. $B_1, \ldots B_n$ form a partition of the sample space $s$ such that $B_1 \cup B_2 \cup \ldots \cup B_n = s$ and $B_i \cap B_j = 0$, for all $i$ and $j$. If the probabilities of $B_1, B_2, \ldots B_n$ and all of the conditional probabilities $P(A|B_i)$ are known, then $P(B_i|A)$ is given by

$$P(B_i) P(A|B_i)$$

The theorem is named for the English mathematician Thomas Bayes (1702–61).

**beam compass** *See* compasses.

**bearing** The horizontal angle between a line and a fixed reference direction, usually measured clockwise from the direction of north. Bearings are expressed in degrees (e.g. 125°); bearings of less than 100° are written with an initial 0 (e.g. 030°). They have important applications in radar, sonar, and surveying.

**bel** *See* decibel.

**Bernoulli trial** An experiment in which there are two possible independent outcomes, for example, tossing a coin. The experiment is named for the Swiss mathe-

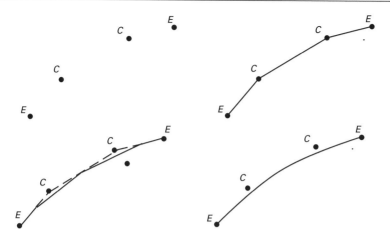

Bezier curve: construction of a curve by successive iterations.

matician Jakob (or Jacques) Bernoulli (1654–1705), who made an important contribution to the discipline of probability theory.

**Bessel functions** /**bess**-ăl/ A set of functions, denoted by the letter $J$, that are solutions to equations involving waves, which are expressed in cylindrical polar coordinates. The solutions form an infinite series and are listed in tables. The set of functions is named for the German astronomer Friedrich Wilhelm Bessel (1784–1846), who used them in his work on planetary movements. *See also* partial differential equation.

**beva-** Symbol: B A prefix used in the USA to denote $10^9$. It is equivalent to the SI prefix giga-.

**Bezier curve** /**bez**-ee-ay/ A type of curve used in computer graphics. A simple Bezier curve is defined by four points. Two of these are the end points of the curve. The other two are control points and lie off the curve. A way of producing the curve is to first join the four points by three straight lines. If the midpoints of these three lines are taken, together with the original two end points, a better approximation is to draw five straight lines joining these

points. The midpoints of these lines can then be used to get an even better approximation, and so on. The Bezier curve is the limit of this recursive process. Bezier curves can be defined mathematically by cubic polynomials. A similar type of curve obtained by using control points is known as a *B-spline*. The curve is named for the French engineer and mathematician Pierre Bezier (1910–99).

**bias** A property of a statistical sample that makes it unrepresentative of the whole population. For example, if medical data is based on a survey of patients in a hospital, then the sample is a biased estimate of the general population, since healthy people will be excluded.

**biconditional** /bÿ-kŏn-**dish**-ŏ-năl/ Symbol: $\leftrightarrow$ or $\equiv$ In logic, the relationship *if and only if* (often abbreviated to *iff*) that holds between a pair of propositions or statements $P$ and $Q$ only when they are both

| $P$ | $Q$ | $P \equiv Q$ |
|-----|-----|--------------|
| T | T | T |
| T | F | F |
| F | T | F |
| F | F | T |

Biconditional

true or both false. It is also the relationship of logical *equivalence*; the truth of *P* is both a necessary and a sufficient CONDITION for the truth of *Q* (and vice versa). The truth-table definition for the biconditional is shown. *See also* truth table.

**bilateral symmetry** A type of symmetry in which a shape is SYMMETRICAL about a plane, i.e. each half of the shape is a mirror image of the other.

**billion** A number equal to $10^9$ (i.e. one thousand million). This has always been the definition in the USA. In the UK a billion was formerly $10^{12}$ (i.e. one million million). In the 1960s, the British Treasury started using the term billion in the American sense of $10^9$, and this usage has now supplanted the original 'million million'.

**bimodal** /bȳ-**moh**-d'l/ Describing distributions of numerical data that have two peaks (modes) in frequency.

**binary** /**bȳ**-nă-ree, -nair-ee/ Denoting or based on two. A binary number is made up using only two different digits, 0 and 1, instead of ten in the decimal system. Each digit represents a unit, twos, fours, eights, sixteens, etc., instead of units, tens, hundreds, etc. For example, 2 is written as 10, 3 is 11, 16 (= $2^4$) is 10 000. Computers calculate using binary numbers. The digits 1 and 0 correspond to on/off conditions in an electronic switching circuit or to presence or absence of magnetization on a disk or tape. *Compare* decimal; duodecimal; hexadecimal; octal.

**binary logarithm** A LOGARITHM to the base two. The binary logarithm of 2 (written $\log_2 2$) is 1.

**binary operation** A mathematical procedure that combines two numbers, quantities, etc., to give a third. For example, multiplication of two numbers in arithmetic is a binary operation.

**binary relation** A relation between two elements *a* and *b* of a set *S*. For example, the symbol > for 'greater than' indicates a

binary relation between two real numbers. In general, the existence of a relation between *a* and *b* can be indicated by *a*R*b*.

**binomial** /bȳ-**moh**-mee-ăl/ An algebraic expression with two variables in it. For example, $2x + y$ and $4a + b = 0$ are binomials. *Compare* trinomial.

**binomial coefficient** The factor multiplying the variables in a term of a BINOMIAL EXPANSION. For example, in $(x + y)^2 = x^2 + 2xy + y^2$ the binomial coefficients are 1, 2, and 1 respectively.

**binomial distribution** The distribution of the number of successes in an experiment in which there are two possible outcomes, i.e. success and failure. The probability of *k* successes is
$$b(k,n,p) = n!/k!(n - k)! \times p^n \times q^{n-k}$$
where *p* is the probability of success and *q* (= 1 − *p*) the probability of failure on each trial. These probabilities are given by the terms in the binomial theorem expansion of $(p + q)^n$. The distribution has a mean *np* and variance *npq*. If *n* is large and *p* small it can be approximated by a Poisson distribution with mean *np*. If *n* is large and *p* is not near 0 or 1, it can be approximated by a normal distribution with mean *np* and variance *npq*

**binomial expansion** A rule for the expansion of an expression of the form $(x + y)^n$. *x* and *y* can be any real numbers and *n* is an integer. The general formula, sometimes called the *binomial theorem*, is
$$(x + y)^n = x^n + nx^{n-1}y + [n(n - 1)/2!]x^{n-2}y^2 + \dots + y^n$$
This can be written in the form:
$$x^n + {}_nC_1 x^{n-1}y + {}_nC_2 x^{n-2}y^2 + \dots + {}_nC_r x^{n-r}y^r + \dots$$
The coefficients ${}_nC_1$, ${}_nC_2$, etc., are called *binomial coefficients*. In general, the *r*th coefficient, ${}_nC_r$, is given by
$$n!/(n - r)!r! \text{ See also combination.}$$

**binomial theorem** *See* binomial expansion.

**birectangular** /bȳ-rek-**tang**-gyŭ-ler/ Having two right angles. *See* spherical triangle.

**bisector** /bỹ-**sek**-ter, **bỹ**-sek-/ A straight line or a plane that divides a line, a plane, or an angle into two equal parts.

**bistable circuit** /bỹ-**stay**-băl/ An electronic circuit that has two stable states. The circuit will remain in one state until the application of a suitable pulse, which will cause it to assume the other state.

Bistable circuits are used extensively in computer equipment for storing data and for counting. They usually have two input terminals to which pulses can be applied. A pulse on one input causes the circuit to change state; it will remain in that state until a pulse on the other input causes it to assume the alternative state. These circuits are often called *flip-flops*.

**bit** Abbreviation of *binary digit*, i.e. either of the digits 0 or 1 used in binary notation. Bits are the basic units of information in computing as they can represent the states of a two-valued system. For example, the passage of an electric pulse along a wire could be represented by a 1 while a 0 would mean that no pulse had passed. Again, the two states of magnetization of spots on a magnetic tape or disk can be represented by either a 1 or 0. *See also* binary; byte; word.

**bitmap** /**bit**-map/ *See* computer graphics.

**bivariate** /bỹ-**vair**-ee-ayt/ Containing two variable quantities. A plane vector, for example, is bivariate because it has both magnitude and direction.

A bivariate random variable $(X, Y)$ has the joint probability $P(x, y)$; i.e. the probability that $X$ and $Y$ have the values $x$ and $y$ respectively, is equal to $P(x) \times P(y)$, when $X$ and $Y$ are independent.

**Board of Trade unit** (BTU) A unit of energy equivalent to the kilowatt-hour ($3.6 \times 10^6$ joules). It was formerly used in the UK for the sale of electricity.

**Bolzano–Weierstrass theorem** /bohl-**tsah**-noh vỹ-er-shtrahss/ The theorem that any bounded infinite set has an accumulation point. The accumulation point need not be in the set; e.g. the set $\{1, \frac{1}{2}, \frac{1}{4}, \ldots\}$ is bounded and infinite and it does have an accumulation point, namely 0, but that point is not in the set. The theorem is named for the Czech philosopher, mathematician and theologian Bernard Bolzano (1781–1848) and the German mathematician Karl Weierstrass (1815–97).

**Boolean algebra** /**boo**-lee-ăn/ A system of mathematical logic that uses symbols and set theory to represent logic operations in a mathematical form. It was the first system of logic using algebraic methods for combining symbols in proofs and deductions. Various systems have been developed and are used in probability theory and computing. Boolean algebra is named for the British mathematician George Boole (1815–64).

**bound** 1. In a set of numbers, a value beyond which there are no members of the set. A *lower bound* is less than or equal to every number in the set. An *upper bound* is greater than or equal to every number in the set. The *least upper bound* (or *supremum*) is the lowest of its upper bounds and the *greatest lower bound* (or *infimum*) is the largest of its lower bounds. For example, the set $\{0.6, 0.66, 0.666, \ldots\}$ has a least upper bound of 2/3.
2. A bound of a function is a bound of the set of values that the function can take for the range of values of the variable. For example, if the variable $x$ can be any real number, then the function $f(x) = x^2$ has a lower bound of 0.
3. In formal logic a variable is said to be bound if it is within the scope of a quantifier. For example, in the sentence $Fy \rightarrow (\exists x)Fx$, $x$ is a bound variable, whereas $y$ is not. A variable which is not bound is said to be *free*.

**boundary condition** In a DIFFERENTIAL EQUATION, the value of the variables at a certain point, or information about their relationship at a point, that enables the arbitrary constants in the solution to be determined. For example, the equation
$$d^2y/dx^2 - 4dy/dx + 3y = 0$$
has a general solution

$$y = Ae^{-x} + Be^{-3x}$$

where $A$ and $B$ are arbitrary constants. If the boundary conditions are $y = 1$ at $x = 0$ and $dy/dx = 3$ at $x = 0$, the first can be substituted to obtain $B = 1 - A$. Differentiating the general solution for $y$ gives

$$dy/dx = -Ae^{-x} - 3Be^{-3x}$$

and substituting the second boundary condition then gives

$$3 = -A - 3B = 2A - 3$$

That is, $A = 3$ and $B = -2$.

**boundary line** A line on a graph that indicates the boundary at which an inequality in two dimensions holds. A common convention for drawing the boundary line is that if the inequality is of the form 'greater than or equal to' (or 'less than or equal to'), the line is solid, to indicate that points on the line are included. If the inequality is of the type 'greater than' (or 'less than') the line is dotted to show that points on the line are not included.

**bounded function** A real function $f(x)$ defined on a domain $D$ for which there is a number $M$ such that $f(x) \leq M$ for all $x$ in the domain $D$. It is an important result in mathematical analysis that if $f(x)$ is a continuous function on a closed interval between $a$ and $b$ then it is also a bounded function on that interval.

**bounded set** A set that is bounded both above and below; i.e. the set has both an upper BOUND and a lower bound.

**brackets** In mathematical expressions, brackets are used to indicate the order in which operations are to be carried out. For example:

$$9 + (3 \times 4) = 9 + 12$$
$$= 21$$
$$(9 + 3) \times 4 = 12 \times 4$$
$$= 48$$

Brackets are also used in FACTORIZATION. For example:

$$4x^3y^2 - 10x^2y^2 = 2x^2y(2xy - 5y)$$

**branch 1.** A section of a curve separated from the remainder of the curve by discontinuities or special points such as vertices, maximum points, minimum points, or cusps.
**2.** A departure from the normal sequential execution of instructions in a computer program. Control is thus transferred to another part of the program rather than passing in strict sequence from one instruction to the next. The branch will be either *unconditional*, i.e. it will always occur, or it will be *conditional*, i.e. the transfer of control will depend on the result of some arithmetical or logical test. *See also* loop.

**breadth** A horizontal distance, usually taken at right angles to a length.

**Briggsian logarithm** /**brig**-zee-ăn/ *See* logarithm.

**British thermal unit** (**Btu**) A unit of energy equal to $1.055\,06 \times 10^3$ joules. It was formerly defined by the heat needed to raise the temperature of one pound of air-free water by one degree Fahrenheit at standard presure. Slightly different versions of the unit were in use depending on the temperatures between which the degree rise was measured.

**B-spline** *See* Bezier curve.

**BTU** *See* Board of Trade unit.

**Btu** *See* British thermal unit.

**buffer store** (**buffer**) A small area of the main STORE of a computer in which information can be stored temporarily before, during, and after processing. A buffer can be used, for instance, between a peripheral device and the CENTRAL PROCESSOR, which operate at very different speeds.

**bug** An error or fault in a computer program. *See* debug.

**buoyancy** // The tendency of an object to float. The term is sometimes also used for the upward force (UPTHRUST) on a body. *See* center of buoyancy. *See also* Archimedes' principle.

**bushel** A unit of capacity usually used for solid substances. In the USA it is equal to 64 US dry pints or 2150.42 cubic inches. In the UK it is equal to 8 UK gallons.

**butterfly effect** Any effect in which a small change to a system results in a disproportionately large disturbance. The term comes from the idea that the Earth's atmosphere is so sensitive to initial conditions that a butterfly flapping its wings in one part of the world may be the cause of a tornado in another part of the world. *See* chaos theory.

**byte** /bÿt/ A subdivision of a WORD in computing, often being the number of BITS used to represent a single letter, number, or other CHARACTER. In most computers a byte consists of a fixed number of bits, usually eight. In some computers bytes can have their own individual addresses in the store.

**calculus** /**kal**-kyŭ-lŭs/ (infinitesimal calculus) (*pl.* calculuses) The branch of mathematics that deals with the DIFFERENTIATION and INTEGRATION of functions. By treating continuous changes as if they consisted of infinitely small step changes, *differential calculus* can, for example, be used to find the rate at which the velocity of a body is changing with time (acceleration) at a particular instant.

*Integral calculus* is the reverse process, that is finding the end result of known continuous change. For example, if a car's acceleration $a$ varies with time in a known way between times $t_1$ and $t_2$, then the total change in velocity is calculated by the integration of $a$ over the time interval $t_1$ to $t_2$. Integral calculus is also used to find the area under a curve. *See* differentiation; integration.

**calculus of variations** The branch of mathematics that uses calculus to find minimum or maximum values for a system. It has many applications in physical science and engineering. For example, FERMAT'S PRINCIPLE in optics can be regarded as an application of the calculus of variations. *See also* variational principle.

**calibration** /kal-ă-**bray**-shŏn/ The marking of a scale on a measuring instrument. For example, a thermometer can be calibrated in degrees Celsius by marking the freezing point of water (0°C) and the boiling point of water (100°C).

**calorie** /**kal**-ŏ-ree/ Symbol: cal A unit of energy approximately equal to 4.2 joules. It was formerly defined as the energy needed to raise the temperature of one gram of water by one degree Celsius. Because the specific thermal capacity of water changes with temperature, this definition is not precise. The mean or thermochemical calorie ($cal_{TH}$) is defined as 4.184 joules. The international table calorie ($cal_{IT}$) is defined as 4.186 8 joules. Formerly the mean calorie was defined as one hundredth of the heat needed to raise one gram of water from 0°C to 100°C, and the 15°C calorie as the heat needed to raise it from 14.5°C to 15.5°C.

**cancellation** Removing a common factor in a numerator and denominator, or removing the same quantity from both sides of an algebraic equation. For example, $xy/yz$ can be simplified, by the cancellation

$$\begin{pmatrix} 1 & 0 & 0 \\ 3 & 2 & 0 \\ 0 & 0 & 2 \end{pmatrix} \longrightarrow \begin{pmatrix} -3 & 0 & 0 \\ 3 & 2 & 0 \\ 0 & 0 & 2 \end{pmatrix} \begin{matrix} \text{multiply} \\ \text{row 1} \\ \text{by } -3 \end{matrix}$$

$$\begin{pmatrix} -3 & 0 & 0 \\ 3 & 2 & 0 \\ 0 & 0 & 2 \end{pmatrix} \longrightarrow \begin{pmatrix} -3 & 0 & 0 \\ 0 & 2 & 0 \\ 0 & 0 & 2 \end{pmatrix} \begin{matrix} \text{add row 1} \\ \text{to row 2} \end{matrix}$$

Reduction of a matrix to canonical form.

of $y$, to $x/z$. The equation $z + x = 2 + x$ is simplified to $z = 2$ by canceling (subtracting) $x$ from both sides.

**candela** Symbol: cd The SI base unit of luminous intensity, defined as the intensity (in the perpendicular direction) of the black-body radiation from a surface of 1/600 000 square meter at the temperature of freezing platinum and at a pressure of 101 325 pascals.

**canonical form** (**normal form**) In matrix algebra, the DIAGONAL MATRIX derived by a series of transformations on another SQUARE MATRIX of the same order.

**Cantor's diagonal argument** /kan-terz/ An argument to show that the real numbers, unlike the rationals, are not countable, and hence that there are more real numbers than rational numbers. The argument proceeds by assuming that the reals are countable and showing that this leads to a contradiction.

Every real number can be expressed as an infinite decimal expansion. We suppose that all reals are expressed in this way and since they are countable they can be arranged in order in a list as shown. We now define a real number $b_1 . b_2 b_3 b_4$ … by saying that $b_1$ must be any number different from $a_1$. Hence our new number will not be equal to the first real number in our list. $b_2$ must be any number different

from $a_{21}$, and hence the new number will not be equal to the second number on the list. Continuing in this way we have a method of producing an infinite decimal that must define a real number, but is not equal to any of the real numbers in our list. But we assumed that all real numbers occurred somewhere in the list; hence there is a contradiction, and so it cannot be true that the real numbers are countable.

Cantor's diagonal argument is named for the Russian mathematician Georg Cantor (1845–1918). *See also* Cantor set.

**Cantor set** /kan-ter/ The set obtained by taking a closed interval in the real line, e.g. [0,1], and removing the middle third, i.e. the open interval (1/3, 2/3), and then doing the same to the two remaining closed intervals [0,1/3] and [2/3,1], and so on *ad infinitum*. The set generated in this way has the remarkable property of containing uncountably many points – i.e. the same number of points as the whole real line – yet being nowhere dense – i.e. for any point in the set one can always find a point not in the set arbitrarily close to it. This set is also sometimes known as the *Cantor discontinuum*

**capital 1.** The total sum of all the assets of a person or company, including cash, investments, household goods, land, buildings, machinery, and finished or unfinished goods.
**2.** A sum of money borrowed or lent on which interest is payable or received. *See* compound interest; simple interest.
**3.** The total amount of money contributed by the shareholders when a company is formed, or the amount contributed to a partnership by the partners.

**carat** /ka-răt/ (**metric**) A unit of mass used for precious stones. It is equal to 200 milligrams.

**cardinality** /kar-dă-**nal**-ă-tee/ Symbol n($A$). The number of elements in a finite set $A$. The symbols |$A$| and #$A$ are sometimes used to denote the cardinality of $A$. The cardinality is sometimes called the *cardinal number* of the set. If $A$, $B$, and $C$ are all fi-

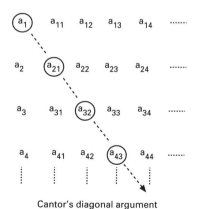

Cantor's diagonal argument

nite subsets of a set $S$ then the following relations between cardinal numbers hold:
$$n(A \cup B) = n(A) + n(B) - n(A \cap B)$$
and
$$n(A \cup B \cup C) = n(A) + n(B) + n(C) - n(A \cap B) - n(A \cap C) - n(B \cap C) + n(A \cap B \cap C).$$

**cardinal numbers** Whole numbers that are used for counting or for specifying a total number of items, but not the order in which they are arranged. For example, when one says 'three books', the three is a cardinal number.

Two sets are said to have the same cardinal number if their elements can be put into one-to-one correspondence with each other. *Compare* ordinal numbers. *See also* aleph.

**cardioid** /**kar**-dee-oid/ An EPICYCLOID that has only one loop, formed by the path of a point on a circle rolling round the circumference of another that has the same radius.

**Cartesian coordinates** /kar-**tee**-zhăn/ A method of defining the position of a point by its distance from a fixed point (origin) in the direction of two or more straight lines. On a flat surface, two straight lines, called the $x$-axis and the $y$-axis, form the basis of a two-dimensional Cartesian coordinate system. The point at which they cross is the origin (O). An imaginary grid is formed by lines parallel to the axes and one unit length apart. The point (2,3), for example, is the point at which the line parallel to the $y$-axis two units in the direction of the $x$-axis, crosses the line parallel to the $x$-axis three units in the direction of the $y$-axis. Usually the $x$-axis is horizontal and the $y$-axis is perpendicular to it. These are known as *rectangular coordinates*. If the axs are not at right angles, they are *oblique coordinates*

In three dimensions, a third axis, the $z$-axis, is added to define the height or depth of a point. The COORDINATES of a point are then three numbers $(x,y,z)$. A right-handed system is one for which if the thumb of the right hand points along the $x$-axis, the fingers of the hand fold in the direction in which the $y$-axis would have to rotate to

point in the same direction as the $z$-axis. A left-handed system is the mirror image of this. In a rectangular system, all three axs are mutually at right-angles. The Cartesian coordinate system is named for the French mathematician, philosopher, and scientist René du Perron Descartes (1596–1650). *See also* cylindrical polar coordinates; polar coordinates.

**Cartesian product** The Cartesian product of two sets $A$ and $B$, which is written $A \times B$, is the set of ordered pairs $\langle x,y \rangle$ where $x$ belongs to $A$ and $y$ belongs to $B$.
$$A \times B = \{\langle x,y \rangle \mid x \in A \text{ and } y \in B\}$$

**catastrophe theory** The study of the ways in which discontinuities appear, both in mathematics and its physical applications. Catastrophe theory deals with problems that cannot be analyzed using calculus since calculus is largely concerned with continuity. Catastrophe theory can be regarded as a branch of TOPOLOGY. It has been applied to several physical problems, including problems in the theory of waves, especially optics. It has also been applied to the evolution of complex systems such as biological systems and (far more dubiously) to social and economic changes.

**category** A category consists of two classes: a class of objects and a class of *morphisms* – i.e. mappings that are in some sense structure-preserving. Associated with each pair of objects are a set of the morphisms and a law of composition for these morphisms. *Category theory* is the study of such entities. It provides a model for many situations where sets with certain structures are studied along with a class of mappings that preserve these structures. Examples of categories are sets with functions and groups with homomorphisms.

**catenary** /**kat**-ĕ-nair-ee/ The plane curve of a flexible uniform line suspended from two points. For example, an empty washing line attached to two poles and hanging freely between them follows a catenary. The catenary is symmetrical about an axis perpendicular to the line joining the two points of suspension. In Cartesian coordi-

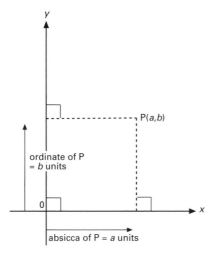

Two-dimensional rectangular Cartesian coordinates showing a point P($a,b$).

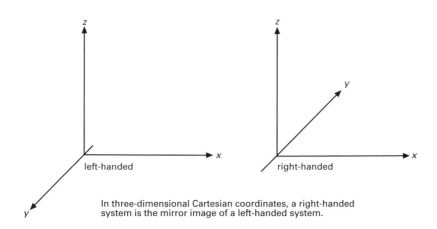

In three-dimensional Cartesian coordinates, a right-handed system is the mirror image of a left-handed system.

nates, the equation of a catenary that has its axis of symmetry lying along the $y$-axis at $y = a$, is

$$y = (a/2)(e^{x/a} + e^{-x/a})$$

**catenoid** /**kat**-ĕ-noid/ The curved surface formed by rotating a catenary about its axis of symmetry.

**Cauchy's inequality** /**koh**-sheez, koh-**sheez**/ For sums, the inequality has the form that if $a_1, a_2, ..., an$ and $b_1, b_2, ..., bn$ are two sets of real numbers then the in-equality $(\Sigma a_i b_i)^2 \leq (\Sigma a_i^2)(\Sigma b_i^2)$. This in-equality only becomes an equality in the case when there are constants $k$ and $l$ that satisfy $ka_i = lb_i$ for all $i$, i.e. if the $a_i$ and the $b_i$ are all proportional. It is also called the *Cauchy–Schwarz inequality*.

**CD-ROM** A type of compact disc with a read-only memory, which provides read-only access to up to 640 megabytes of memory for use on a computer. The disk can carry data as text, images (video) and sound (audio).

**Celsius degree** /**sel**-see-ŭs/ Symbol: °C  A unit of temperature difference equal to one hundredth of the difference between the temperatures of freezing and boiling water at one atmosphere pressure. It was formerly known as the degree centigrade and is equivalent to 1 K. On the Celsius scale water freezes at 0°C and boils at 100°C. It is named for the Swedish astronomer Anders Celsius (1701–44).

**center**  A point about which a geometric figure is symmetrical.

**center of buoyancy**  For an object in a fluid, the center of mass of the displaced volume of fluid. In order for a floating object to be stable the center of mass of the object must lie below the center of buoyancy; when the object is in equilibrium, the two lie on a vertical line. *See also* Archimedes' principle.

**center of curvature**  *See* curvature.

**center of gravity**  *See* center of mass.

**center of mass** (barycentre)  A point in a body (or system) at which the whole mass of the body may be considered to act. Often the term *center of gravity* is used. This is, strictly, not the same unless the body is in a constant gravitational field. The center of gravity is the point at which the weight may be considered to act.

The center of mass coincides with the center of symmetry if the symmetrical body has a uniform density throughout. In other cases the principle of moments may be used to locate the point. For instance, two masses $m_1$ and $m_2$ a distance $d$ apart have a center of mass on the line between them. If this is a distance $d_1$ from $m_1$ and $d_2$ from $m_2$ then $m_1d_1 = m_2d_2$ or:
$$m_1d_1 = m_2(d - d_1)$$
$$d_1 = m_2d/(m_1 + m_2)$$
A more general relationship can be applied to a number of masses $m_1$, $m_2$,..., $m_i$ that are respectively distances $r_1$, $r_2$,..., $r_i$ from an origin. The distance $r$ from the origin to the center of mass is given by:
$$r = \Sigma r_i m_i / \Sigma m_i$$

In the case of a body having a uniform density an integration must be used to obtain the position of the center of mass, which coincides with the CENTROID.

**center of pressure**  For a body or surface of a fluid, the point at which the resultant of pressure forces acts. If a surface lies horizontally in a fluid, the pressure at all points will be the same. The resultant force will then act through the centroid of the surface. If the surface is not horizontal, the pressure on it will vary with depth. The resultant force will now act through a different point and the center of pressure is not at the centroid.

**center of projection**  The point at which all the lines forming a central projection meet. *See* central projection.

**centi-** Symbol: c  A prefix denoting $10^{-2}$. For example, 1 centimeter (cm) = $10^{-2}$ meter (m).

**central conic**  A conic with a center of symmetry; e.g. an ellipse or hyperbola.

**central enlargement**  A central projection. *See also* scale factor.

**central force**  A force that acts on any affected object along a line to an origin. For instance, the motion of electric forces between charged particles are central; frictional forces are not.

**central limit theorem**  A result of probability theory that if $X_1$, $X_2$, ... $Xn$ is a random sample of size $n$ from a distribution, which is not necessarily a normal distribution, in which the finite mean is $\mu$ and the finite variance is $\sigma^2$, then as the sample size increases the distribution is approximately a normal distribution. This theorem can also be stated in the forms that $\bar{X} = (\Sigma Xi)/n$ is approximately equal to $N(\mu,\sigma^2/n)$ and $\Sigma(\bar{X} - \mu)/(\sigma/\sqrt{n})$ is approximately equal to $N(0,1)$. The size that $n$ needs to be for the theorem to be useful depends on how close the distribution is to a normal distribution. If the distribution is close to a normal distribution then for $n = 10$ it is a reasonably good approximation. If it is a long way from being a normal distribution it is not a good approximation until well after $n = 100$.

**central processor** (**central processing unit**; CPU) A highly complex electronic device that is the nerve center of a COMPUTER. It consists of the *control unit* and the *arithmetic and logic unit* (ALU). Also sometimes considered part of the central processor is the main store, or memory, where a program or a section of a program is stored in binary form.

The control unit supervises all activities within the computer, interpreting the instructions that make up the program. Each instruction is automatically brought, in turn, from the main store and kept temporarily in a small store called a *register*. Electronic circuits analyze the instruction and determine the operation to be carried out and the exact location or locations in store of the data on which the operation is to be performed. The operation is actually performed by the ALU, again using electronic circuitry and a set of registers. It may be an arithmetical calculation, such as the addition of two numbers, or a logical operation, such as selecting or comparing data. This process of fetching, analyzing, and executing instructions is repeated in the required order until an instruction to stop is executed.

The size of central processors has diminished considerably with advances in technology. It is now possible to form a central processor on a single silicon chip a few millimeters square in area. This tiny device is known as a *microprocessor*.

**central projection** (**conical projection**) A geometrical transformation in which a straight line from a point (called the *center of projection*) to each point in the figure is continued to the point at which it passes through a second (image) plane. These points form the image of the original figure. When a photographic image is created from a film using an enlarger, this is the kind of PROJECTION that takes place. The light source is at the center of projection, the light rays are the straight lines, the film

is the first plane, and the screen or print is the second. In this case the two planes are usually parallel, but this is not always so in central projection.

**central quadric** A quadric surface that is not a DEGENERATE QUADRIC and has a center of symmetry. This definition means that a central quadric has to be either an ellipsoid or a hyperboloid. The hyperboloid can have either one sheet or two. *See* conicoid.

**centrifugal force** A force supposed to act radially outward on a body moving in a curve. In fact there is no real force acting; centrifugal force is said to be a 'fictitious' force, and the use is best avoided. The idea arises from the effect of inertia on an object moving in a curve. If a car is moving around a bend, for instance, it is forced in a curved path by friction between the wheels and the road. Without this friction (which is directed toward the center of the curve) the car would continue in a straight line. The driver also moves in the curve, constrained by friction with the seat, restraint from a seat belt, or a 'push' from the door. To the driver it appears that there is a force radially outward pushing his or her body out – the centrifugal force. In fact this is not the case; if the driver fell out of the car he or she would move straight forward at a tangent to the curve. It is sometimes said that the centrifugal force is a 'reaction' to the CENTRIPETAL FORCE – this is not true. (The 'reaction' to the centripetal force is an outward push on the road surface by the tires of the car.)

**centripetal force** A force that causes an object to move in a curved path rather than continuing in a straight line. The force is provided by, for instance:
– the tension of the string, for an object whirled on the end of a string;
– gravity, for an object in orbit round a planet or a star;
– electric force, for an electron in the shell of an atom.

The centripetal force for an object of mass $m$ with constant speed $v$ and path radius $r$ is $mv^2/r$, or $m\omega^2 r$, where $\omega$ is angular speed. A body moving in a curved path has an acceleration because the direction of the velocity changes, even though the magnitude of the velocity may remain constant. This acceleration, which is directed toward the center of the curve, is the *centripetal acceleration*. It is given by $v^2/r$ or $\omega^2 r$.

**centroid** /sen-troid/ (mean center) The point in a figure or solid at which the CENTER OF MASS would be if the figure or body were of uniform-density material. The centroid of a symmetrical figure is at the center of symmetry; thus, the centroid of a circle is at its center. The centroid of a triangle is the point at which the medians meet.

For non-symmetrical figures or bodies integration is used to find the centroid. The centroid of a line, figure, or solid is the point that has coordinates that are the mean values of the coordinates of all the points in the line, figure, etc. For a surface, the coordinates of the centroid are given by:
$$\bar{x} = [\textstyle\iint x \, dx \, dy]/A, \text{ etc.}$$
the integration being over the surface, and $A$ being the area. For a volume, a triple integral is used to obtain the coordinates of the centroid:
$$\bar{x} = [\textstyle\iiint x \, dx \, dy \, dz]/V, \text{ etc.}$$

**c.g.s. system** A system of units that uses the centimeter, the gram, and the second as the base mechanical units. Much early scientific work used this system, but it has now almost been abandoned.

**chain** A former unit of length equal to 22 yards. It is equivalent to 20.116 8 m.

**chain rule** A rule for expressing the derivative of a function $z = f(x)$ in terms of another function of the same variable, $u(x)$, where $z$ is also a function of u. That is:
$$dz/dx = (dz/du)(du/dx)$$
This is often called the 'function of a function' rule.

For a function $z = f(x_1, x_2, x_3, ...)$ of several variables, in which each of the variables $x_1$, $x_2$, $x_3$, ... is itself a function of a single variable, $t$, the derivative $dz/dt$, called the *total derivative*, is given by the chain rule for partial differentiation, which is:
$$dz/dt = (\partial z/\partial x_1)(dx_1/dt) + (\partial z/\partial x_2)(dx_2/dt) + ...$$

**changing the subject of a formula** Rearranging a formula so that the single term on the left-hand side (the subject) is replaced by another term from the formula. For example, the formula for the area of a sphere is $A = 4\pi r^2$ (where $A$ is the area and $r$ is the radius). Changing the subject of the formula to $r$ gives $r = \frac{1}{2}\sqrt{(A/\pi)}$.

**channel** A path along which information can travel in a computer system or communications system.

**chaos theory** The theory of systems that exhibit apparently random unpredictable behavior. The theory originated in studies of the Earth's atmosphere and the weather. In such a system there are a number of variables involved and the equations describing them are nonlinear. As a result, the state of the system as it changes with time is extremely sensitive to the original conditions. A small difference in starting conditions may be magnified and produce a large variation in possible future states of the system. As a result, the system appears to behave in an unpredictable way and may exhibit seemingly random fluctuations (chaotic behavior). The study of such nonlinear systems has been applied in a number of fields, including studies of fluid dynamics and turbulence, random electrical oscillations, and certain types of chemical reaction. *See also* attractor.

**character** One of a set of symbols that can be represented in a computer. It can be a letter, number, punctuation mark, or a special symbol. A character is stored or manipulated in the computer as a group of BITS (i.e. binary digits). *See also* byte; word; store.

**characteristic 1.** *See* logarithm.
**2.** *See* eliminant.

**chi-square distribution** /kȳ-skwair/ ($\chi^2$ distribution) The distribution of the sum of the squares of random variables with standard normal distributions. For example, if $x_1$, $x_2$,...$x_i$,... are independent variables with standard normal distribution, then

$$\chi^2 = \Sigma x_i^2$$

has a chi-square distribution with $n$ degrees of freedom, written $\chi_n^2$. The mean and variance are $n$ and $2n$ respectively. The values $\chi_n^2(\alpha)$ for which $P(\chi^2 \leq \chi_n^2(\alpha)) = \alpha$ are tabulated for various values of $n$.

**chi-square test** A measure of how well a theoretical probability distribution fits a set of data. For $i = 1, 2, \ldots m$ the value $x_i$ occurs $o_i$ times in the data and the theory predicts that it will occur $e_i$ times. Provided $e_i \geq 5$ for all values of $i$ (otherwise values must be combined), then

$$\chi^2 = \Sigma\, (o_i - e_i)^2/e_i$$

has a chi-square distribution with $n$ degrees of freedom. *See also* chi-square distribution.

**choice, axiom of** The axiom states that, given any collection of sets, one can form a new set by *choosing* one element from each. This axiom may seem intuitively obvious and it was presupposed in many classical mathematical works. However, it has been a point of debate and controversy since many of its consequences appeared to be paradoxical. An example is the *Banach-Tarski theorem*, which proves that it is possible to cut a solid sphere into a finite number of pieces and to reassemble these pieces to form two solid spheres the same size as the original sphere. Despite these apparent paradoxes the axiom is widely accepted. It has many equivalents including the well-ordering principle and ZORN'S LEMMA.

**chord** A straight line joining two points on a curve, for example, the line segment joining two points on the circumference of a circle.

**chord of contact** The line that connects the points of contact between two tangents to a curve from a point P $(p,q)$. For a conic the chord of contact is also known as the *polar* of P with respect to the conic and the point P is known as the *pole*. For a general conic section described by the equation

$$ax^2 + 2hxy + by^2 + 2gx + 2fy + c = 0,$$

the equation of the chord of contact for P $(p,q)$ is

$$apx + h(py + qx) + bqy + g(p + x) + f(q + y) + c = 0.$$

**circle** The plane figure formed by a closed curve consisting of all the points that are at a fixed distance (the radius, $r$) from a particular point in the plane; the point is the center of the circle. The diameter of a circle is twice its radius; the circumference is $2\pi r$; and its area is $\pi r^2$. In Cartesian coordinates, the equation of a circle centred at the origin is

$$x^2 + y^2 = r^2$$

The circle is the curve that encloses the largest possible area within any given perimeter length. It is a special case of an ellipse with eccentricity 0.

**circular argument** An argument that, tacitly or explicitly, assumes what it is trying to prove, and is consequently invalid.

**circular cone** A CONE that has a circular base.

**circular cylinder** A CYLINDER in which the base is circular.

**circular functions** *See* trigonometry.

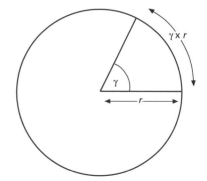

Circular measure: in a circle of radius $r$ and circumference $2\pi r$, the angle $\gamma$ radians subtends an arc length $\gamma \times r$.

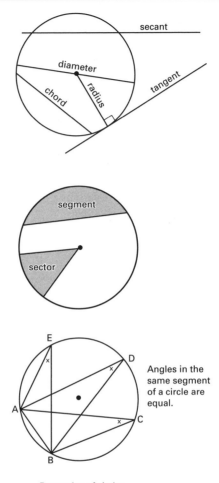

Properties of circles

**circular measure** The measurement of an angle in radians.

**circular mil** *See* mil.

**circular motion** A form of periodic (or cyclic) motion; that of an object moving in a circular path. For this to be possible, a positive central force must act. If the object has a uniform speed $v$ and the radius of the circle is $r$, the angular velocity ($\omega$) is $v/r$. There is an acceleration toward the center of the circle (the centripetal acceleration) equal to $v^2/r$ or $\omega^2 r$. *See also* centripetal force; rotational motion.

**circumcenter** /ser-kŭm-**sen**-ter/ *See* circumcircle.

**circumcircle** /ser-kŭm-**ser**-kăl/ (**circumscribed circle**) The circle that passes through all three vertices of a triangle or through the vertices of any other cyclic polygon. The figure inside the circle is said to be *inscribed*. The point in the figure that is the center of the circle is called the *circumcenter*. For a triangle with side lengths $a$, $b$, and $c$ the radius $r$ of the circumcircle is given by:
$$r = abc/\{4\sqrt{[s(s - a)(s - b)(s - c)]}\}$$
where $s$ is $(a + b + c)/2$.

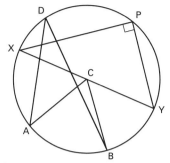

An angle that an arc subtends at the center of a circle is twice the angle that it subtends at the circumference: $A\hat{C}B = 2A\hat{D}B$

An angle in a semicircle is a right angle: $X\hat{P}Y$ $(=\frac{1}{2}X\hat{C}Y) = 90°$

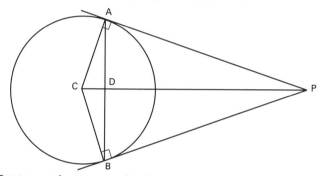

Two tangents from an external point:
(1) are equal, PA = PB
(2) subtend equal angles at the center, $P\hat{C}A = P\hat{C}B$
(3) the line from the point to the center bisects the line AB, AD = DB

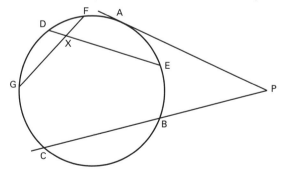

A tangent and a secant from an external point: PC.PB = PA²

Two intersecting chords: FX.GX = DX.XE

Properties of circles

**circumference** The boundary, or length of the boundary, of a closed curve, usually a circle. The circumference of a circle is equal to $2\pi r$, where $r$ is the radius of the circle.

**circumscribed** Describing a geometric figure that is drawn around and enclosing another geometrical figure. For example, in a square, a circle can be drawn through the vertices. This is called the *circumscribed circle*, and the square is called the *inscribed square* of the circle. Similarly, a regular polyhedron might have a circumscribed sphere, and a rectangular pyramid a circumscribed cone. *Compare* inscribed.

**cissoid** /sis-oid/ A curve that is defined by the equation $y^2 = x^3/2(a-x)$. It has the line $x = 2a$ as an ASYMPTOTE and a CUSP at the origin.

**class 1.** A grouping of data that is taken as one item in a FREQUENCY TABLE or HISTOGRAM.
**2.** Often used simply as a synonym for *set*. However, in set theory it is sometimes desirable, to avoid paradoxes, to allow the existence of collections that are not sets. Such collections are known as *classes* or *proper classes*. For example, the collection of all sets is a proper class not a set.

**classical mechanics** A system of mechanics that is based on Newton's laws of motion. Relativity effects and quantum theory are not taken into account in classical mechanics.

**class mark** *See* frequency table.

**clock pulse** One of a series of regular pulses produced by an electronic device called a *clock* and are used to synchronize operations in a computer. Every instruction in a computer program causes a number of operations to be done by the CENTRAL PROCESSOR of the computer. Each of these operations, performed by the control unit or arithmetic and logic unit, are triggered by one clock pulse and must be completed before the next clock pulse. The interval at which the pulses occur is usually a few microseconds (millionths of a second).

**clockwise** Rotating in the same sense as the hands of a clock. For example, the head of an ordinary screw is turned clockwise (looking at the head of the screw) to drive it in. Looking at the other end the rotation appears to be anticlockwise (counterclockwise).

**closed** Describing a set for which a given operation gives results in the same set. For example, the set of positive integers is closed with respect to addition and multiplication. Adding or multiplying any two members gives another positive integer. The set is not closed with respect to division since dividing certain integers does not give a positive integer (e.g. 4/5). The set of positive integers is also not closed with respect to subtraction (e.g. 5−7 = −2). *See also* closed interval; closed set.

**closed curve** (**closed contour**) A curve, such as a circle or an ellipse, that forms a complete loop. It has no end points. A *simple* closed curve is a closed curve that does not cross itself. *Compare* open curve.

**closed interval** A set consisting of the numbers between two given numbers (end points), including the end points. For example, all the real numbers greater than or equal to 2 and less than or equal to 5 constitute a closed interval. The closed interval between two real numbers $a$ and $b$ is written $[a,b]$. On a number line the end points are marked by a blacked-in circle. *Compare* open interval. *See also* interval.

**closed set** A set in which the limits that define the set are included. The set of rational numbers greater than or equal to 0 and less than or equal to ten, written $\{x: 0 \leq x \leq 10; \ x \in R\}$, and the set of points on and within a circle are examples of closed sets. *Compare* open sets.

**closed surface** A surface that has no boundary lines or curves, for example a sphere or an ellipsoid.

**closed system** (isolated system) A set of one or more objects that may interact with each other, but do not interact with the world outside the system. This means that there is no net force from outside or energy transfer. Because of this the system's angular momentum, energy, mass, and linear momentum remain constant.

**closure** *See* group.

**cluster point** *See* accumulation point.

**coaxial** /koh-**aks**-ee-ǎl/ **1.** *Coaxial circles* are circles such that all pairs of the circles have the same RADICAL AXIS.
**2.** *Coaxial planes* are planes that pass through the same straight line (the *axis*).

**COBOL** /**koh**-bôl/ *See* program.

**coding** The writing of instructions in a computer programming language. The person doing the coding starts with a written description or a diagram representing the task to be carried out by the computer. This is then converted into a precise and ordered sequence of instructions in the language selected. *See also* flowchart; program.

**coefficient** /koh-i-**fish**-ĕnt/ A multiplying factor. For example, in the equation $2x^2 + 3x = 0$, where $x$ is a variable, the coefficient of $x^2$ is 2 and the coefficient of $x$ is 3. Sometimes the value of the coefficients is not known, although they are known to stay constant as $x$ changes, for example, $ax^2 + bx = 0$. In this case $a$ and $b$ are *constant coefficients*. *See also* constant.

**coefficient, binomial** *See* binomial coefficient.

**coefficient of friction** *See* friction.

**coefficient of restitution** *See* restitution, coefficient of.

**cofactor** /koh-**fak**-ter/ The DETERMINANT of the matrix obtained by removing the row and column containing the element. The matrix formed by all the cofactors of

the elements in a matrix is called the *adjoint* of the matrix.

$$A = \begin{pmatrix} a & b & c \\ d & e & f \\ g & h & i \end{pmatrix}$$

$$a' = \begin{vmatrix} e & f \\ h & i \end{vmatrix} = ei - hf$$

$$b' = \begin{vmatrix} d & f \\ g & i \end{vmatrix} = di - gf$$

$$c' = \begin{vmatrix} d & e \\ g & h \end{vmatrix} = dh - ge$$

The cofactors $a'$, $b'$, and $c'$ of the elements $a$, $b$, and $c$ in a 3 x 3 matrix A.

The adjoint of A.
$$\begin{pmatrix} a' & b' & c' \\ d' & e' & f' \\ g' & h' & i' \end{pmatrix}$$

Cofactor

**coherent units** A system or subset of units (e.g. SI units) in which the derived units are obtained by multiplying or dividing together base units, with no numerical factor involved.

**colatitude** /koh-**lat**-ă-tewd/ *See* spherical polar coordinates.

**collinear** /kŏ-**lin**-ee-er/ Lying on the same straight line. Any two points, for example, could be said to be collinear because there is a straight line that passes through both. Similarly, two vectors are collinear if they are parallel and both act through the same point.

**cologarithm** /koh-**lôg**-ă-ri*th*-<m/ The LOGARITHM of the reciprocal of a given number; i.e. the negative of the logarithm. It is sometimes used in logarithmic computation to avoid the use of negative mantissas or of subtraction of logarithms.

**column matrix** *See* column vector.

**column vector** (**column matrix**) A number (*m*) of quantities arranged in a single column; i.e. an $m \times 1$ matrix. For example, the vector that defines the displacement of the point (*x*,*y*,*z*) from the origin of a Cartesian coordinate system is usually written as a column vector.

The column vector that defines the displacement of a point (*x*, *y*, *z*) from the origin of a Cartesian coordinate system.

Column vector

**combination** Any subset of a given set of objects regardless of the order in which they are selected. If *r* objects are selected from *n*, and each object can only be chosen once, the number of different combinations is

$$n!/[r!(n-r)!]$$

written as $_nC_r$ or $C(n,r)$. For instance, if there are 15 students in a class and only 5 books, then each book has to be shared by 3 students. The number of ways in which this can happen – i.e. the number of combinations of 3 from 15 – is 15!/3!12!, or 455. If each object can be selected more than once the number of different combinations is $_{n+r-1}C_r$. *See also* factorial, permutation.

**combinatorics** /kŏm-bў-nă-**tor**-iks, -**toh**-riks/ (**combinatorial analysis**) The branch of mathematics that studies the number of possible configurations or arrangements of a certain type. It forms the basis for the theory of probability since we have to know how to calculate the total number of different ways an event *can* happen before we can hope to predict how it is *likely* to happen. There are many unsolved problems in combinatorics that at first appear simple. For example, a rectangular grid of some fixed dimension, $m \times n$, is made up of unit squares each of one of two colors. How many different color patterns are there if the number of boundary edges between the two colors is a certain fixed number? This problem was completely solved in two dimensions in the 1960s, but the solution for the three-dimensional problem is still unknown.

**commensurable** /kŏ-**men**-shŭ-ră-băl/ Able to be measured in the same way and in terms of the same units. For example, a 30-centimeter rule is commensurable with a 1-meter length of rope, because both can be measured in centimeters. Neither is commensurable with an area.

**common denominator** A whole number that is a common multiple of the denominators of two or more fractions. For example, 6 and 12 are both common denominators of 1/2 and 1/3. The *lowest* (or *least*) *common denominator* (LCD) is the smallest number that is a common multiple of the denominators of two or more fractions. For example, the LCD of 1/2, 1/3, and 1/4 is 12. Fractions are put in terms of the LCDs when they are to be added or subtracted:

$$1/2 + 1/3 + 1/4 = 6/12 + 4/12 + 3/12$$
$$= 13/12$$

**common difference** The difference between successive terms in an arithmetic sequence or arithmetic series.

**common factor** 1. A whole number that divides exactly into two or more given numbers. For example, 7 is a common factor of 14, 49, and 84. Since 7 is the largest number that divides into all three exactly, it is the *highest common factor* (HCF). *See also* factor.
2. A number or variable by which several parts of an expression are multiplied. For

example, in $4x^2 + 4y^2$, 4 is a common factor of $x^2$ and $y^2$, and from the distributive law for multiplication and addition,
$$4x^2 + 4y^2 = 4(x^2 + y^2)$$

**common fraction** *See* fraction.

**common logarithm** *See* logarithm.

**common multiple** A whole number that is a multiple of each of a group of numbers. For example, 100 is a common multiple of 5, 25, and 50. The *lowest* (or *least*) *common multiple* (LCM) is the smallest number that is a common multiple; in this case it is 50.

**common ratio** The ratio of successive terms in a geometric sequence or geometric series.

**common tangent** A single line that forms a tangent to two or more separate curves. The term is also used for the length of the line joining the two tangential points.

**commutative** /kŏ-**myoo**-tă-tiv, **kom**-yŭ-tay-tiv/ Denoting an operation that is independent of the order of combination. A binary operation • is commutative if $a•b = b•a$ for all values of $a$ and $b$. In ordinary arithmetic, multiplication and addition are commutative operations. This is sometimes referred to as the *commutative law of multiplication* and the *commutative law of addition*. Subtraction and division are not commutative operations. *See also* associative; distributive.

**commutative group** *See* Abelian group.

**compact** A set $S$ of real numbers is compact if, given any collection of open sets whose union contains $S$, we can find a finite subcollection of those open sets whose union also contains $S$. The concept can be generalized to any topological space, and also to mathematical logic. In logic a formal system is said to be compact if it is such that, when a given sentence is a logical consequence of a given set of sentences, it is a consequence of some finite subset of them.

**compasses** An instrument used for drawing circles. It consists of two rigid arms joined by a hinge. At one end is a sharp point, which is placed at the center of the circle. At the other end is a pencil or other marker, which traces out the circumference when the compasses are pivoted around the point. In a *beam compass*, used for drawing large circles, the sharp point and the marker are attached to opposite ends of a horizontal beam.

**compiler** *See* program.

**complement** The set of all the elements that are not in a particular set. If the set $A = \{1, 2, 3\}$, and the universal set, $E$, is taken as containing all the natural numbers, then the complement of $A$, written $A'$ or , is $\{4, 5, 6, ...\}$. *See* Venn diagram.

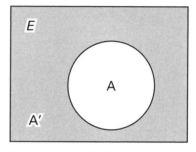

The shaded area in the Venn diagram is the complement A′ of the set A.

**complementary angles** A pair of angles that add together to make a right angle (90° or $\pi/2$ radians). *Compare* conjugate angles; supplementary angles.

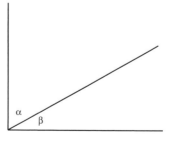

Complementary angles: $\alpha + \beta = 90°$

**complementary function** *See* auxiliary equation.

**complete** In mathematical logic a formal system is said to be complete if every true sentence in the system is also provable within the system. Not all logical systems have this property. *See* Gödel's incompleteness theorem.

**completing the square** A way of solving a QUADRATIC EQUATION, by dividing both sides by the coefficient of the square term and adding a constant, in order to express the equation as a single squared term. For example, to solve $3x^2 + 6x + 2 = 0$:

$$x^2 + 2x + \tfrac{2}{3} = 0$$
$$(x + 1)^2 - 1 + \tfrac{2}{3} = 0$$
$$x + 1 = +\sqrt{\tfrac{1}{3}} \text{ or } -\sqrt{\tfrac{1}{3}}$$
$$x = -1 + \sqrt{\tfrac{1}{3}} \text{ or } -1 - \sqrt{\tfrac{1}{3}}$$

**complex analysis** The branch of analysis that is specifically concerned with COMPLEX FUNCTIONS. Some of the results and techniques of real analysis can be extended to complex analysis but many results arise specifically because of the mathematical structure imposed by dealing with complex functions. Complex analysis is used in the theory of ASYMPTOTIC SERIES and has many applications in the physical sciences and in engineering.

**complex fraction** *See* fraction.

**complex function** A FUNCTION in which the variable quantities are complex numbers. *See* complex analysis.

**complexity theory** The mathematical theory of systems that can exhibit complex behavior. It includes such topics as chaos theory, ideas about how organization and order can emerge in systems, and how emergent laws of phenomena arise at different levels of description. Complexity theory is a subject of great interest to mathematicians, physicists, chemists, biologists, and economists, amongst others. Computers are used extensively in complexity theory since the systems being analyzed do not, in general, have exact solutions. At the time of writing, there is not yet a consensus as to whether there are general 'laws of complexity', which are applicable to all complex systems.

**complex number** A number that has both a real part and an imaginary part. The imaginary part is a multiple of the square root of minus one (i). Some algebraic equations cannot be solved with real numbers. For example, $x^2 + 4x + 6 = 0$ has the solutions $x = -2 + \sqrt{(-2)}$ and $x = -2 - \sqrt{(-2)}$. If the number system is extended to include i $= \sqrt{-1}$, all algebraic equations can be solved. In this case the solutions are $x = -2 + i\sqrt{2}$ and $x = -2 - i\sqrt{2}$. The real part is $-2$ and the imaginary part is $+i\sqrt{2}$ or $-i\sqrt{2}$.

Complex numbers are sometimes represented on an *Argand diagram*, which is similar to a graph in Cartesian coordinates, but with the horizontal axis representing the real part of the number and the vertical axis the imaginary part. The diagram is named for the Swiss mathematician Jean Argand (1768–1822).

Any complex number can also be written as a function of an angle $\theta$, just as Cartesian coordinates can be converted into polar coordinates. Thus $r(\cos\theta + i\sin\theta)$ is equivalent to $x + iy$, where $x = r\cos\theta$ and $y = r\sin\theta$. Here, $r$ is the *modulus* of the complex number and $\theta$ is the *argument* (or *amplitude*). This can also be written in the exponential form $r = e^{i\theta}$.

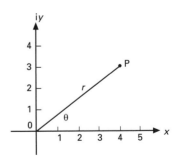

The point P(4,3) on an Argand diagram represents the complex number $z = 4 + 3i$. In the polar form $z = r(\cos\theta + i\sin\theta)$.

**component** /kŏm-**poh**-nĕnt/ The resolved part of a VECTOR in a particular direction, often one of two components at right angles.

**component forces** *See* component vectors.

**component vectors** The components of a given VECTOR (such as a force or velocity) are two or more vectors with the same effect as the given vector. In other words the given vector is the resultant of the components. Any vector has an infinite number of sets of components. Some sets are more use than others in a given case, especially pairs at 90°. The component of a given vector ($V$) in a given direction is the projection of the vector on to that direction; i.e. $V\cos\theta$, where $\theta$ is the angle between the vector and the direction.

**component velocities** *See* component vectors.

**composite function** A function formed by combining two functions f and g. There are several notations for such a composite function: for example, $g[f(x)]$, gof, g.f, gf. For a composite function to exist it is necessary that the domain of g contains the range of f. The order of the functions f and g in a composite function has to be specified since, in general, $g[f(x)] \neq f[g(x)]$. This means that to find the composite function $g[f(x)]$, $f(x)$ is found first and then $g[f(x)]$ is found. The non-commuting nature of composite functions can be seen by considering the functions $f(x) = \sin x$ and $g(x) = x^2$ for which $g[f(x)] = (\sin x)^2$ and $f[g(x)] = \sin(x^2)$.

**composite number** An integer that has more than one prime factor. For example, $4 (= 2 \times 2)$, $6 (= 2 \times 3)$, $10 (= 2 \times 5)$ are composite numbers. The prime numbers and $\pm 1$ are not composite.

**composition** The process of combining two or more functions to obtain a new one. For example the composition of $f(x)$ and $g(x)$, which is written f•g, is obtained by applying $g(x)$ and then applying $f(x)$ to the result. If $f(x) = x - 2$ and $g(x) = x^3 + 1$ then $f•g(x) = f(x^3+1) = x^3 - 1$, whereas $g•f(x) = g(x-2) = (x - 2)^3 + 1$. As can be seen, these two resulting functions are not the same. In general, composing functions in a different order will produce different results.

**compound growth** The growth of money that is associated with COMPOUND INTEREST. Compound growth can be tabulated in tables called *compound growth tables*, which give the multiplying factors for different periods of time (usually years) and different percentage rates. These tables can be drawn in the form of a matrix, with the number of years increasing from left to right and the percentage rates increasing from top to bottom. Compound growth tables can also be used to record the growth in population with time for different population growth rates.

**compound interest** The interest earned on capital, when the interest in each period is added to the original capital as it is earned. Thus the capital, and therefore the interest on it, increases year by year. If $P$ is the principal (the original amount of money invested), $R$ percent the interest rate per annum, and $n$ the number of interest periods, then the compound interest is

$$P(1 + R/100)^n$$

This formula is a geometric progression whose first term is $P$ (when $n = 0$) and whose common ratio is $(1 + R/100)$. *Compare* simple interest.

**compound proposition** *See* proposition.

**compound transformation** A transformation on a figure that can be regarded as one transformation followed by another transformation. For example, a compound transformation could consist of a rotation followed by a translation through space. Another example is a translation followed by a reflection. If the first and second transformations on the object O are denoted $T_1$ and $T_2$ respectively, then the compound transformation is denoted by $T_2T_1(O)$. In general, performing the two transformations in different order leads to two different compound transformations. This is consistent with the fact that transformations can be described by matrices.

**computability** /kŏm-pyoo-tă-**bil**-ă-tee/ Intuitively, a problem or function is computable if it is capable of being solved by an ideal machine (computer) in a finite time. In the 1930s it was discovered that some problems had no algorithmic solution and could not be solved by computers. This led many mathematicians to try to formulate a precise definition of the intuitive concept of computability, and Turing, Gödel, and Church independently came up with three very different abstract definitions, which all turned out to define exactly the same set of functions. The definition given by Turing is that a function f is computable if for each element $x$ of its domain, when some representation of $x$ is placed on the tape of a TURING MACHINE, the machine stops in a finite time with a representation of $f(x)$ on the tape.

**computer** Any automatic device or machine that can perform calculations and other operations on data. The data must be received in an acceptable form and is processed according to instructions. The most versatile and most widely used computer is the *digital computer*, which is usually referred to simply as a computer. *See also* analog computer; hybrid computer.

A digital computer is an automatically controlled calculating machine in which information, generally known as data, is represented by combinations of discrete electrical pulses denoted by the binary digits 0 and 1. Various operations, both arithmetical and logical, are performed on the data according to a set of instructions (a program). Instructions and data are fed into the main store or memory of the computer, where they are held until required. The instructions, coded like the data in binary form, are analyzed and carried out by the central processor of the computer. The result of this processing is then delivered to the user.

The technology used in digital computers is so highly advanced that they can operate at extremely high speeds and can store a huge amount of information. The tube valves used in early computers were replaced by transistors; transistors, resistors, etc., were subsequently packed into integrated circuits, which have become more and more complicated. As the electronic circuits used in the various devices in a computer system have decreased in size and increased in complexity, so the computers themselves have grown smaller, faster, and more powerful. The *microcomputer* has been developed as a somewhat simpler version of the full-size *mainframe* computer. Computers now have an immense range of uses in science, technology, industry, commerce, education, and many other fields.

**computer graphics** The creation and reproduction of pictures, photographs, and diagrams using a computer. There are many different formats for storing images but they fall into two main classes. In *raster graphics* the picture is stored as a series of dots (or pixels). The information in the computer file is a stream of data indicating the presence or absence of a dot and the color if present. Images of this type are sometimes known as *bitmaps*. This format is used for high-quality artwork and for photographs. Diagrams are more conveniently stored using *vector graphics*, in which the information is stored as mathematical instructions. For example, it is possible to specify a circle by its center, its radius, and the thickness of the line forming the circumference. More complicated curves are usually drawn using BEZIER CURVES. Vector images are easier to change and take up less storage space than raster images.

**computer modeling** The development of a description or mathematical representation (i.e. a *model*) of a complicated process or system, using a computer. This model can then be used to study the behavior or control of the process or system by varying the conditions in it, again with the aid of a computer.

**concave** /kong-**kayv**, **kong**-kayv/ Curved inwards. For example, the inner surface of a hollow sphere is concave. Similarly in two dimensions, the inside edge of the circumference of a circle is concave. A *concave polygon* is a polygon that has one (or

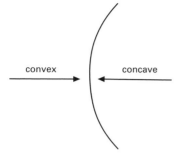

convex          concave

Concave and convex curvatures

more) interior angles greater than 180°. *Compare* convex.

**concentric** Denoting circles or spheres that have the same center. For example, a hollowed out sphere consists of two concentric spherical surfaces. *Compare* eccentric.

**conclusion** The proposition that is asserted at the end of an argument; i.e. what the argument sets out to prove.

**condition** In logic, a proposition or statement, *P*, that is required to be true in order that another proposition *Q* be true. If *P* is a *necessary condition* then *Q* could not be true without *P*. If *P* is a *sufficient condition*, then whenever *P* is true *Q* is also true, but not vice versa. For example, for a quadrilateral to be a rectangle it must satisfy the necessary condition that two of its sides be parallel, but this is not a sufficient condition. A sufficient condition for a quadrilateral to be a rhombus is that all its sides have a length of 5 centimeters, but this is not a necessary condition. For a rectangle to be a square it is both a necessary and a sufficient condition that all its sides are of equal length.

In formal terms, if *P* is a necessary condition for *Q*, then $Q \rightarrow P$. If *P* is a sufficient condition, then $P \rightarrow Q$. If *P* is a necessary and sufficient condition for *Q* then $P \equiv Q$. *See also* biconditional, symbolic logic.

**conditional** (conditional statement; conditional proposition) An *if... then...* statement.

**conditional convergence** *See* absolute convergence.

**conditional equation** *See* equation.

**conditional probability** *See* probability.

**cone** A solid defined by a closed plane curve (forming the base) and a point outside the plane (the vertex). A line segment from the vertex to a point on the plane curve generates a curved lateral surface as the point moves around the plane curve. The line is the *generator* of the cone and the plane curve is its *directrix*. Any line segment from the vertex to the directrix is an *element* of the cone.

If the directrix is a circle the cone is a *circular cone*. If the base has a center, a line from the vertex to this is an *axis* of the cone. If the axis is at right angles to the base the cone is a *right cone*; otherwise it is an *oblique cone*. The volume of a cone is one third of the base area multiplied by the altitude (the perpendicular distance from the vertex to the base). For a right circular cone

$$V = \pi r^2 h / 3$$

where *r* is the radius of the base and *h* the altitude. The area of the curved (lateral) surface of a right circular cone is $\pi rs$, where *s* is the length of an element (the *slant height*).

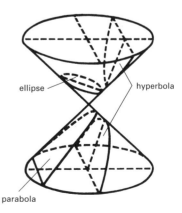

ellipse          hyperbola

parabola

The three sections of a cone – the ellipse, the parabola, and the hyperbola.

If an extended line is used to generate the curved surface (i.e. extending beyond the directrix and beyond the vertex), an extended surface is produced with two parts (*nappes*) each side of the vertex. More strictly, this is called a *conical surface*

**confidence interval** An interval that is thought, with a preselected degree of confidence, to contain the value of a parameter being estimated. For example, in a binomial experiment the $\alpha$% confidence interval for the probability of success $P$ lies between $P - a$ and $P + a$, where
$$a = z\sqrt{[P(1 - P)/N]}$$
$N$ is the sample size, $P$ the proportion of successes in the sample, and $z$ is given by a table of area under the standard normal curve. $P$ will lie in this interval $\alpha$ times out of every 100.

**confocal conics** /kon-**foh**-kăl/ Two or more conics that have the same focus.

**conformable matrices** *See* matrix.

**conformal mapping** A geometrical transformation that does not change the angles of intersection between two lines or curves. For example, Mercator's projection is a conformal mapping in which any angle between a line on the spherical surface and a line of latitude or longitude will be the same on the map.

**congruence** /**kong**-groo-ĕns/ The property of being congruent.

**congruent** /**kong**-groo-ĕnt/ **1.** Denoting two or more figures that are identical in size and shape. Two congruent plane figures will fit into the area occupied by each other; i.e. one could be brought into coincidence with the other by moving it without change of size. Two circles are congruent if they have the same radius. The conditions for two triangles to be congruent are: **1.** Two sides and the included angle of one are equal to two sides and the included angle of the other.
**2.** Two angles and the included side of one are equal to two angles and the included side of the other.

**3.** Three sides of one are equal to three sides of the other.

In solid geometry, two figures are congruent if they can be brought into coincidence in space.

Sometimes the term *directly congruent* is used to describe identical figures; *indirectly congruent* figures are ones that are mirror images of each other. *Compare* similar.
**2.** Two elements $a$ and $b$ of a ring are *congruent modulo d* if there exist elements in the ring, $p$, $q$, and $r$, such that $a = dp + r$, $b = dq + r$. Intuitively, this means that they both leave the same remainder when divided by $d$.
**3.** Two square matrices $A$ and $B$ are congruent if $A$ can be transformed into $B$ by a *congruent transformation*; i.e. there exists a nonsingular matrix $C$ such that $B = C^T A C$, where $C^T$ is the transpose of $C$.

**conic** /**kon**-ik/ A type of plane curve defined so that for all points on the curve the distance from a fixed point (the *focus*) has a constant ratio to the perpendicular distance from a fixed straight line (the *directrix*). The ratio is the eccentricity of the conic, $e$; i.e. the eccentricity is the distance from curve to focus divided by distance from curve to directrix.

The type of conic depends on the value of $e$: when $e$ is less than 1 it is an ELLIPSE; when $e$ equals 1 it is a PARABOLA; when $e$ is greater than 1 it is a HYPERBOLA. A circle is a special case of an ellipse with eccentricity 0.

The original definition of conics was as plane sections of a conical surface – hence the name *conic section*. In a conical surface having an apex angle of $2\theta$, the cross-section on a plane that makes an angle $\theta$ with the axis of the cone, (i.e. a plane parallel to the slanting edge of the cone) is a parabola. A cross-section in a plane that makes an angle greater than $\theta$ with the axis is an ellipse. A cross-section in a plane making an angle less than $\theta$ with the axis is a hyperbola and because this plane cuts both halves (nappes) of the cone, the hyperbola has two arms.

There are various ways of writing the equation of a conic. In Cartesian coordinates:

$$(1 - e^2)x^2 + 2e^2qx + y^2 = e^2q$$

where the focus is at the origin and the directrix is the line $x = q$ (a line parallel to the $y$-axis a distance $q$ from the origin). The general equation of a conic (i.e. the *general conic*) is:

$$ax^2 + bxy + cy^2 + dx + ey + f = 0$$

where $a$, $b$, $c$, $d$, $e$, and $f$ are constants (here $e$ is *not* the eccentricity). This includes degenerate cases (*degenerate conics*), such as a point, a straight line, and a pair of intersecting straight lines. A point, for example, is a section through the vertex of the conical surface. A pair of intersecting straight lines is a section down the axis of the surface. The tangent to the general conic at the point $(x_1, y_1)$ is:

$$ax_1x + b(xy_1 + x_1y) + cy_1y + d(x + x_1) + e(y + y_1) + f = 0$$

**conical helix** *See* helix.

**conical projection** *See* central projection.

**conical surface** *See* cone.

**conicoid** A type of surface in which sections of the surface are conics. Conicoids can have equations in three-dimensional Cartesian coordinates as follows:
elliptic paraboloid
$$x^2/a^2 + y^2/b^2 = 2z/c$$
hyperbolic paraboloid
$$x^2/a^2 - y^2/b^2 = 2z/c$$
hyperboloid
$$x^2/a^2 + y^2/b^2 - z^2/c^2 = 1$$
ellipsoid
$$x^2/a^2 + y^2/b^2 + z^2/c^2 = 1$$

Conicoids are also known as *quadric surfaces* or *quadrics*. The simplest types are conicoids of revolution, formed by rotating a conic about an axis. A sphere is a special case of an ELLIPSOID in which all three axes are equal (just as a circle is a special case of an ellipse). *See also* paraboloid. *See illustration overleaf.*

**conic sections** *See* conic.

**conjugate angles** A pair of angles that add together to make a complete revolution (360° or $2\pi$ radians). *Compare* complementary angles; supplementary angles.

Conjugate angles: $\alpha + \beta = 360°$

**conjugate axis** *See* hyperbola.

**conjugate complex numbers** Two complex numbers of the form $x + iy$ and $x - iy$, which when multiplied together have a real product $x^2 + y^2$. If $z = x + iy$, the complex conjugate of $z$ is $\bar{z} = x - iy$.

**conjugate diameter** Two diameters AB and CD of an ellipse are said to be conjugate diameters if CD is the diameter that contains the mid-points of the set of chords parallel to AB, where a diameter of an ellipse is defined to be a locus of the mid-points of a set of parallel chords. For an equation for an ellipse of the form $x^2/a^2 + y^2/b^2 = 1$ the product of the gradients of two conjugate diameters is $-b^2/a^2$. This means that, in general, conjugate diameters are not perpendicular (except in the case of $a = b$; i.e. a circle).

**conjugate hyperbola** *See* hyperbola.

**conjunction** /kŏn-**junk**-shŏn/ Symbol: ∧ In logic, the relationship *and* between two

| $P$ | $Q$ | $P \wedge Q$ |
|-----|-----|-----|
| T | T | T |
| T | F | F |
| F | T | F |
| F | F | F |

Conjunction

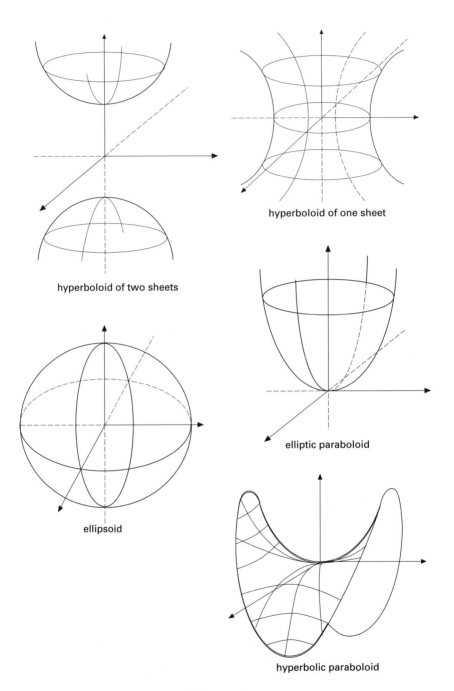

hyperboloid of one sheet

hyperboloid of two sheets

ellipsoid

elliptic paraboloid

hyperbolic paraboloid

Conicoid surfaces

or more propositions or statements. The conjunction of $P$ and $Q$ is true when $P$ is true and $Q$ is true, and false otherwise. The truth table definition for conjunction is shown in the illustration. *Compare* disjunction. *See also* truth table.

**connected** Intuitively, a connected set is a set with only one piece. More rigorously, a set in a topological space is said to be connected if it is not the union of two non-empty disjoint closed sets. For example, the set of rational numbers is not connected since the set of all rational numbers less than $\sqrt{3}$ and the set of all rational numbers greater than $\sqrt{3}$ are both closed in the set of all rational numbers. However, the set of all real numbers is connected since no such decomposition is possible. A subset $S$ of a topological space is *path-connected* if any two points in $S$ can be joined by a *path* in $S$, where a path from $a$ to $b$ in $S$ is a continuous map $f:[0,1] \rightarrow S$ such that $f(0)=a$ and $f(1)=b$.

A *simply connected* set is a path-connected set such that any closed curve within it can be deformed continuously to a point of the set without leaving the set. A path-connected set that is not simply connected is *multiply connected* and the *connectivity* of the set is one plus the maximum number of points that can be deleted if the set is a curve, or one plus the maximum number of closed cuts that can be made if the set is a surface, without separating the set so that it is no longer path-connected. For example, the region between two concentric circles has *connectivity two*, or is *doubly connected*, since one closed cut can be made which still leaves a connected region.

**connected particles** Particles that are connected by a light inextensible string. If particles are connected then their motions are not independent. An example is the motion of two particles joined by a light inextensible string which passes over a fixed pulley. If the two particles have masses $m_1$ and $m_2$ then the heavier particle moves up. If $m_2 > m_1$, the acceleration down of $m_2$ (and the acceleration up of $m_1$) = $a$, the tension in the string is denoted $T$ and the ac-celeration due to gravity is denoted $g$ then: $a = [(m_2 - m_1)g]/(m_1 + m_2)$ and $T = 2m_1m_2g/(m_1 + m_2)$.

**connectivity** The number of cuts needed to break a shape in two parts. For example, a rectangle, a circle, and a sphere, all have a connectivity of one. A flat disk with a hole in it or a torus has a connectivity of two. *See also* topology.

**consequent** In logic, the second part of a conditional statement; a proposition or statement that is said to follow from or be implied by another. For example, in the statement 'if Jill is happy, then Jack is happy', 'Jack is happy' is the consequent. *Compare* antecedent. *See also* implication.

**conservation law** A law stating that the total value of some physical quantity is conserved (i.e. remains constant) throughout any changes in a closed system. The conservation laws applying in mechanics are the laws of constant mass, constant energy, constant linear momentum, and constant angular momentum.

**conservation of angular momentum, law of** *See* constant angular momentum; law of.

**conservation of energy, law of** *See* constant energy; law of.

**conservation of linear momentum, law of** *See* constant linear momentum; law of.

**conservation of mass, law of** *See* constant mass; law of.

**conservation of mass and energy** The law that the total energy (rest mass energy + kinetic energy + potential energy) of a closed system is constant. In most chemical and physical interactions the mass change is undetectably small, so that the measurable rest-mass energy does not change (it is regarded as 'passive'). The law then becomes the classical *law of conservation of energy*. In practice, the inclusion of mass in the calculation is necessary only in the case of nuclear changes or systems involving very high speeds. *See also* mass-energy equation; rest mass.

**conservation of momentum, law of**
*See* constant linear momentum; law of.

**conservative field** A field such that the work done in moving an object between two points in the field is independent of the path taken. *See* conservative force.

**conservative force** A force such that, if it moves an object between two points, the energy transfer (work done) does not depend on the path between the points. It must then be true that if a conservative force moves an object in a closed path (back to the starting point), the energy transfer is zero. Gravitation is an example of a conservative force; friction is a non-conservative force.

**consistent** Describing a theory, system, or set of propositions giving rise to no contradictions. Arithmetic, for example, is thought to be a consistent logical system because none of its axioms nor any of the theorems that are derived from these by the rules, are believed to be contradictory. *See* contradiction.

**consistent equations** A set of equations that can be satisfied by at least one set of values for the variables. For example, the equations $x + y = 2$ and $x + 4y = 6$ are satisfied by $x = 2/3$ and $y = 4/3$ and they are therefore consistent. The equations $x + y = 4$ and $x + y = 9$ are inconsistent.

**constant** A quantity that does not change its value in a general relationship between variables. For example, in the equation $y = 2x + 3$, where $x$ and $y$ are variables, the numbers 2 and 3 are constants. In this case they are *absolute constants* because their values never change. Sometimes a constant can take any one of a number of values in different applications of a general formula. In the general quadratic equation
$$ax^2 + bx + c = 0$$
$a$, $b$, and $c$ are *arbitrary constants* because no values are specified for them. An INDEF-

INITE INTEGRAL includes an arbitrary constant (the *constant of integration*), which depends on the limits chosen.

**constant angular momentum, law of** (**law of conservation of angular momentum**) The principle that the total angular momentum of a system cannot change unless a net outside torque acts on the system. *See also* constant linear momentum; law of.

**constant energy, law of** (**law of conservation of energy**) The principle that the total energy of a system cannot change unless energy is taken from or given to the outside. *See also* mass-energy equation.

**constant linear momentum, law of** (**law of conservation of linear** momentum) The principle that the total linear momentum of a system cannot change unless a net outside force acts.

**constant mass, law of** (**law of conservation of mass**) The principle that the total mass of a system cannot change unless mass is taken from or given to the outside. *See also* mass-energy equation.

**constant momentum, law of** *See* constant linear momentum; law of.

**construct** In geometry, to draw a figure, line, point, etc., meeting certain conditions; e.g. a line that bisects a given line. Usually certain specific restrictions are imposed on the method used; e.g. using only a straight edge and compasses. There is an important class of problems concerning questions of whether certain things can be constructed using given methods. Examples are two celebrated problems of whether it is possible to construct two lines that trisect a given angle, and to construct a square equal in area to a given circle – in both cases using only a straight edge and compasses. Both these constructions have been shown to be impossible.

**constructive proof** A proof that not only shows that a certain mathematical entity, such as a root of an equation or a fixed-point of a transformation, exists, but also explicitly produces it. Constructive proofs are usually considerably longer and more complicated and harder to find than nonconstructive proofs of the same results. Many results that have been proved nonconstructively have yet to be given constructive proofs.

**constructivist mathematics** An approach to mathematics that insists that only constructive proofs are acceptable and rejects as meaningless nonconstructive proofs. Constructivist mathematics is considerably more restricted than classical mathematics and rejects many of its theorems. Different varieties of constructivism differ over what exactly counts as an acceptably constructive proof. One of the best known examples of mathematical constructivism is the intuitionism of Brouwer.

**contact force** A force between bodies in which the bodies are in contact. By contrast, in a *non-contact force* such as the gravitational force between the Earth and the Sun, the two bodies are separated by 'empty' space. Examples of contact forces include *tension* in which a particle is hanging in equilibrium at the end of a string and *thrust* in which a particle is suspended by a spring underneath it. In the first case the tension in the string acts upwards on the particle, as does the thrust in the spring in the second case.

**continued fraction** An expression that can be written in the form:

$$b_0 + \cfrac{a_1}{b_1 + \cfrac{a_2}{b_2 + \cfrac{a_3}{b_3 + \cdots}}}$$

where all the $a$'s and $b$'s are numbers (which are usually positive integers). If this fraction terminates after a finite number of terms it is said to be *terminating* or *finite*. If it does not terminate it is said to be *nonterminating* or *infinite*. The values of a continued fraction that are obtained by truncating the fraction at a finite point

such as $b_0$, $b_1$, $b_2$, etc., are called *convergents*.

**continued product** Symbol: $\Pi$  The product of a number of related terms. For example, $2 \times 4 \times 6 \times 8...$ is a continued product, written:

$$\prod^{k} a_n$$

This means the product of $k$ terms, with the $n$th term, $a_n = 2n$.

$$\prod_{1}^{\infty} a_n$$

has an infinite number of terms.

**continuous function** A function that has no sudden changes in values as the variable increases or decreases smoothly. More precisely, a function $f(x)$ is continuous at a point $x = a$ if the limit of $f(x)$ as $x$ approaches $a$ is $f(a)$. When a function does not satisfy this condition at a point, it is said to be *discontinuous*, or to have a *discontinuity*, at that point. For example, $\tan\theta$ has discontinuities at $\theta = \pi/2, 3\pi/2, 5\pi/2,...$  A function is continuous in an interval of $x$ if there are no points of discontinuity in that interval.

**continuous stationery** A length of fanfolded paper with sprocket holes along each side for transporting it through the printer of a computer. It may be perforated for tearing it into separate sheets after printing; there may also be perforations along the sides to tear off the sprocket holes.

**continuum** /kŏn-**tin**-yoo-ŭm/  (*pl.* continua)  A compact connected set with at least two points. The conditions that the set has at least two points and is connected imply that the set has an infinite number of points. Any closed interval of the real numbers is a continuum and the set of all real numbers is called the *real continuum*.

The *continuum hypothesis* is the conjecture that every infinite subset of the real continuum has the CARDINAL NUMBER either of the positive integers or of the entire set of real numbers. This is equivalent to

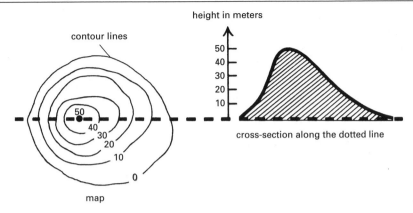

A hill shown as contour lines on a map and as a cross-section.

the statement that $2\aleph_0$ is the lowest cardinal number greater than $\aleph$. *See* aleph.

**contour integral** *See* line integral.

**contour line** A line on a map joining points of equal height. Contour lines are usually drawn for equal intervals of height, so that the steeper a slope, the closer together the contour lines. *See illustration overleaf.*

**contradiction** In LOGIC, a proposition, statement, or sentence that both asserts something *and* denies it. It is a form of words or symbols that cannot possibly be true; for example, 'if I can read the book then I cannot read the book' and 'he is coming and he is not coming'. *Compare* tautology.

**contradiction, law of** *See* laws of thought.

**contrapositive** /kon-tră-**poz**-ă-tiv/ In logic, a statement in which the antecedent and consequent of a conditional are reversed and negated. The contrapositive of $A \rightarrow B$ is $\sim B \rightarrow \sim A$ (not B implies not A), and the two statements are logically equivalent. *See* implication. *See also* biconditional.

**control theory** A branch of applied mathematics that is concerned with trying to obtain a specific type of desired dynamical behavior from a physical system. Some device is added to the system to affect its behavior, and the concept of feedback is applied very widely in control systems. There are many important applications of control theory, including industrial machinery, robots, chemical processes, and the stability of cars, trains, and aircraft. *See also* cybernetics.

**control unit** *See* central processor.

**convergent sequence** A sequence in which the difference between each term and the one following it becomes smaller throughout the SEQUENCE; i.e. the difference between the $n$th term and the $(n + 1)$th term decreases as $n$ increases. For example, $\{1, \frac{1}{2}, \frac{1}{4}, \frac{1}{8}, ...\}$ is a convergent sequence, but $\{1, 2, 4, 8, ...\}$ is not. A convergent sequence has a limit; i.e. a value towards which the $n$th term tends as $n$ becomes infinitely large. In the first example here the limit is 0. *Compare* divergent sequence. *See also* convergent series; geometric sequence.

**convergent series** An infinite SERIES $a_1 + a_2 + ...$ is *convergent* if the partial sums $a_1 + a_2 + ... + a_n$ tend to a limit value as $n$ tends to infinity. For example, the series $S = 1 + \frac{1}{2} + \frac{1}{4} + \frac{1}{8} + ...$ is a convergent series with sum 2, since 2 is the limit approached by the sum of the first $n$ terms, namely $1 - (1/2^n)$ as $n$ tends to infinity. The series 1

+ (−1) + 1 + (−1) + 1 + ... is not convergent. *Compare* divergent series. *See also* convergent sequence; geometric series.

**converse** A logical IMPLICATION taken in the reverse order. For example, the converse of
if I am under 16, then I go to school
is
if I go to school, then I am under 16.
The converse of an implication is not always true if the implication itself is true. There are a number of theorems in mathematics for which both the statement and the converse are true. For example, the theorem:
if two chords of a circle are equidistant from the center, then they are equal
has a true converse:
if two chords of a circle are equal, then they are equidistant from the center of the circle.

**conversion factor** The ratio of a measurement in one set of units to the equivalent numerical value in other units. For example, the conversion factor from inches to centimeters is 2.54 because 1 inch = 2.54 centimeters (to two decimal places).

**conversion graph** A graph showing a relationship between two variable quantities. If one quantity is known, the corresponding value of the other can be read directly from the graph. For example, air pressure depends on height above sea level. A standard curve of altitude against air pressure may be plotted on a graph. An air-pressure measurement can then be converted to an indication of height by reading the appropriate value from the graph.

**convex** /kon-**veks**, **kon**-veks/ Curved outwards. For example, the outer surface of a sphere is convex. Similarly, in two dimensions, the outside of a circle is its convex side. A *convex polygon* is one in which no interior angle is greater than 180°. *Compare* concave.

**coordinate geometry** *See* analytical geometry.

**coordinates** Numbers that define the position of a point, or set of points. A fixed point, called the *origin*, and fixed lines, called *axs*, are used as a reference. For example, a horizontal line and a vertical line drawn on a page might be defined as the $x$-axis and the $y$-axis respectively, and the point at which they cross as the origin (O). Any point on the page can then be given two numbers – its distance from O along the $x$-axis and its distance from O along the $y$-axis. Distances to the right of the origin for $x$ and above the origin for $y$ are positive; distances to the left of the origin for $x$ and below the origin for $y$ are negative. These two numbers would be the $x$ and $y$ coordinates of the point. This type of coordinate system is known as a rectangular Cartesian coordinate system. It can have two axs, as on a flat surface, such as a map, or three axs, when depth or height also have to be specified. Another type of coordinate system (POLAR COORDINATES) expresses the position of a point as radial distance from the origin (the *pole*), with its direction expressed as an angle or angles (positive when anticlockwise) between the radius and a fixed axis (the *polar axis*). *See also* Cartesian coordinates.

height in meters above sea level

4000

3000

2000

1000

0

0.06  0.07  0.08  0.09  0.10
air pressure in megapascals (MPa)

A conversion graph for finding altitude from air pressure measurements. (Standard air pressure at sea level is 0.101325 megapascals.)

**coplanar** /koh-**play**-ner/ Lying in the same plane. Any set of three points, for example, could be said to be coplanar because there is a plane in which they all lie. Two vectors are coplanar if there is a plane that contains both.

**coplanar forces** Forces in a single plane. If only two forces act through a point, they must be coplanar. So too are two parallel forces. However, nonparallel forces that do not act through a point cannot be coplanar. Three or more nonparallel forces acting through a point may not be coplanar. If a set of coplanar forces act on a body, their algebraic sum must be zero for the body to be in equilibrium (i.e. the resultant in one direction must equal the resultant in the opposite direction). In addition there must be no couple on the body (the moment of the forces about a point must be zero).

**Coriolis force** /kor-ee-**oh**-lis, ko-ree-/ A 'fictitious' force used to describe the motion of an object in a rotating system. For instance, air moving from north to south over the surface of the Earth would, to an observer outside the Earth, be moving in a straight line. To an observer on the Earth the path would appear to be curved, as the Earth rotates. Such systems can be described by introducing a tangential Coriolis 'force'. The idea is used in meteorology to explain wind directions. It is named for the French physicist Gustave-Gaspard Coriolis (1792–1843).

**corollary** /kŏ-**rol**-ă-ree/ A result that follows easily from a given theorem, so that it is not necessary to prove it as a separate theorem.

**correction** A quantity added to a previously obtained approximation to yield a better approximation. When using logarithmic or trigonometric tables the correction is the number added to a logarithm or to a trigonometric function in the table to give the logarithm or trigonometric function of a number or angle that is not in the table.

**correlation** In statistics, the *correlation coefficient* of two random variables $X$ and $Y$ is defined by

$$r(X,Y) = cov(X,Y)/\sqrt{(var(X)var(Y))}$$

where cov and var denote covariance and variance respectively. It satisfies $-1 \leq r \leq 1$ and is a measurement of the interdependence between random variables, or their tendency to vary together. If $r \neq 0$ then $X$ and $Y$ are said to be *correlated*: they are correlated *positively* if $0 < r \leq 1$ and *negatively* if $-1 \leq r < 0$. If $r = 0$ then $X$ and $Y$ are said to be *uncorrelated*.

For two sets of numbers $(x_1,...x_n)$ and $(y_1,...y_n)$ the correlation coefficient is

$$r = \frac{\sum_{i=1}^{n} (x_i - \bar{x})(y_i - \bar{y})}{\sqrt{[\sum_{i=1}^{n} (x_i - \bar{x})^2 \sum_{i=1}^{n} (y_i - \bar{y})^2]}}$$

where $\bar{x}$ and $\bar{y}$ are the corresponding means. It measures how near the points $(x_1,y_1)...(x_n,y_n)$ are to lying on a straight line. If $r = 1$ the points lie on a line and the two sets of data are said to be in *perfect correlation*.

**correspondence** *See* function.

**corresponding angles** A pair of angles on the same side of a line (the transversal) that intersects two other lines; they are between the transversal and the other lines. If the intersected lines are parallel, the corresponding angles are equal. *Compare* alternate angles.

**cos** /koz/ *See* cosine.

**cosec** /koh-sek/ *See* cosecant.

**cosecant** /koh-**see**-kănt, -kant/ (**cosec; csc**) A trigonometric function of an angle equal to the reciprocal of its sine; i.e. cosecα = 1/sinα. *See also* trigonometry.

**cosech** /**koh**-sech, -sek/ A hyperbolic cosecant. *See* hyperbolic functions.

**cosh** /kosh, kos-**aych**/ A hyperbolic cosine. *See* hyperbolic functions.

**cosine** /**koh**-sÿn/ (**cos**) A trigonometric function of an angle. The cosine of an angle α (cosα) in a right-angled triangle is the ratio of the side adjacent to it, to the hy-

Cosine: The graph of $y = \cos x$.

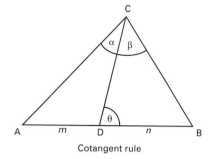

Cotangent rule

potenuse. This definition applies only to angles between 0° and 90° (0 and π/2 radians). More generally, in rectangular Cartesian coordinates, the $x$-coordinate of any point on the circumference of a circle of radius $r$ centred on the origin is $r\cos\alpha$, where $\alpha$ is the angle between the $x$-axis and the radius to that point. In other words, the cosine function is the horizontal component of a point on a circle. Cosα varies periodically in the same way as sinα, but 90° ahead. That is: cosα is 1 when α is 0°, falls to zero when α = 90° (π/2) and then to –1 when α = 180° (π), returning to zero at α = 270° (3π/2) and then to +1 again at α = 360° (2π). This cycle is repeated every complete revolution. The cosine function has the following properties:

$$\cos\alpha = \cos(\alpha + 360°) = \sin(\alpha + 90°)$$
$$\cos\alpha = \cos(-\alpha)$$
$$\cos(90° + \alpha) = -\cos\alpha$$

The cosine function can also be defined as an infinite series. In the range from +1 to –1:

$$\cos x = 1 - x^2/2! + x^4/4! - x^6/6! + \ldots$$

**cosine rule** In any triangle, if $a$, $b$, and $c$ are the side lengths and γ is the angle opposite the side of length $c$, then
$$c^2 = a^2 + b^2 - 2ab\cos\gamma$$

**cot** /kot/ *See* cotangent.

**cotangent** /koh-**tan**-jĕnt/ (**cot**) A trigonometric function of an angle equal to the reciprocal of its tangent; i.e. cotα = 1/tanα. *See also* trigonometry.

**cotangent rule** A rule for triangles that states that if the side AB of a triangle is divided into the ratio $m{:}n$ by a point D then $(m + n) \cot\theta = m\cot\alpha - n\cot\beta$, where θ, α, and β are defined in the diagram. The

cotangent rule can also be expressed in the form: $(m + n)\cot\theta = n\cot A - m\cot B$, where $A$ is the angle at A and $B$ the angle at B.

An important special case of the cotangent rule is that in which D is the mid-point of AB. In this case the rule becomes $2\cot\theta = \cot\alpha - \cot\beta$.

The cotangent rule can be used to analyze some problems in mechanics involving equilibrium, particularly in the case when D is the mid-point of AB.

**coth** /koth/ A hyperbolic cotangent. *See* hyperbolic functions.

**coulomb** /koo-**lom**/ Symbol: C The SI unit of electric charge, equal to the charge transported by an electric current of one ampere flowing for one second. 1 C = 1 A s. The unit is named for the French physicist Charles Augustin de Coulomb (1736–1806).

**countable** (**denumerable**) A set is countable if it can be put in one-one correspondence with the integers. The set of rational numbers, for example, is countable whereas the set of real numbers is not (*see* Cantor's diagonal argument). To show that the rationals are countable we need to show how they can be arranged in a series such that every rational number will be included somewhere.

If we consider the array in the diagram it is clear that every rational number will occur in it somewhere. But by starting with 1/1 and following the path indicated we can enumerate every number in the array. If we reduce each fraction to its lowest terms and then remove any that has al-

| | 1 | 2 | 3 | 4 | 5 | ... |
|---|---|---|---|---|---|---|
| 1 | 1/1 | 1/2 | 1/3 | 1/4 | 1/5 | ... |
| 2 | 2/1 | 2/2 | 2/3 | 2/4 | 2/5 | ... |
| 3 | 3/1 | 3/2 | 3/3 | 3/4 | 3/5 | ... |
| 4 | 4/1 | 4/2 | 4/3 | 4/4 | 4/5 | ... |
| 5 | 5/1 | 5/2 | 5/3 | 5/4 | 5/5 | ... |
| | : | : | : | : | : | |

Countable

ready occurred in the list it is clear that this will give a list in which each rational number occurs once and only once.

**counterclockwise** *See* anticlockwise.

**couple** A pair of equal parallel forces in opposite directions and not acting through a single point. Their linear resultant is zero, but there is a net turning effect (moment). The net turning effect $T$ (the torque) is given by:
$$T = Fd_1 + Fd_2$$
$F$ being the magnitude of each force and $d_1$ and $d_2$ the distances from any point to the lines of action of each force. This is equivalent to:
$$T = Fd$$
where $d$ is the distance between the forces.

**covariance** /koh-**vair**-ee-ăns/ A statistic that measures the association between two variables. If for $x$ and $y$ there are $n$ pairs of values $(x_n, y_i)$, then the covariance is defined as
$$[1/(n-1)]\Sigma(x_i - x')(y_i - y')$$
where $x'$ and $y'$ are the mean values.

**CPU** /see-pee-**yoo**/ *See* central processor.

**Cramer's rule** /**kray**-merz/ A rule that gives the solution of a set of linear equations in terms of a matrix. If a set of $n$ linear equations for $n$ unknown variables $x_1$, $x_2$, ... $x_n$ can be written in the form of a matrix equation $Ax = b$, where $A$ is an invertible matrix the solution for the equations can be written uniquely as $x = A^{-1} = b$. Cramer's rule states that the $x_i$ can be written as:
$$x_i = (b_1A_{1i} + b_2A_{2i} + ... b_nA_{ni})/(\det A),$$
for all $i$ between 1 and $n$, where the $b_j$ are the entries in the $b$ column and the $A_{ji}$ are the cofactors of $A$. This is the case because $A^{-1} = \text{adj}A/\det A$, where adj$A$ is the *adjoint* of the matrix $A$. In turn, this means that $x = (\text{adj } A) b/\det A$. In the case of two linear equations with two unknowns $x_1$, and $x_2$ of the form
$$ax_1 + bx_2 = l,$$
$$cx_1 + dx_2 = m,$$
with $ad - bc \neq 0$, Cramer's rule leads to the result:
$$x_1 = (ld-bm)/(ad-bc),$$
$$x_2 = (am-lc)/(ad-bc).$$

**critical damping** *See* damping.

**critical path** The sequence of operations that should be followed in order to complete a complicated process, task, etc., in the minimum time. It is usually determined by using a computer.

**critical region** *See* acceptance region.

**cross multiplication** A way of simplifying an equation in which one or both terms are fractions. The product of the numerator on the left-hand side of the equation and the denominator on the right-hand side equals the product of the denominator on the left-hand side and the numerator on the right-hand side. For example, cross multiplying the equation
$$\frac{4x}{3} = \frac{3y}{2}$$
gives
$$4x \times 2 = 3y \times 3$$
or
$$8x - 9y$$

**cross product** *See* vector product.

**cross-section** (section) A plane cutting through a solid figure or the plane figure produced by such a cut. For example, the cross-section through the middle of a sphere is a circle. A vertical cross-section through an upright cone and off the axis is a hyperbola.

**crystallographic group** /kris-tă-lŏ-**graf**-ik/ A POINT GROUP that is compatible with the symmetry of a crystal. This restriction means that there are 32 possible point groups. Crystallographic groups can only have twofold, threefold, fourfold, or sixfold rotational symmetry since any other rotational symmetry would be incompatible with the translational symmetry existing in a crystal. However, fivefold and icosahedral symmetries are possible in QUASICRYSTALLINE SYMMETRY. The SPACE GROUP of a crystal is characterized by the combination of the crystallographic point group and the translational symmetry of the crystal lattice.

**crystallographic symmetry** The symmetry that is associated with the regular three-dimensional structure of a crystal. The set of crystallographic symmetry operations of a crystal makes up the SPACE GROUP of the crystal. This consists of the symmetry operations of the CRYSTALLOGRAPHIC GROUP and the symmetry operations associated with the translational symmetry of the crystal.

**csc** *See* cosecant.

**cube 1.** The third power of a number or variable. The cube of $x$ is $x \times x \times x = x^3$ ($x$ cubed).
**2.** In geometry, a solid figure that has six square faces. The volume of a cube is $l^3$, where $l$ is the length of a side.

**cube root** An expression that has a third power equal to a given number. The cube root of 27 is 3, since $3^3 = 27$.

**cube root of unity** A complex number $z$ for which $z^3 = 1$. There are three cube roots of unity, which can be denoted as 1, $\omega$, and $\omega^2$, where $\omega$ and $\omega^2$ are defined by
$$\omega = \exp(2\pi i/3) = \cos(2\pi/3) + i \sin (2\pi/3)$$
$$= + 1/2 - \sqrt{3}i/2,$$
$$\omega^2 = \exp(4\pi i/3) = \cos(4\pi/3) + i \sin (4\pi/3) = -1/2 - \sqrt{3}i/2.$$

The cube roots of unity satisfy the following relations:
$$\bar{\omega} = \omega^2$$
$$\omega^2 + \omega + 1 = 0$$

**cubic equation** A polynomial equation in which the highest power of the unknown variable is three. The general form of a cubic equation in a variable $x$ is
$$ax^3 + bx^2 + cx + d = 0$$
where $a$, $b$, $c$, and $d$ are constants. It is also sometimes written in the reduced form
$$x^3 + bx^2/a + cx/a + d/a = 0$$
In general, there are three values of $x$ that satisfy a cubic equation. For example,
$$2x^3 - 3x^2 - 5x + 6 = 0$$
can be factorized to
$$(2x + 3)(x - 1)(x - 2) = 0$$
and its solutions (or roots) are $-3/2$, 1, and 2. On a Cartesian coordinate graph, the curve
$$y = 2x^3 - 3x^2 - 5x + 6$$
crosses the $x$-axis at $x = -3/2$, $x = +1$, and $x = +2$.

**cubic graph** A graph of the equation
$$y = ax^3 + bx^2 + cx + d,$$
where $a$, $b$, $c$, and $d$ are constants. As $x$ becomes very large $y \to ax^3$. This means that if $a > 0$ then $y \to \infty$ as $x \to \infty$ and $y \to -\infty$ as $x \to -\infty$ and that if $a < 0$ then $y \to -\infty$ as $x \to \infty$ and $y \to \infty$ as $x \to -\infty$. Since $dy/dx = 3ax^2 + 2bx + c$ the nature of the stationary points of the cubic graph are determined by the nature of the solutions of the quadratic equation $3ax^2 + 2bx + c = 0$. If the quadratic equation has two real roots that are distinct then the curve has two turning points that are distinct, one of which is a maximum and one of which is a minimum. If the quadratic equation has two real roots that are equal the curve has a point of inflexion. If the quadratic equation does not have any real roots the curve does not have any real stationary points. The curve is continuous since $y$ does not tend to infinity for any finite value of $x$. A cubic curve can either: (1) cross the $x$-axis three times; (2) cross and touch the $x$-axis; (3) cross the $x$-axis once; or (4) touch the $x$-axis at the point of inflexion. This means that there has to be at least one point of intersection with the $x$-axis. This result means, in turn,

that a CUBIC EQUATION $ax^3 + bx^2 + cx + d = 0$ must have at least one real root.

**cuboid** A box-shaped solid figure bounded by six rectangular faces. The opposite faces are congruent and parallel. At each of the eight vertices, three faces meet at right angles to each other. The volume of a cuboid is its length, $l$, times its breadth, $b$, times its height, $h$. The surface area is the sum of the areas of the faces, that is
$$2(l \times b) + 2(b \times h) + 2(l \times h)$$
In the special case in which $l = b = h$, all the faces are square and the cuboid is a cube of volume $l^3$ and surface area $6l^2$.

**cumulative distribution** *See* distribution function.

**cumulative frequency** The total frequency of all values up to and including the upper boundary of the class interval under consideration. *See also* frequency table.

**curl** (Symbol $\nabla$) A vector operator on a vector function that, for a three-dimensional function, is equal to the sum of the vector product (cross product) of the unit vectors and partial derivatives in each of the component directions. That is:
$$\text{curl } F = \nabla F = i \times \partial F/\partial x + j \times \partial F/\partial y + k \times \partial F/\partial z$$
where $i$, $j$, and $k$ are the unit vectors in the $x$, $y$, and $z$ directions respectively. In physics, the curl of a vector arises in the relationship between electric current and magnetic flux, and in the relationship between the velocity and angular momentum of a moving fluid. *See also* div; grad.

**curvature** The rate of change of the slope of the tangent to a curve, with respect to distance along the curve. For each point on a smooth curve there is a circle that has the same tangent and the same curvature at that point. The radius of this circle, called the *radius of curvature*, is the reciprocal of the curvature, and its center is known as the *center of curvature*. If the graph of a function $y = f(x)$ is a continuous curve, the slope of the tangent at any point is given by the derivative $dy/dx$ and the curvature is given by:

$$(d^2y/dx^2)/[1 + (dy/dx)^2]^{3/2}$$

**curve** A set of points forming a continuous line. For example, in a graph plotted in Cartesian coordinates, the curve of the equation $y = x^2$ is a parabola. A curved surface may similarly represent a function of two variables in three-dimensional coordinates.

**curve sketching** Sketching the graph of a function $y = f(x)$ in such a way as to indicate the main features of interest of that curve. This usually involves determining the general shape of the curve and investigating how it behaves at points of special interest. To be more specific, symmetry of the curve about the $x$-axis and the $y$-axis are investigated, as is symmetry about the origin. The behavior as $x$ and $y$ become very large is investigated in both the positive and negative directions. The points at which the curve cross the $x$-axis and the $y$-axis are established. The problems of whether there are any values of $x$ for which $y$ is infinite and any values of $y$ for which $x$ is infinite are investigated. The nature of any stationary points is also investigated. There are also a number of other features that can be looked at, including: establishing the intervals for which the function is always decreasing or always increasing; the concave or convex nature of the curve; and all the asymptotes of the curve. Usually curve sketching is done using Cartesian coordinates. It is a way of visualizing how functions behave without calculating the exact values.

**curvilinear integral** /ker-vă-**lin**-ee-er/ *See* line integral.

**cusp** A sharp point formed by a discontinuity in a curve. For example, two semicircles placed side by side and touching form a cusp at which they touch.

**cut** *See* Dedekind cut.

**cybernetics** /sÿ-ber-**net**-iks/ The branch of science concerned with control systems, especially with regard to the comparisons between those of machines and those of human beings and other animals. In a series of operations, information gained at one stage can be used to modify later performances of that operation. This is known as *feedback* and enables a control system to check and possibly adjust its actions when required.

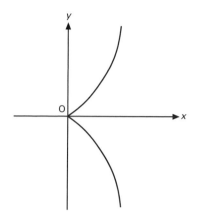

In this graph a cusp occurs at the origin O.

**cycle** A series of events that is regularly repeated (e.g. a single orbit, rotation, vibration, oscillation, or wave). A cycle is a complete single set of changes, starting from one point and returning to the same point in the same way.

**cyclic function** *See* periodic function.

**cyclic group** A group in which each element can be expressed as a power of any other element. For example, the set of all numbers that are powers of 3 could be written as $\{\ldots 3^{1/3},\ 3^{1/2},\ 3,\ 3^2,\ 3^3,\ \ldots\}$ or $\{\ldots 9^{1/6},\ 9^{1/4},\ 9^{1/2},\ 9,\ 9^{3/2},\ \ldots\}$, etc. *See also* Abelian group.

**cyclic polygon** A polygon for which there is a circle on which all the vertices lie. All triangles are cyclic. All regular polygons are cyclic. All squares and rectangles are cyclic quadrilaterals. However, not all quadrilaterals are cyclic. Convex quadrilaterals are cyclic if the opposite angles are supplementary. For a cyclic quadrilateral with sides of length *a, b, c,* and *d* (in order) the expression $(ac + bd)$ is equal to the product of the diagonals. This is known as *Ptolemy's theorem*, named for the Egyptian astronomer Ptolemy (*or* Claudius Ptolemaeus) (*fl.* 2nd century AD).

**cyclic quadrilateral** A four-sides figure whose corners (vertices) lie on a circumscribed circle. The opposite angles are supplementary, i.e. they add to 180°. *See* circumcircle; supplementary angles.

**cycloid** /sȳ-kloid/ The curve traced out by a point on a circle rolling along a straight line, for example, a point on the rim of a wheel rolling along the ground. For a circle

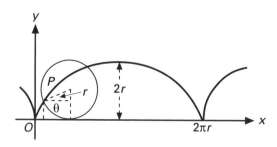

Cycloid

of radius $r$ along a horizontal axis, the cycloid produced is a series of continuous arcs that rises from the axis to a height $2\pi$ and fall to touch the axis again at a cusp point, where the next arc begins. The horizontal distance between successive cusps is $2\pi r$, the circumference of the circle. The length of the cycloid between adjacent cusps is $8r$. If $\theta$ is the angle formed by the radius to a point P $(x,y)$ on the cycloid and the radius to the point of contact with the $x$-axis, the parametric equations of the cycloid are:

$$x = r(\theta - \sin\theta)$$
$$y = r(1 - \cos\theta)$$

**cylinder** A solid defined by a closed plane curve (forming a base) with an identical curve parallel to it. Any line segment from a point on one curve to a corresponding point on the other curve is an *element* of the cylinder. If one of these elements moves parallel to itself round the base it sweeps out a curved lateral surface. The line is a *generator* of the cylinder and the plane closed curve forming the base is called the *directrix*

If the bases are circles the cylinder is a *circular cylinder*. If the bases have centers the line joining them is an axis of the cylinder. A *right cylinder* is one with its axis at right angles to the base; otherwise it is an *oblique cylinder*. The volume of a cylinder is $Ah$, where $A$ is the base area and $h$ the altitude (the perpendicular distance between the bases). For a right circular cylinder, the curved lateral surface area is $2\pi rh$, where $r$ is the radius.

If the generator is an indefinitely extended line it sweeps out an extended surface – a *cylindrical surface*

**cylindrical helix** *See* helix.

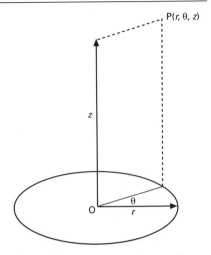

A point P($r,\theta,z$) in cylindrical polar coordinates.

**cylindrical polar coordinates** A method of defining the position of a point in space by its horizontal radius $r$ from a fixed vertical axis, the angular direction $\theta$ of the radius from an axis, and the height $z$ above a fixed horizontal reference plane. Starting at the origin O of the coordinate system, the point P($r,\theta,z$) is reached by moving out along a fixed horizontal axis to a distance $r$, following the circumference of the horizontal circle radius $r$ centred at O through an angle $\theta$, and then moving vertically upward by a distance $z$. For a point P($r,\theta,z$), the corresponding rectangular CARTESIAN COORDINATES $(x,y,z)$ are:

$$x = r\cos\theta$$
$$y = r\sin\theta$$
$$z = z$$

*Compare* spherical polar coordinates. *See also* coordinates; polar coordinates.

**cylindrical surface** *See* cylinder.

# D

**d'Alembertian** /**dal**-ahm-bair-ti-ăn/ Symbol $\Box^2$. An operator that acts on the function $u(x,y,z,t)$ for the displacement of a wave. It is related to the LAPLACIAN operator $\nabla^2$ by:
$$\Box^2 = \partial^2/\partial x^2 + \partial^2/\partial y^2 + \partial^2/\partial z^2 - (1/c^2)\partial^2/\partial t^2$$
$$= \nabla^2 - (1/c^2)\, \partial^2/\partial t^2,$$
where $c$ is the speed of the wave. Using this operator the WAVE EQUATION can be written very concisely as $\Box^2 u = 0$. Sometimes the d'Alembertian is taken to refer specifically to the wave equation for electromagnetic waves, with $c$ being the speed of light in that case. It is named for the French mathematician, encyclopedist, and philosopher Jean Le Rond d'Alembert (1717–83).

**d'Alembert's principle** /dal-ahm-**bairz**/ A principle that combines Newton's second and third laws of motion into the equation $F–ma = 0$. Using this principle, all problems involving forces can be treated as equilibrium problems, with the $–ma$ being regarded as an inertial reaction force. For bodies in motion this type of equilibrium is called *dynamic equilibrium* or *kinetic equilibrium* to distinguish it from *static equilibrium*. The application of d'Alembert's principle can be used to simplify many problems in Newtonian mechanics.

**d'Alembert's ratio test** /dal-ahm-**bairz**/ (**generalized ratio test**) A method of showing whether a series is convergent or divergent. The absolute value of the ratio of each term to the one before it is taken:
$$|u_{n+1}/u_n|$$
If the LIMIT of this is $l$ as $n$ tends to infinity and $l$ is less than 1, then the series is convergent. If $l$ is greater than 1, the series is divergent. If $l$ is equal to 1, the test fails and some other method has to be used.

**damped oscillation** An oscillation with an amplitude that progressively decreases with time. *See* damping.

**damping** The reduction in amplitude of a vibration with time by some form of resistance. A swinging pendulum will at last come to rest; a plucked string will not vibrate for long – in both cases internal and/or external resistive forces progressively reduce the amplitude and bring the system to equilibrium.

In many cases the damping force(s) will be proportional to the object's speed. In any event, energy must be transferred from the vibrating system to overcome the resistance. Where damping is an asset (as in bringing the pointer of a measuring instrument to rest), the optimum situation occurs when the motion comes to zero in the shortest time possible, without vibration. This is *critical damping*. If the resistive force is such that the time taken is longer than this, *overdamping* occurs. Conversely, *underdamping* involves a longer time with vibrations of decreasing amplitude.

**data** /**day**-tă/ (**now often used as a singular noun**) (*sing.* **datum**) The facts that refer to or describe an object, idea, condition, situation, etc. In computing, data can be regarded as the facts on which a PROGRAM operates as opposed to the instructions in the program. It can only be accepted and processed by the computer in binary form. Data is sometimes considered to be numerical information only.

**data bank** A large collection of organized computer data, from which particular pieces of information can be readily extracted. *See also* database.

**database** A large collection of organized data providing a common pool of information for users, say, in the various sections of a large organization. Information can be added, deleted, and updated as required. The management of a database is very complicated and costly so that computer programs have been developed for this purpose. These programs allow the information to be extracted in many different ways. For example, a request could be put in for an alphabetical list of people over a certain age and living in a specified area, in which their employment and income should be given. Alternatively the request could be for an alphabetical list of people over a certain age and income level in which their address and form of employment should be given.

**data processing** The sequence of operations performed on data in order to extract information or to achieve some form of order. The term usually means the processing of data by computers but can also include its observation and collection.

**data transmission** Any method of transferring data from one computer to another or from an outstation (such as a cash-point machine) to a central computer.

**debug** To detect, locate, and correct errors or faults (bugs) that occur in computer programs or in pieces of computer equipment. Since PROGRAMS and equipment are often highly complicated, debugging can be a tedious and lengthy job. Programming errors may result from the incorrect coding of an instruction (known as a *syntax error*) or from using instructions that will not give the required solution to a problem (a *logic error*). Syntax errors can usually be detected and located by the compiler; logic errors can be more difficult to find.

**deca-** Symbol: da  A prefix denoting 10. For example, 1 decameter (dam) = 10 meters (m).

**decagon** /**dek**-ă-gon/ A plane figure with ten straight sides. A *regular decagon* has ten equal sides and ten equal angles of 36°.

**decahedron** /dek-ă-**hee**-drŏn/ A polyhedron that has ten faces. *See* polyhedron.

**deci-** Symbol: d  A prefix denoting $10^{-1}$. For example, 1 decimeter (dm) = $10^{-1}$ meter (m).

**decibel** /**des**-ă-bel/ Sybol: dB A unit of power level, usually of a sound wave or electrical signal, measured on a logarithmic scale. The threshold of hearing is taken as 0 dB in sound measurement. Ten times this power level is 10 dB. The fundamental unit is the *bel*, but the decibel is almost exclusively used (1 dB = 0.1 bel).

A power $P$ has a power level in decibels given by:

$$10 \log_{10}(P/P_0)$$

where $P_0$ is the reference power.

**decimal** Denoting or based on the number ten. The numbers in common use for counting form a decimal number system. A *decimal fraction* is a rational number written as units, tenths, hundredths, thousandths, and so on. For example, ¼ is 0.25 in decimal notation. This type of decimal fraction (or decimal) is a *finite decimal* because the third and subsequent digits after the decimal point are 0. Some rational numbers, such as 5/27 (= 0.185 185 185...) cannot be written as an exact decimal, but result in a number of digits that repeat indefinitely. These are called *repeating decimals*. All rational numbers can be written as either finite decimals or repeating decimals. A decimal that is not finite and does not repeat is an irrational number and can be quoted to any number of decimal places, but never exactly. For example, π to an accuracy of six decimal places is 3.141 593 and to seven decimal places is 3.141 592 7.

A decimal measure is any measuring system in which larger and smaller units are derived by multiplying and dividing the basic unit by powers of ten. *See also* metric system.

**decision box** *See* flowchart.

**decomposition** 1. The process of breaking a fraction up into partial fractions.

**2.** Decomposition of a vector $v$ is the process of writing it in the form
$$v = \text{grad}(\phi) + \text{curl}(A)$$
where $\phi$ is a scalar and $A$ is a vector. Every vector may be decomposed in this form.

**decreasing function** A real function $f(x)$ for which $f(x_1) \geq f(x_2)$ for all $x_1$ and $x_2$ in an interval $I$ when $x_1 < x_2$. If the stronger inequality $f(x_1) > f(x_2)$ holds when $x_1 < x_2$ then $f(x)$ is said to be a *strictly decreasing function*.

**decreasing sequence** A sequence $a_1$, $a_2$, $a_3$, ... for which $a_i \geq a_{i+1}$ for all $i$ in the sequence. If $a_i > a_{i+1}$ for all $i$ in the sequence the sequence is said to be a *strictly decreasing sequence*.

**Dedekind cut** /**day**-dĕ-kint/ A method of defining the real numbers, starting from the rational numbers. A Dedekind cut is a division of the rational numbers into two disjoint sets, $A$ and $B$, which are nonempty and which satisfy the conditions: (1) if $x \in A$ and $y \in B$ then $x<y$, and (2) $A$ has no largest member (or, equivalently, $B$ has no smallest member). The real numbers can be defined as the set of all Dedekind cuts and can be shown to have all the requisite properties. The method is named for the German mathematician Richard Dedekind (1831–1916).

**deduction** A series of logical steps in which a conclusion is reached directly from a set of initial statements (premises). A deduction is valid if a sentence or statement that asserts the premises and denies the conclusion is a contradiction. *Compare* induction. *See* contradiction.

**definite integral** (**Riemann integral**) The result of integrating any function of a single variable, $f(x)$, between two specified values of $x$: $x_1$ and $x_2$. The definite integral of $f(x)$ is written
$$\int_{x_1}^{x_2} f(x)\mathrm{d}x$$
If the general expression for the integral of $f(x)$ (its indefinite integral) is another function of $x$, $g(x)$, the definite integral is given by:

$$g(x_1) - g(x_2)$$
*Compare* indefinite integral. *See also* integration.

**definition** In a measurement, the ACCURACY with which the instrument reading reflects the true value of the quantity being measured.

**deformation** A geometrical transformation that stretches, shrinks, or twists a shape but does not break up any of its lines or surfaces. It is often called, more precisely, a *continuous deformation*. *See also* topology; transformation.

**degeneracy** The occurrence of two different EIGENFUNCTIONS of an eigenvalue problem that have the same eigenvalue. An important example of degeneracy is given in quantum mechanical systems such as nuclei, atoms and molecules, where degeneracy occurs when different quantum states have the same energy. The degeneracy of a system is closely associated with its symmetry.

**degenerate conic** *See* conic.

**degenerate conicoid** A quadric surface described by the equation
$$ax^2 + by^2 + cz^2 + 2fyz + 2gxz + 2hxy + 2ux + 2vy + 2wz + d = 0$$

$$\Delta = \begin{vmatrix} a & h & g & u \\ h & b & f & v \\ g & f & c & w \\ u & v & w & d \end{vmatrix}$$

for which $\Delta = 0$, where $\Delta$ is the determinant defined by
A quadric that is not a degenerate quadric is called a *non-degenerate quadric*. The non-degenerate quadrics can be listed: the *ellipsoid*, the *hyperboloid* (both of one sheet and two sheets), the *elliptic paraboloid* and the *hyperbolic paraboloid*. *See* concoid.

**degree 1.** Symbol: ° A unit of plane angle equal to one ninetieth of a right angle.

**2.** A unit of temperature. *See* Celsius degree; Fahrenheit degree; kelvin.
**3.** The exponent of a variable. For instance $3x^3$ has a degree of 3. If there are several terms, the sum of the exponents is used; $5xy^2z^3$ has a degree of 6.
**4.** The highest power of a variable in a polynomial. For example in $x^3 + 2x + 1$, the degree is 3.
**5.** The highest power of an equation. For example, the degree of the equation $x^4 + 2x = 0$ is 4.
**6.** The highest power to which the highest-order derivative is raised in a DIFFERENTIAL EQUATION. For example, the degree of
$$(d^2y/dx^2)^3 + dy/dx = 0$$
is three. The degree of
$$d^3y/dx^3 + 2y(d^2y/dx^2)^2 = 0$$
is one.

**degrees of freedom** The number of independent parameters that are needed to specify the configuration of a system. For example, an atom in space has three independent coordinates needed to specify its position. A molecule with two atoms (e.g. $O_2$) has additional degrees of freedom because it can also vibrate and rotate. In fact, a diatomic molecule of this type has six degrees of freedom. It is usual in physics to interpret the number of degrees of freedom as the number of independent ways in which the system can store energy. *See also* phase space.

**De l'Hôpital's rule** /dě-loh-pee-**tahlz**/ The rule stating that the limit of the ratio of two functions of the same variable $(x)$ as $x$ approaches a value $a$, is equal to the limit of the ratios of their derivatives with respect to $x$. That is, the limit of $f(x)/g(x)$ as $x→a$ is the limit of $f'(x)/g'(x)$ as $x→a$.

De l'Hôpital's rule can be used to find the limits of $f(x)/g(x)$ at points at which both $f(x)$ and $g(x)$ are zero and the ratio is therefore indeterminate. Any function that gives rise to an indeterminate form and that can be expressed as a ratio of two functions, can be dealt with in this way. For example, in
$$F(x) = (x^2 - 3)/(x - 3)$$
writing
$$f(x) = (x^2 - 3)$$

and
$$g(x) = (x - 3)$$
gives
$$F(x) = f(x)/g(x)$$
The limit of $F(x)$ as $x→3$ is indeterminate (since $x - 3 = 0$). It can be obtained by using the limit of
$$f'(x)/g'(x) = 2x$$
as $x→3$. Thus the limit is 6.

If $f'(x)/g'(x)$ also gives an indeterminate form at $x = a$, De l'Hôpital's rule can be applied again, differentiating as many times as is necessary.

The rule is named for the French mathematician Guillaume François Antoine, Marquis De l'Hôpital (1661–1704).

**delta function** *See* Dirac delta function.

**De Moivre's theorem** /dě-**mwahvrz**/ A formula for calculating a power of a complex number. If the number is in the polar form
$$z = r(\cosθ + i \sinθ)$$
then $z^n = r^n(\cos nθ + i \sin nθ)$
The theorem is named for the French mathematician Abraham De Moivre (1667–1754).

**De Morgan's laws** /dě-**mor**-gănz/ Two laws governing the relation between complementation, intersection, and union of sets. If represents the complement of the set $A$ (i.e. the set of all things not in $A$) then De Morgan's laws state that:
(1) $(\overline{A \cup B}) = \bar{A} \cap \bar{B}$ and
(2) $(A \cap B) = \overline{A \cup B}$.
Analogous laws are true for any finite intersection or union of sets. Parallel laws exist in other areas, e.g. in propositional logic the equivalences
$$\sim(p \,\&\, q) \equiv \sim p \vee \sim q \text{ and}$$
$$\sim(p \vee q) \equiv \sim p \,\&\, \sim q$$
are also known as De Morgan's laws. The laws are named for the English mathematician Augustus De Morgan (1806–71).

**denominator** The bottom part of a fraction. For example, in the fraction ¾, 4 is the denominator and 3 is the numerator. The denominator is the divisor.

**dense** A set $S$ is said to be dense in another set $T$ if every point of $T$ either belongs to $S$ or is a limit point of $S$. If it is obvious what the other set is, then a set is sometimes simply said to be dense; e.g. if we are considering sets of real numbers. For example, the set of rational numbers is dense (on the real line) because every point of the real line is either a rational number or is the limit of a sequence of rational numbers. Another way of putting this is to say that any interval on the real line will always contain infinitely many rationals.

**density** The amount of matter per unit volume; mass divided by volume.

**denumerable set** /di-**new**-mĕ-ră-băl/ A SET in which the elements can be counted. For example, the set of prime numbers, although infinite, can be counted, as can the set of positive even integers. These are known as denumerably infinite sets. The set of real numbers, on the other hand, is not denumerable because between any two elements there can always be found a third. *See* countable.

**dependent** 1. An equation is *dependent* on a set of equations if every set of values of the unknowns that satisfy the set of equations also satisfies this equation. Otherwise the equation is *independent* of the set of equations.
2. Two events are *independent* if the occurrence or nonoccurrence of one of them does not affect the probability of the occurrence of the other. Otherwise the events are *dependent*. If $A$ and $B$ are events whose probabilities are P($A$) and P($B$), then $A$ and $B$ are independent if and only if P($A$ and $B$) = P($A$)P($B$). For example, each toss of a coin is an independent event.
3. A set of functions $\{f_1, \ldots f_n\}$ are *dependent* if one can be expressed as a function of the others or equivalently there exists an expression F($f_1, \ldots f_n$) ≡ 0 with not all $\partial F / \partial f_i$ = 0. Otherwise they are *independent*. For example, the functions $2x + y$ and $4x + 2y + 6$ are dependent since $4x + 2y + 6 = 2(2x + y) + 6$.
4. *See* variable.

5. A set of vectors, matrices, or other objects $\{x_1, \ldots x_n\}$ is said to be *linearly dependent* if there exists a linear relation
$$a_1 x_1 + a_2 x_2 + \ldots + a_n x_n = 0$$
with at least one of the coefficients nonzero. A set of objects is *linearly independent* if it is not linearly dependent. It should be noted that the dependence is relative to the set from which we may pick the coefficients $a_1, \ldots a_n$. For example, 2 and π are linearly independent with respect to the rational numbers but linearly dependent with respect to the real numbers. This is the case since a relation of the form $a_1 2 + a_2 \pi$ = 0 does not exist if $a_1$ and $a_2$ are rational numbers but if $a_1$ and $a_2$ are allowed to be irrational numbers we may take $a_1 = \pi$, $a_2$ = 2.

**dependent variable** *See* variable.

**deposit** A sum of money paid by a buyer, either to reserve goods or property that he wishes to buy at a later date or as the first of a series of installments in an installment plan. If the buyer fails to complete the installments the deposit is normally forfeited.

**depth** The distance downward from a reference level or backward from a reference plane. For example, the distance below a water surface and the distance between a wall surface and the back of an alcove in the wall, are both called depths.

**derivative** The result of DIFFERENTIATION.

**derived unit** A unit defined in terms of base units, and not directly from a standard value of the quantity it measures. For example, the newton is a unit of force defined as a kilogram meter seconds$^{-2}$ (kg m s$^{-2}$). *See also* SI units.

**desktop publishing** A system that uses a microcomputer with word-processing facilities linked to a laser printer to produce multiple copies of a document. The word processor can provide various type fonts and a scanner can be included to add illustrations (graphics). The result can rival the quality of conventional printing.

$$\begin{vmatrix} a_1 & b_1 \\ a_2 & b_2 \end{vmatrix} = a_1b_2 - a_2b_1$$

The second-order determinant of a 2 x 2 matrix.

$$\begin{vmatrix} a_1 & b_1 & c_1 \\ a_2 & b_2 & c_2 \\ a_3 & b_3 & c_3 \end{vmatrix} = \begin{aligned} & a_1b_2c_3 - a_1b_3c_2 + a_2b_3c_1 - a_2b_1c_3 \\ & + a_3b_1c_2 - a_3b_2c_1 \end{aligned}$$

The third-order determinant of a 3 x 3 matrix.

$$\begin{vmatrix} a_1 & b_1 & c_1 \\ a_2 & b_2 & c_2 \\ a_3 & b_3 & c_3 \end{vmatrix} = a_1 \begin{vmatrix} b_2 & c_2 \\ b_3 & c_3 \end{vmatrix} - b_1 \begin{vmatrix} a_2 & c_2 \\ a_3 & c_3 \end{vmatrix} + c_1 \begin{vmatrix} a_2 & b_2 \\ a_3 & b_3 \end{vmatrix}$$

$$= a_1a_1' - b_1b_1' + c_1c_1'$$
$$= a_1a_1' - a_2a_2' + a_3a_3'$$

$$\begin{array}{ccc} + & - & + \\ - & + & - \\ + & - & + \end{array}$$

Determinants: A third-order determinant is equal to the sum along any row, or down any column, of the product of each element with its cofactor. The cofactors are given alternate positive and negative signs in the pattern shown. Fourth- and higher order determinants can be calculated in a similar way.

**determinant** /di-**ter**-mă-nănt/ A function of a square matrix derived by multiplying and adding the elements together to obtain a single number. For example, in a $2 \times 2$ matrix the determinant is $a_1b_2 - a_2b_1$. This is written as a square array in vertical lines, symbol $D_2$, and is called a *second-order determinant*. Determinants occur in simultaneous equations. The solution of
$$a_1x + b_1y + c_1 = 0$$
and
$$a_2x + b_2y + c_2 = 0$$
is
$$x = (b_1c_2 - c_1b_2)/D_2$$
and
$$y = (c_1a_2 - a_1c_2)/D_2$$
If $a_1$, $a_2$, $b_1$, $b_2$, $c_1$, and $c_2$ are 1, 2, 3, 4, 5, and 6 respectively, then $D_2 = -2$ and
$$x = [(3 \times 6) - (5 \times 4)]/-2 = 1$$
and

$$y = [(5 \times 2) - (1 \times 6)]/-2 = -2$$
A *third-order determinant* has three rows and columns and arises in a similar way in sets of three simultaneous equations in three variables.

The determinant of a transpose of a matrix, $|\bar{A}|$, is equal to the determinant of the matrix, $|A|$. If the position of any of the rows or columns in the matrix is changed, the determinant remains the same.

**determinism** The idea that if the present state of a system is known exactly and the law governing the evolution of that system with time is known then the subsequent evolution of the system can be determined exactly. Classical mechanics is governed by determinism. Two developments in the twentieth century undermined this simple picture. In quantum mechanics, the uncertainty principle of the German physicist Werner Heisenberg (1901–76) showed that it is not possible to know the state of a system exactly since the position and the momentum of a particle cannot be determined exactly simultaneously. Even within classical mechanics itself the development of CHAOS THEORY led to the realization that the word *exactly* is very important both in principle and in practice because there are many systems for which slightly different initial conditions result in widely different results for the state of the system over a period of time. Consequently, even systems that are described in a deterministic way can, in practice, behave in ways that appear random or chaotic.

**developable surface** A surface that can be rolled out flat onto a plane. The lateral surface of a cone, for example, is developable. A spherical surface is not.

**deviation** *See* mean deviation; standard deviation.

**diagonal** Joining opposite corners. A diagonal of a square, for example, cuts it into two congruent right angled triangles. In a solid figure, usually a polyhedron, a diagonal plane is one that passes through two edges that are not adjacent.

**diagonal matrix** A square matrix in which all the elements are zero except those on the leading diagonal, that is, the first element in the first row, the second element in the second row, and so on. Diagonal matrices, unlike most others, are commutative in matrix multiplication.

$$\begin{pmatrix} a_{11} & 0 & 0 \\ 0 & a_{22} & 0 \\ 0 & 0 & a_{33} \end{pmatrix}$$

A 3 × 3 diagonal matrix.

**diameter** The distance across a plane figure or a solid at its widest point. The di-ameter of a circle or a sphere is twice the radius.

**diametral** /dÿ-**am**-ĕ-trăl/ Denoting a line or plane that forms a diameter of a figure. For example, a cross-section through the center of a sphere is a diametral plane.

**dichotomy, principle of** In logic, the principle that a proposition is either true or false, but not both. For example, for two numbers $x$ and $y$ either $x = y$ or $x \neq y$, but not both.

**diffeomorphism** /diff-ee-oh-**mor**-fiz-ăm/ A continuous transformation of a space that moves the points of a space around but preserves those relationships between them that are used to define which points are close to one another. The set of diffeomorphisms of a space is called the *diffeomorphism group* of that space. The diffeomorphism group underlies discussions of invariance in RIEMANNIAN GEOMETRY and general relativity theory.

**difference** The result of subtracting one quantity or expression from another.

**difference between two squares** A special case of factorization involving two numbers or expressions that are squares. In general terms,
$$x^2 - y^2 = (x + y)(x - y)$$

**difference equation** An equation that expresses a relation between finite differences $\Delta$, where $\Delta$ is a *difference operator* which operates on the $r$th term $U_r$ of a sequence $U_1, U_2, \dots U_n$ to produce the $(r + 1)$th term $U_{r+1}$. The definition of the difference operator means that $\Delta U_r = U_{r+1}$, $\Delta U_1 = U_2 - U_1$, $\Delta U_2 = U_3 - U_2$, etc. The expressions just given are *first finite differences* of $U_1$, $U_2, \dots U_n$ since $\Delta$ only operates on them once. A *second finite difference* is defined by operating with $\Delta$ a second time. This leads to the results $\Delta^2 U_1 = U_3 - 2U_2 + U_1$, $\Delta^2 U_2 = U_4 - 2U_3 + U_2$, etc. Higher-order differences $\Delta^3 U_r, \Delta^4 U_r, \dots$ can be defined in an analogous way.

The *order* of a difference equation is the order of the highest-order finite difference in it. For example, $\Delta U_r - aU_r = b$, where $a$ and $b$ are constants is a *first-order difference equation*, while $\Delta^2 U_r - a\Delta U_r + bU_r = f(r)$, where $a$ and $b$ are constants and $f(r)$ is a function of $r$, is a *second-order difference equation*. There are analogies between difference equations and differential equations.

**difference set** The set that is made up of all the elements of set A that are not elements of set B. In terms of a VENN DIAGRAM the difference set can be shown as a shaded region for two overlapping sets.

**differentiable function** /dif-ĕ-ren-shă-băl/ A function f for which the derivative of f exists. More formally, if f is a real function of one variable $x$ then the function is differentiable in some interval if the limit of $[f(x + \delta x]/\delta x$ exists as $\delta x \to 0$ for all values of $x$ in the interval. Thus, a function is differentiable at a point if the gradient of the graph $y = f(x)$ can be defined as that point, and therefore that the tangent to the curve can be defined at that point. In the case of a function of a complex variable the function $f(z)$ is differentiable at a particular point if the limit $[f(z + \delta z) - f(z)]/\delta z$ exists as $\delta z \to 0$ for that point and is independent of the way in which that point is approached.

**differential** /dif-ĕ-ren-shăl/ An infinitesimal change in a function of one or more variables, resulting from a small change in the variables. For example, if $f(x)$ is a function of $x$, and f changes by $\Delta f$ as a result of a change $\Delta x$ in $x$, the differential df, is defined as the limit of $\Delta f$ as $\Delta x$ becomes infinitely small. That is, $df = f'(x)dx$, where $f'(x)$ is the derivative of f with respect to $x$. This is a *total differential*, because it takes into account changes in all of the variables, just one in this case.

For a function of two variables, $f(x,y)$ the rate of change of f with respect to $x$ is the partial derivative $\partial f/\partial x$. The change in f resulting from changing $x$ by $dx$ and keeping $y$ constant is the *partial differential*, $(\partial f/\partial y).dy$. For any function, the total differential is the sum of all the partial differentials. For $f(x,y)$:
$$df = (\partial f/\partial x).dx + (\partial f/\partial y).dy$$

*See also* differentiation.

**differential equation** An equation that contains derivatives. An example of a simple differential equation is:
$$dy/dx + 4x + 6 = 0$$
To solve such equations it is necessary to use integration. The equation above can be rearranged to give:
$$dy = -(4x + 6)dx$$
integrating both sides:
$$\int dy = \int -(4x + 6)dx$$
which gives:
$$y = -2x^2 - 6x + C$$
where $C$ is a constant of integration. The value of $C$ can be found if particular values of $x$ and $y$ are known: for example, if $y = 1$ when $x = 0$ then $C = 1$, and the full solution is
$$y = -2x^2 - 6x + 1$$
Note that the solution to a differential equation is itself an equation. Differentiating the solution gives the original equation. Equations like that above, which contain only first derivatives $(dy/dx)$ are said to be *first order*, if they contain second derivatives they are *second order*; in general, the *order* of a differential equation is the highest derivative in the equation. The *degree* of a differential equation is the highest power of the highest order derivative.

The differential equation in the example given is a first order and first degree equation. It is an example of a type of equation solvable by separating the variables onto both sides of the equation, so that each can be integrated (the *variables separable* method of solution). Another type of first-order first-degree equation is one of the form:
$$dy/dx = f(y/x)$$
Such equations are known as homogeneous differential equations. An example is the equation:
$$dy/dx = (x^2 + y^2)/x^2$$
To solve homogeneous equations a substitution is made, $y = mx$, where $m$ is a function of $x$. Then:
$$dy/dx = m + xdm/dx$$
and
$$(x^2 + y^2)/x^2 = (x^2 + m^2x^2)/x^2$$
So the equation becomes:
$$m + xdm/dx = (x^2 + m^2x^2)/x^2$$

or:
$$x\,dm/dx = 1 + m^2 - m$$
The equation can now be solved by separating the variables.

An equation of the form:
$$dy/dx + P(x)y = Q(x)$$
where $P(x)$ and $Q(x)$ are functions of $x$ (but not $y$), is a *linear differential* equation. Equations of this type can be put in a solvable form by multiplying both sides by the expression:
$$\exp(\textstyle\int P(x)dx)$$
This is known as an *integrating factor*. For example, the differential equation
$$dy/dx + y/x = x^2$$
is a linear first-order differential equation. The function $P(x)$ is $1/x$, so the integrating factor is:
$$\exp(\textstyle\int dx/x)$$
which is $\exp(\log x)$; i.e. $x$. Multiplying both sides of the equation by $x$ gives:
$$x\,dy/dx + y = x^3$$
The left-hand side of the equation is equal to $d(xy)/dx$, so the equation becomes
$$d(xy)/dx = x^3$$
Integrating both sides gives:
$$xy = x^4/4 + C$$
where $C$ is a constant.

**differential form** Any entity that is under an integral sign. For example, the integral $\int Adx$ has a *one-form* $Adx$ associated with it. Similarly, a volume integral $\iiint Bdxdydz$ has a three-form $Bdxdydz$ associated with

it. More generally, if an integral is over $n$ dimensions then there is an *n-form* associated with that integral. There are many problems in mathematics and its applications to physical sciences and engineering that can be analyzed using differential forms.

**differentiation** /dif-ĕ-ren-shee-**ay**-shŏn/ A process for finding the rate at which one variable quantity changes with respect to another. For example, a car might travel along a road from position $x_1$ to position $x_2$ in a time interval $t_1$ to $t_2$. Its average speed is $(x_2 - x_1)/(t_2 - t_1)$, which can be written $\Delta x/\Delta t$, where $\Delta x$ represents the change in $x$ in the time $\Delta t$. However, the car might accelerate or decelerate in this interval and it may be necessary to know the speed at a particular instant, say $t_1$. In this case the time interval $\Delta t$ is made infinitely small, i.e. $t_2$ can be as close as necessary to $t_1$. The limit of $\Delta x/\Delta t$ as $\Delta t$ approaches zero is the instantaneous velocity at $t_1$. The result of differentiation (i.e. the *derivative*) of a function $y = f(x)$ is written $dy/dx$ or $f'(x)$. On a graph of $f(x)$, $dy/dx$ at any point is the slope of the tangent to the curve $y = f(x)$ at that point. *See also* integration.

**differentiator** /dif-ĕ-**ren**-shee-ay-ter/ An analog computer device whose output (which is variable) is proportional to the

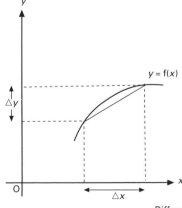

Differentiation of a function $y = f(x)$. The derivative $dy/dx$ is the limit of $\triangle y/\triangle x$ as $\triangle y$ and $\triangle y$ become infinitely small.

Differentiation

time differential of the input (also variable). *See* differential.

**digit** A symbol that forms part of a number. For example, in the number 3121 there are four digits. The ordinary (decimal) number system has ten digits (0–9), whereas the binary (base two) system needs only two, 0 and 1.

**digital** Using numerical digits. For example, a digital watch shows the time in numbers of hours and minutes and not as the position of hands on a dial. In general, digital devices work by some kind of counting process, either mechanical or electronic. The abacus is a very simple example. Early calculating machines counted with mechanical relays. Modern calculators use electronic switching circuits.

**digital/analog converter** A device that converts digital signals, usually from a digital computer, into continuously varying electrical signals for inputting to an ANALOG COMPUTER. *See* computer.

**digital computer** *See* computer.

**dihedral** /dÿ-**hee**-drǎl/ Formed by two intersecting planes. Two planes intersect along a straight line (edge). The *dihedral angle* (or *dihedron*) between the planes is the angle between two lines (one in each plane) drawn perpendicular to the edge from a point on the edge. The dihedral angle of a polyhedron is the angle between two faces.

**dihedron** /dÿ-**hee**-drŏn/ *See* dihedral.

**dilatation** /dil-ǎ-**tay**-shŏn, dÿ-lǎ-/ A geometrical mapping or projection in which a figure is 'stretched', not necessarily by the same amount in each direction. A square, for example, may be mapped into a rectangle by dilatation, or a cube into a cuboid.

**dimension 1.** The number of coordinates needed to represent the points on a line, shape, or solid. A plane figure is said to be two-dimensional; a solid is three-dimensional. In more abstract studies $n$-dimensional spaces can be used.
**2.** The size of a plane figure or solid. The dimensions of a rectangle are its length and width; the dimensions of a rectangular parallelepiped are its length, width, and height.
**3.** One of the fundamental physical quantities that can be used to express other quantities. Usually, mass [M], length [L], and time [T] are chosen. Velocity, for example, has dimensions of $[L][T]^{-1}$ (distance divided by time). Force, as defined by the equation:

$$F = ma$$

where $m$ is mass and $a$ acceleration, has dimensions $[M][L][T]^{-2}$. *See also* dimensional analysis.
**4.** Of a matrix, the number of rows or number of columns. A matrix with 4 rows and 5 columns is $4 \times 5$ matrix.

**dimensional analysis** The use of the dimensions of physical quantities to check relationships between them. For instance, Einstein's equation $E = mc^2$ can be checked. The dimensions of speed$^2$ are $([L][T]^{-1})^2$, i.e. $[L]^2[T]^{-2}$, so $mc^2$ has dimensions of $[M][L]^2[T]^{-2}$. Energy also has these dimensions since it is force $[M][L][T]^{-2}$ multiplied by distance [L]. Dimensional analysis is also used to obtain the units of a quantity and to suggest new equations.

**Diophantine equation** /dÿ-ŏ-**fan**-tÿn, -teen, -tin/ *See* indeterminate equation.

**Dirac delta function** /**di**-rak/ Symbol $\delta(x)$. A mathematical symbol which is used to represent a sudden pulse. Strictly speaking, the Dirac delta function is not a proper mathematical function. It can be defined as the generalization to continuous variables of the *Kronecker delta* $\delta_{ij}$ which is defined for discrete variables $i$ and $j$ by: $\delta_{ij} = 1$, if $i = j$, and $=0$ if $i \neq j$. The Direct delta function can also be defined by the properties: $\delta(x) = 0$, if $x \neq 0$, $\int_{-\infty}^{\infty} \delta(x)dx = 1$, $\int_{-\infty}^{\infty} f(x)\delta(x) = f(0)$. It is named for the British physicist Paul Dirac (1902–84).

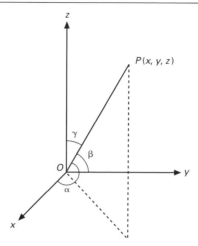

Direction angles: α, β, and γ are made by the line *OP* with the *x*-, *y*-, and *z*-axes respectively.

**directed** Having a specified positive or negative sign, or a definite direction. A directed number usually has one of the signs + or − written in front of it. A directed angle is measured from one specified line to the other. If the direction were reversed, the size of the angle would be a negative number.

**directed line** A straight line in which a direction along the line is specified. The direction specified is called the *positive direction*: the opposite direction is called the *negative direction*. It is also possible to indicate the direction on a directed line by specifying a point 0 on the line to be the origin, with the positive direction from this origin being directed towards the end *x* of the line. In this notation the directed line is given by 0*x*.

**directed line-segment** The combination of a straight line and a specific direction along that line. For example if P and Q are two points on a straight line then PQ is the directed line-segment from P to Q, and QP is the directed line-segment from Q to P. The concept of a directed line-segment is closely related to that of a VECTOR.

**direction** A property of vector quantities, usually defined in reference to a fixed origin and axes. The direction of a curve at a point is the angle from the *x* axis to the tangent at the point.

**directional derivative** The rate of change of a function with respect to distance *s* in a particular direction, or along a specified curve. Going from a point $P(x,y,z)$ in the direction that makes angles α, β, and γ with the *x*, *y*, and *z* axs respectively, the directional derivative of a function $f(x,y,z)$ is

$$df/ds = (\partial f/\partial x)\cos\alpha + (\partial f/\partial y)\cos\beta$$
$$+ (\partial f/\partial z)\cos\gamma$$

If there is a direction for which the directional derivative is a maximum, then this derivative is the gradient of f (grad f) at point P. *See also* grad.

**direction angle** The angle between a line and one of the axs in a rectangular Cartesian coordinate system. In a plane system, it is the angle, α, that the line makes with the positive direction of the *x*-axis. In three dimensions, there are three direction angles, α, β, and γ, for the *x*, *y*, and *z* axes respectively. If two direction angles are known, the third can be calculated by the relationship:

$$\cos^2\alpha + \cos^2\beta + \cos^2\gamma = 1$$

Cosα, cosβ, and cosγ are called the *direction cosines* of the line, sometimes given the symbols $l$, $m$, and $n$. Any three numbers in the ratio $l:m:n$ are called the *direction numbers* or the *direction ratio* of the line. If a line joins the point A $(x_1, y_1, z_1)$ and the point B $(x_2, y_2, z_2)$ and the distance between A and B is $D$, then

$$l = (x_2 - x_1)/D$$
$$m = (y_2 - y_1)/D$$
$$n = (z_2 - z_1)/D$$

**direction cosines** *See* direction angle.

**direction numbers** *See* direction angle.

**direction ratio** *See* direction angle.

**director circle** The locus of points P$(x,y)$ from which two perpendicular tangents are drawn to an ellipse of the form $x^2/a^2 + y^2/b^2 = 1$ as the points of contact vary. The equation of a director circle defined in this way is $x^2 = a^2 + b^2$, as can be shown by applying coordinate geometry to the equations of the two tangents from P$(x,y)$ to the ellipse.

**direct proof** A logical argument in which the theorem or proposition being proved is the conclusion of a step-by-step process based on a set of initial statements that are known or assumed to be true. *Compare* indirect proof.

**directrix** /dă-**rek**-triks/ **1.** A straight line associated with a conic, from which the shortest distance to any point on the conic maintains a constant ratio with the distance from that point to the focus. *See also* conic.
**2.** A plane curve defining the base of a cone or cylinder.

**discontinuity** *See* continuous function.

**discontinuous** *See* continuous function.

**discount 1.** The difference between the issue price of a stock or share and its nominal value when the issue price is less than the nominal value. *Compare* premium.
**2.** A reduction in the price of an article or commodity for payment in cash (*cash discount*), or for a large order (*bulk discount*), or for a retailer who will be selling the goods on to members of the public (*trade discount*).

**discrete** Denoting a set of events or numbers in which there are no intermediate levels. The set of integers, for example, is discrete but the set of rational numbers is not. Between any two rational numbers, no matter how close, there can always be found another rational number. The results of tossing dice form a discrete set of events, since a die has to land on one of its six faces. Putting the shot, on the other hand, does not have a discrete set of outcomes, since it may travel for any distance in a continuous range of lengths.

**discriminant** /dis-**krim**-ă-nănt/ The expression $(b^2 - 4ac)$ in a QUADRATIC EQUATION of the form $ax^2 + bx + c = 0$. If the roots of the equation are equal, the discriminant is zero. For example, in
$$x^2 - 4x + 4 = 0$$
$b^2 - 4ac = 0$ and the only root is 2. If the discriminant is positive, the roots are different and real. For example, in
$$x^2 + x - 6 = 0$$
$b^2 - 4ac = 25$ and the roots are 2 and $-3$. If the discriminant is negative, the roots of the equation are complex numbers. For example, the equation:
$$x^2 + x + 1 = 0$$
has roots $[-\frac{1}{2} - (\sqrt{3}/2)]i$ and $[-\frac{1}{2} + (\sqrt{3}/2)]i$.

| P Q | P ∨ Q |
|-----|-------|
| T T | F |
| T F | T |
| F T | T |
| F F | F |

Disjunction (exclusive)

| P Q | P ∨ Q |
|-----|-------|
| T T | T |
| T F | T |
| F T | T |
| F F | F |

Disjunction (inclusive)

**disjoint** Two sets are said to be disjoint if they have no members in common; i.e. if $A \cap B = 0$ then $A$ and $B$ are disjoint.

**disjunction** Symbol: $\vee$ In logic, the relationship *or* between two propositions or statements. Disjunction can be either inclusive or exclusive. *Inclusive disjunction* (sometimes called *alternation*) is the one most commonly used in mathematical logic, and can be interpreted as 'one or the other or both'. For two propositions $P$ and $Q$, $P \vee Q$ is false if $P$ and $Q$ are both false, and true in all other cases. The more rarely used *exclusive disjunction* can be interpreted as 'either one or the other but not both'. With this definition $P \vee Q$ is false when $P$ and $Q$ are both true, as well as when they are both false. The truth-table definitions for both types of disjunction are shown in the illustrations. *Compare* conjunction. *See also* truth table.

**disk** A device that is widely used in computer systems to store information. It is a flat circular metal plate coated usually on both sides with a magnetizable substance. Information is stored in the form of small magnetized spots, which are closely packed in concentric *tracks* on the coated surfaces of the disk. The spots are magnetized in one of two directions so that the information is in binary form. The magnetization pattern of a group of spots represents a letter, digit (0–9), or some other character. One disk can store several million characters. Information can be altered or deleted as necessary by magnetic means. Disks are usually stacked on a common spindle in a single unit known as a *disk pack*. Disk packs storing 200 million characters are common.

Information can be recorded on a disk using a special typewriter; this method is known as *key-to-disk*. The information is fed into a computer using a complex device called a *disk unit*. The disk pack is rotated at very great speed in the disk unit. Small electromagnets, known as *read–write heads*, move radially in and out over the surfaces of the rotating disks. They extract (read) or record (write) items of information at specified locations on a track, following instructions from the central processor. The time to reach a specified location is very short. This factor, together with the immense storage capacity, makes the disk unit a major backing store in a computer system. *Compare* drum; magnetic tape. *See also* floppy disk; hard disk.

**diskette** *See* floppy disk.

**dispersion** A measure of the extent to which data are spread about an average. The range, the difference between the largest and smallest results, is one measure. If $P_r$ is the value below which $r\%$ of the results occur, then the range can be written as $(P_{100} - P_0)$. The interquartile range is $(P_{75} - P_{25})$. The semi-interquartile range is $(P_{75} - P_{25})/2$. The mean deviation of $X_1$, $X_2$, ..., $X_n$ measures the spread about the MEAN $X$ and is

$$\sum_{n} |x_j - x|/n$$

If values $X_1$, $X_2$, ..., $X_k$ occur with frequencies $f_1, f_2, ...f_k$ it becomes

$$\sum_{n} f_j |x_j - x|/\Sigma f_j$$

**displacement** Symbol: $s$ The vector form of distance, measured in meters (m) and involving direction as well as magnitude.

**dissipation** The removal of energy from a system to overcome some form of resistive force. Without resistance (as in motion in a vacuum) there can be no dissipation. Dissipated energy normally appears as thermal energy.

**distance** Symbol: $d$ The length of the path between two points. The SI unit is the meter (m). Distance may or may not be measured in a straight line. It is a scalar; the vector form is displacement.

**distance formula** The formula for the distance between two points $(x_1, y_1)$ and $(x_2, y_2)$ in Cartesian coordinates. It is:

$$\sqrt{[(x_1 - x_2)^2 + (y_1 - y_2)^2]}$$

**distance ratio** (**velocity ratio**) For a MACHINE, the ratio of the distance moved by the effort in a given time to the distance moved by the load in the same time.

**distribution function** For a random variable $x$, the function $f(x)$ that is equal to the probability of each value of $x$ occurring. If all values of $x$ between $a$ and $b$ are equally likely, $x$ has a *uniform distribution* in this interval and a graph of the distribution function $f(x)$ against $x$ is a horizontal line. For example, the probability of the results 1 to 6 when throwing dice is a uniform distribution. Continuous random variables usually have a varying distribution function with a maximum value $x_m$ and in which the probability of $x$ decreases as $x$ moves away from $x_m$. The *cumulative distribution* function $F(x)$ is the probability of a value less than or equal to $x$. For the dice example, $F(x)$ is a step function that increases from zero to one in six equal steps. For continuous functions, $F(x)$ is often an s-shaped curve. In both cases $F(x)$ is the area under the curve of $f(x)$ to the left of $x$.

**distributive** Denoting an operation that is independent of being carried out before or after another operation. For two operations $\bullet$ and $\circ$, $\bullet$ is distributive with respect to $\circ$ if $a \bullet (b \circ c) = (a \bullet b) \circ (a \bullet c)$ for all values of $a$, $b$, and $c$. In ordinary arithmetic, multiplication is distributive with respect to addition $[a(b + c) = ab + ac]$ and to subtraction.

In set theory intersection ($\cap$) is distributive with respect to union ($\cup$):
$$[A \cap (B \cup C) = (A \cap B) \cup (A \cap C)]$$
*See also* associative; commutative.

**div** /div/ (**divergence**) Symbol: $\nabla$. A scalar operator that, for a three-dimensional vector function $F(x,y,z)$, is the sum of the scalar products of the unit vectors and the partial derivatives in each of the three component directions. That is:
$$\text{div } F = \nabla.F = i.\partial F/\partial x + j.\partial F/\partial y + k.\partial F/\partial z$$
In physics, div $F$ is used to describe the excess flux leaving an element of volume in space. This may be a flow of liquid, a flow of heat in a field of varying temperature, or an electric or magnetic flux in an electric or magnetic field. If there is no source of flux (heat source, electric charge, etc.) within the volume, then div $F = 0$ and the total flux entering the volume equals the total flux leaving. *See also* grad.

**divergence theorem** (**Gauss's theorem; Ostrogradsky's theorem**) A basic theorem in vector calculus that states that the integral of the divergence of a vector over a volume enclosed by a surface is equal to the integral of the normal component of the vector over the closed surface. This can be expressed by the equation
$$\iint F.n \, dS = \iiint \text{div} F \, dV,$$
where $F$ is the vector, $n$ is the normal to $S$, and $V$ is the volume. The divergence theorem has important physical applications, particularly in electrostatics (where it is known as *Gauss's law*).

**divergent sequence** A SEQUENCE in which the difference between the $n$th term and the one after it is constant or increases as $n$ increases. $\{1,2,4,8,...\}$ is divergent. A divergent sequence has no limit. *Compare* convergent sequence. *See also* divergent series; geometric sequence.

**divergent series** A SERIES in which the sum of all the terms after the $n$th term does not decrease as $n$ increases. A divergent series, unlike a convergent series, has no sum to infinity. An infinite series $a_1 + a_2 + ...$ is *divergent* if the partial sums $a_1 + a_2 + ... + a_n$ tend to infinity as $n$ tends to infinity. For example, the series $1 + 2 + 3 + 4 + ...$ is divergent. *Compare* convergent series. *See also* divergent sequence; geometric series.

**dividend** 1. The number into which another number (the divisor) is divided to give a quotient. For example, in $16 \div 3$, 16 is the dividend and 3 is the divisor.
2. A share of the profits of a corporation paid to shareholders. The rate of dividend paid will depend on profits in the preceding year. It is expressed as a percentage of the nominal value of the shares. For example, a 10% dividend on a 75c share will pay 7.5c per share (independent of the market price of the share). *See also* yield.

**dividers** A drawing instrument, similar to compasses, but with sharp points on both ends. Dividers are used for measuring lengths on a drawing, or for dividing straight lines.

**division** Symbol: ÷ The binary operation of finding the quotient of two quantities. Division is the inverse operation to MULTI-PLICATION. In arithmetic, the division of two numbers is not commutative ($2 \div 3 \neq 3 \div 2$), nor associative [$(2 \div 3) \div 4 \neq 2 \div (3 \div 4)$]. The identity element for division is one only when it comes on the right hand side ($5 \div 1 = 5$ but $1 \div 5 \neq 5$).

**division of fractions** *See* fractions.

**divisor** The number by which another number (the dividend) is divided to give a quotient. For example, in $16 \div 3$, 16 is the dividend and 3 is the divisor. *See also* factor.

**documentation** Written instructions and comments that give a full description of a computer program. The documentation describes the purposes for which the program can be used, how it operates, the exact form of the inputs and outputs, and how the computer must be operated. It allows the program to be amended when necessary or to be converted for use on different types of machines.

**dodecagon** /doh-**dek**-ă-gon/ A plane figure with 12 sides and 12 interior angles. The sum of the interior angles is 1800°.

**dodecahedron** /doh-dek-ă- **hee**-drŏn/ A solid figure with 12 faces. The faces of a regular dodecahedron are all regular pentagons.

**domain** A set of numbers or quantities on which a mapping is, or may be, carried out. In algebra, the domain of the function $f(x)$ is the set of values that the independent variable $x$ can take. If, for example, $f(x)$ represents taking the square root of $x$, then the domain might be defined as all the positive rational numbers. *See also* range.

**D operator** The differential operator $d/dx$. The derivative $df/dx$ of a function $f(x)$ is often written as $Df$. This notation is used in solving DIFFERENTIAL EQUATIONS. A second derivative, $d^2f/dx^2$, is written as $D^2f$, a third derivative, $d^3f/dx^3$, as $D^3f$, and so on. In some ways, the D operator can be treated like an ordinary algebraic quantity, despite the fact that it has no numerical value. For example, the differential equation

$$d^2y/dx^2 + 2x\,dy/dx + dy/dx + 2x = 0$$

or

$$D^2y + 2xD + D + 2x = 0$$

can be factorized to $(D + 2x)(D + 1) = 0$. In this case, $(D + 2x)$ then operates on the function $(D + 1)$.

**dot product** *See* scalar product.

**double-angle formulae** *See* addition formulae.

**double integral** The result of integrating the same function twice, first with respect to one variable, holding a second variable constant, and then with respect to the second variable, holding the first variable constant. For example, if $f(x,y)$ is a function of the variables $x$ and $y$, then the double integral, first with respect to $x$ and then with respect to $y$, is:

$$\iint f(x,y)\,dy\,dx$$

This is equivalent to summing $f(x,y)$ over intervals of both $x$ and $y$, or to finding the volume bounded by the surface represent-

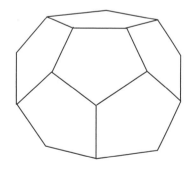

Dodecahedron: a regular dodecahedron has regular pentagonal faces.

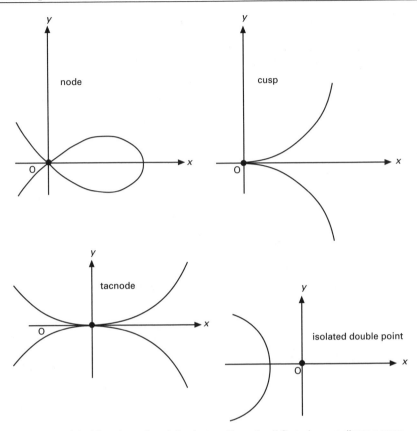

Four types of double point at the origin of a two-dimensional Cartesian coordinate system.

ing $f(x,y)$. The integral is not affected by the order in which the integrations are carried out if they are definite integrals. Another kind of double integral is the result of integrating twice with respect to the same variable. For example, if a car's acceleration $a$ increases with time $t$ in a known way, then the integral

$$\int a\,dt$$

is the velocity ($v$) expressed as a function of time; the double integral

$$\iint a\,dt^2 = \int v.dt = x$$

where $x$ is the distance traveled as a function of time.

**double point** A singular point on a curve at which the curve crosses itself or is tan-

gential to itself. There are several types of double point. At a *node* the curve crosses over itself forming a loop. In this case it has two distinct tangents. At a *cusp* it double back on itself and has only one tangent. At an *acnode* two arcs of a curve touch each other and have the same tangent but, unlike a cusp, the arcs continue through the singular point to form four arms. An *isolated double point* may also occur. This satisfies the equation of the curve but does not lie on the main arc of the curve. *See also* isolated point; multiple point.

**drum** A metal cylinder coated with a magnetizable substance and used in a computer system to store information.

**dual** *See* duality.

**duality** The principle in mathematics whereby one true theorem can be obtained from another merely by substituting certain words in a systematic way. In a plane, 'point' and 'line' are *dual elements* and, for example, drawing a line thrugh a point and marking a point on a line are *dual operations*. Theorems that can be obtained from one another by replacing each element and operation by its dual are *dual theorems*. Dual theorems feature prominently in PROJECTIVE GEOMETRY. In logic, 'implied' and 'is implied by' are *dual relations* and may be interchanged together with the logical connectives 'and' and 'or'. In set theory, the relations 'is contained in' and 'contains' are dual relations and can be interchanged along with the 'union' and 'intersection'. In general, the principle of duality is found where the structure under consideration is a LATTICE.

**dummy variable** A symbol that can be replaced by any other symbol without changing the meaning of the expression in which it occurs.

**duodecahedron** /dew-ŏ-dek-ă-**hee**-drŏn/ *See* dodecahedron.

**duodecimal** /dew-ŏ-**dess**-ă-măl/ Denoting or based on twelve. In a duodecimal number system there are twelve different digits instead of ten. If, for example, ten and eleven were given the symbols *A* and *B* re-spectively, 12 would be written as 10 and 22 as 1A. Duodecimal numbers are of little use, but some duodecimal units (1 foot = 12 inches) are still in use. *Compare* binary; decimal; hexadecimal; octal.

**duplication of the cube** A classic problem of ancient Greek geometry, to find a way, using only a straight edge and compass, to find the side of a cube the volume of which is exactly double that of a given cube. It is now known that this cannot be done. This is the case because it is equivalent to the problem of finding the cube root of 2 using a ruler and compass. This is impossible because geometrical methods involving a ruler and compass can only lead to length in the cases of addition, subtraction, multiplication, division and square roots but definitely not cube roots.

**duty** A tax levied on certain kinds of transactions. Examples include taxes on importing and exporting goods and the tax levied on alcohol and tobacco products.

**dynamic friction** *See* friction.

**dynamics** The study of how objects move under the action of forces. This includes the relation between force and motion and the description of motion itself. *See also* mechanics.

**dyne** /dỹn/ Symbol: dyn The former unit of force used in the c.g.s. system. It is equal to $10^{-5}$ N.

**e** A fundamental mathematical constant defined by the series:

$$e = \lim(1 + 1/n) \, n \to \infty$$

or by

$$e = 1 + 1/1! + 1/2! \ldots + 1/n! + \ldots$$

e is the base of natural logarithms. It is both irrational and transcendental. It is related to $\pi$ by the formula $e^{i\pi} = -1$. The function $e^x$ has the property that it is its own derivative, that is, $de^x/dx = e^x$.

**eccentric** Denoting intersecting circles, spheres, etc., that do not have the same center. *Compare* concentric.

**eccentricity** A measure of the shape of a conic. The eccentricity is the ratio of the distance of a point on the curve from a fixed point (the focus) to the distance from a fixed line (the directrix). For a parabola, the eccentricity is 1. For a hyperbola, it is greater than 1, and for an ellipse, it is between 0 and 1. A circle has an eccentricity of 0.

**echelon matrix** A matrix in which all the rows of a matrix in which all the entries are zero come beneath the rows in which not all the entries are zero and in which the first non-zero entry in a row with non-zero values is 1, with this entry being in a column which is to the right of the first 1 in the row above it. It is possible to convert any matrix into an echelon matrix by using the technique of GAUSSIAN ELIMINATION.

$$\begin{pmatrix} 1 & 3 & 8 & -2 & 7 \\ 0 & 1 & 5 & 3 & -4 \\ 0 & 0 & 0 & 1 & 6 \\ 0 & 0 & 0 & 0 & 0 \end{pmatrix}$$

Echelon matrix

**ecliptic** /i-**klip**-tik/ The apparent path along which the Sun moves each year. It is the great circle formed by the intersection of the plane of the Earth's orbit with the celestial sphere.

**edge** A straight line where two faces of a solid meet. A cube has twelve edges.

**efficiency** Symbol: $\eta$ A measure used for processes of energy transfer; the ratio of the useful energy produced by a system or device to the energy input. For example, the efficiency of an electric motor is the ratio of its mechanical power output to the electrical power input. There is no unit of efficiency; however efficiency is often quoted as a percentage. In practical systems some dissipation of energy always occurs (by friction, air resistance, etc.) and the efficiency is less than 1. For a machine, the efficiency is the force ratio divided by the distance ratio.

**effort** The force applied to a MACHINE.

**eigenfunction** /**ÿ**-gĕn-fung-shŏn/ *See* eigenvalue.

**eigenvalue** /**ÿ**-gĕn-val-yoo/ (**from German** *eigen* = 'allowed') An eigenvalue for a linear transformation $L$ on a vector space $V$ is a scalar $\lambda$ for which there is a nonzero solution vector $v$ in $V$ such that $Lv = \lambda v$ and $v$ satisfies any given boundary conditions. The vector $v$ is an *eigenvector* (or *characteristic vector*) belonging to the eigenvalue $\lambda$. An eigenvector for a linear operator on a vector space whose vectors are functions is also called an *eigenfunction*. As an example, consider the equation $-y'' = \lambda y$ with boundary conditions $y = 0$ when $x = a$ or $b$.

A nonzero solution exists only if $\lambda = n^2\pi^2$ where $n$ is an integer. The solution then is
$$y = C[\sin n\pi(x - a)]/(b - a)$$
where $C$ is an arbitrary constant. The eigenvalues of the equation are given by $n^2\pi^2$ (where $n$ is an integer). The corresponding eigenfunctions (or eigenvectors) are given by
$$y = C[\sin(x - a)]/(b - a)$$

**eigenvector** /ȳ-gĕn-vek-ter/ *See* eigenvalue.

**elastic collision** A collision for which the restitution coefficient is equal to one. Kinetic energy is conserved during an elastic collision. In practice, collisions are not perfectly elastic as some energy is transferred to internal energy of the bodies. *See also* restitution; coefficient of.

**electronvolt** /i-lek-tron-**vohlt**/ Symbol: eV A unit of energy equal to $1.602\ 191\ 7 \times 10^{-19}$ joule. It is defined as the energy required to move the charge of an electron across a potential difference of one volt. It is normally used only to measure energies of elementary particles, ions, or states.

**element** /**el**-ĕ-mĕnt/ **1.** A single item that belongs to, or is a member of, a set. 'February', for example, is an element of the set {month in the year}. The number 5 is an element of the set of integers between 2 and 10. In set notation this is written as
$$5 \in \{2,3,4,5,6,7,8,9,10\}$$
**2.** A line segment forming part of the curved surface of a surface, as of a cone or cylinder.
**3.** A small part of a line, surface, or volume summed by integration.
**4.** (of a matrix) *See* matrix.

**elevation** A drawing that shows the appearance of a solid object as viewed from the front, back or side. *See* plan. *See also* angle of elevation.

**eliminant** /i-**lim**-ă-nănt/ (**characteristic; resultant**) The relationship between coefficients that results from eliminating the variable from a set of simultaneous equations. For example, in the equations

$$a_1 x + b_1 y + c_1 = 0$$
$$a_2 x + b_2 y + c_2 = 0$$
$$a_3 x + b_3 y + c_3 = 0$$

the eliminant is given by the matrix of coefficients:

$$\begin{vmatrix} a_1 & b_1 & c_1 \\ a_2 & b_2 & c_2 \\ a_3 & b_3 & c_3 \end{vmatrix}$$

**elimination** Removing one of the unknowns in an algebraic equation, for example, by the substitution of variables or by cancellation.

**ellipse** A CONIC with an eccentricity between 0 and 1. An ellipse has two foci. A line through the foci cuts the ellipse at two *vertices*. The line segment between the vertices is the *major axis* of the ellipse. The point on the major axis mid-way between the vertices is the *center* of the ellipse. A line segment through the center perpendicular to the major axis is the *minor axis*. Either of the chords of the ellipse through a focus parallel to the minor axis is a *latus rectum*. The area of an ellipse is $\pi ab$, where $a$ is half the major axis and $b$ is half the minor axis. (Note that for a circle, in which the eccentricity is zero, $a = b = r$ and the area is $\pi r^2$.)

The *sum property* of an ellipse is that for any point on the ellipse the sum of the distances from the point to each focus is a constant. The ellipse also has a reflection property; for a given point on the ellipse the two lines from each focus to the point make equal angles with a tangent at that point.

In Cartesian coordinates the equation:
$$x^2/a^2 + y^2/b^2 = 1$$
represents an ellipse with its center at the origin. The major axis is on the $x$-axis and the minor axis on the $y$-axis. The major axis is $2a$ and the minor axis is $2b$. The foci of the ellipse are at the points $(+ea,0)$ and $(-ea,0)$, where $e$ is the eccentricity. The two directrices are the lines $x = a/e$ and $x = -a/e$. The length of the latus rectum is $2b^2/a$.

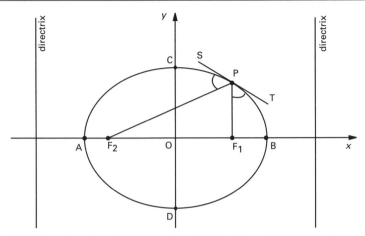

Ellipse: AB is the major axis and CD is the minor axis. $F_1$ and $F_2$ are the foci. Lines from these to any point P make equal angles with a tangent ST.

**ellipsoid** /i-**lip**-soid/ A solid body or curved surface in which every plane cross-section is an ellipse or a circle. An ellipsoid has three axes of symmetry. In three-dimensional Cartesian coordinates, the equation of an ellipsoid with its center at the origin is:

$$x^2/a^2 + y^2/b^2 + z^2/c^2 = 1$$

where $a$, $b$, and $c$ are the points at which it crosses the $x$, $y$, and $z$ axes respectively. In this case the axes of symmetry are the coordinate axes. A *prolate ellipsoid* is one generated by rotating an ellipse about its major axis. An *oblate ellipsoid* is generated by rotation about the minor axis.

**elliptic paraboloid** A CONICOID that is described by the equation $x^2/a^2 + y^2/b^2 = 2z/c$, where $a$, $b$ and $c$ are constants. The $xz$ and $yz$ planes are planes of reflection symmetry. Cross-sections of an elliptic paraboloid formed by planes that satisfy the equation $z = k$, where $k$ is a non-negative number are, in general, ellipses. In the special case in which $a = b$, the ellipses become circles. If $k$ is a negative number then the plane does not intersect the elliptic paraboloid. Cross-sections of an elliptic paraboloid formed by planes that are parallel to either the $xz$ or $yz$ plane are parabolas.

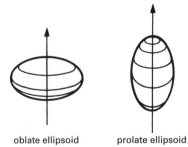

oblate ellipsoid      prolate ellipsoid

An ellipsoid can be generated by rotating an ellipse about one of its axes. An oblate ellipsoid is generated by rotation about the minor axis and a prolate ellipsoid is generated by rotation about a major axis.

**empirical** /em-**pi**-ră-kăl/ Derived directly from experimental results or observations.

**empirical probability** The probability of an event occurring as determined empirically by carrying out a large number of trials in which the event could take place. The number of times the event occurs is counted, as is the number of trials. The value of the number of times the event occurred divided by the number of trials ($n$) is then calculated. The empirical probability is defined to be the limiting value of this ratio as $n \to \infty$. In practice, taking $n \to \infty$ means that the number of trials has to become very large. Exactly how large depends on the nature of the problem being examined. It can also happen that empirical probability gives a value that is different from the value theoretically expected. For example, if a coin is tossed one might expect that the probability of it landing head up is exactly 0.5 but it might be the case that some bias in the coin causes a different empirical probability.

**empty set** Symbol: $\varnothing$ (**null set**) The set that contains no elements. For example, the set of 'natural numbers less than 0' is an empty set. This could be written as $\{m: m \in N; m<0\} = \varnothing$.

**energy** Symbol: $W$ A property of a system – its capacity to do work. Energy and work have the same unit: the joule (J). It is convenient to divide energy into KINETIC ENERGY (energy of motion) and POTENTIAL ENERGY ('stored' energy). Names are given to many different forms of energy (chemical, electrical, nuclear, etc.); the only real difference lies in the system under discussion. For example, chemical energy is the kinetic and potential energies of electrons in a chemical compound. *See also* mass–energy equation.

**engineering notation** *See* standard form.

**enlargement** A geometrical PROJECTION that produces an image larger (smaller if the scale factor is less than 1) than, but similar to, the original shape.

**entailment** In logic, the relationship that holds between two (or more) propositions when one can be deduced from the other. If conclusion $C$ is deducible from premises $A$ and $B$, then $A$ and $B$ are said to *entail C*. *See* deduction.

**entropy** /en-trŏ-pee/ A physical quantity that originated in thermodynamics and statistical mechanics but can be defined for general dynamical systems and also defined in terms of information theory. The definition of entropy in statistical mechanics, which provides a molecular picture for entropy in thermodynamics, means that entropy can be regarded as a measure of the disorder of a system. In thermodynamics, entropy is a measure of a quantity that increases in a system as the ability of the energy of that system to do work decreases. The entropy $S$ of a system in thermodynamics is defined by $dS = dQ/T$, where $dS$ is an infinitesimal change in the entropy of a system, $dQ$ the infinitesimal amount of heat absorbed by the system and $T$ is the thermodynamic temperature. In statistical mechanics entropy is given by $S = k\ln W + B$, where $k$ is a constant of statistical mechanics known as the Boltzmann constant, $W$ is the statistical probability for the system, i.e. the number of distinguishable ways the system can be realized, and $B$ is another constant. Here, $\ln W$ is the natural logarithm of $W$. The definition of entropy used in general dynamical systems leads to a non-zero value of entropy being associated with chaos. In information theory, entropy can be regarded as a measure of the uncertainty in our knowledge about a system.

**enumerable** /i-**new**-mĕ-ră-băl/ *See* countable.

**envelope** Consider a one-parameter family of curves in three dimensions – i.e. a family of curves that can be represented in terms of a common parameter that is constant along each curve, but is changed from curve to curve. The envelope of this family of curves is the surface traced out by these curves. This surface is tangent to every curve of the family. Its equation is obtained

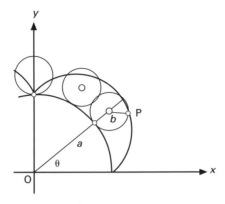

Epicycloid: the epicycloid traced out by a point P on a circle of radius *b* that rolls round a large circle of radius *a*.

by eliminating the parameter between the equation of the curve and the partial derivative of this equation with respect to the parameter. For example, the envelope of the family of paraboloids given by $x^2 + y^2 = 4a(z - a)$ is the equation obtained by eliminating $a$ from the equations $x^2 + y^2 = 4a(z - a)$ and $z - 2a = 0$, i.e. the circular cone $x^2 + y^2 = z^2$.

**epicycle** /**ep**-ă-sÿ-kăl/ A circle that rolls around the circumference of another, tracing out an EPICYCLOID.

**epicycloid** /ep-ă-**sÿ**-kloid/ The plane curve traced out by a point on a circle or *epicycle* rolling along the outside of another fixed circle. For example, if a small cog wheel turns on a larger stationary wheel, then a point on the rim of the smaller wheel traces out an epicycloid. In a two-dimensional Cartesian-coordinate system that has a fixed circle of radius $a$ centred at the origin and another of radius $b$ rolling around the circumference, the epicycloid is a series of continuous arcs that move away from the first circle to a distance $2b$ and then return to touch it again at a cusp point where the next arc begins. The epicycloid has only one arc if $a = b$, two if $a = b/2$, and so on. If the angle between the radius from the origin to the moving point of contact between the two

circles is $\theta$, the epicycloid is defined by the parametric equations:
$x = (a + b) \cos\theta - a \cos[(a + b)\theta/a]$
$y = (a + b) \sin\theta - a \sin[(a + b)\theta/a]$ *See also* hypocycloid.

**equal** Describing quantities that are the same. For example, the quantities on the left-hand side of an equation are equal to the quantities on the right-hand side of the equation; two sides of an isosceles triangle are equal in length. Equality is often denoted by the equals sign =.

**equality** Symbol: = The relationship between two quantities that have the same value or values. If two quantities are not equal, the symbol ≠ is used. For example, $x \neq 0$ means that the variable $x$ cannot take the value zero. When the equality is only approximate, the symbol ≅ is used. For example, if $\Delta x$ is small compared to $x$ then $x + (\Delta x)^2 \cong x$. When two expressions are exactly equivalent the symbol ≡ is used. For example $\sin^2\alpha \equiv 1 - \cos^2\alpha$ because it applies for all values of the variable $\alpha$. *See also* equation; inequality.

**equation** A mathematical statement that one expression is equal to another, that is, two quantities joined by an equals sign. An algebraic equation contains unknown or variable quantities. It may state that two quantities are identical for all values of the

variables, and in this case the identity symbol ≡ is normally used. For example:
$$x^2 - 4 \equiv (x - 2)(x + 2)$$
is an *identity* because it is true for all values that $x$ might have. The other kind of algebraic equation is a *conditional equation*, which is true only for certain values of the variables. To solve such an equation, that is, to find the values of the variables at which it is valid, it often has to be rearranged into a simpler form. In simplifying an equation, the same operation can be carried out on the expressions on both sides of the equals sign. For example,
$$2x - 3 = 4x + 2$$
can be simplified by adding 3 to both sides to give
$$2x = 4x + 5$$
then subtracting $4x$ from both sides to give
$$-2x = 5$$
and finally dividing both sides by –2 to obtain the solution $x = -5/2$.

This kind of equation is called a linear equation, because the highest power of the variable $x$ is one. It could also be written in the form $-2x - 5 = 0$. On a Cartesian coordinate graph,
$$y = -2x - 5$$
is a straight line that crosses the $x$-axis at $x = -5/2$.

Performing the same operation on both sides of an equation does not necessarily give an equation exactly equivalent to the original. For example, starting with $x = y$ and squaring both sides gives $x^2 = y^2$, which means that $x = y$ or $x = -y$. In this case the symbol $\Rightarrow$ is used between the equations, meaning that the first implies the second, but the second does not imply the first. That is,
$$x = y \Rightarrow x^2 = y^2$$
Where the two equations are equivalent, the symbol $\Leftrightarrow$ is used, for example,
$$2x = 2 \Leftrightarrow x = 1$$

**equations of motion** Equations that describe the motion of an object with constant accleration ($a$). They relate the velocity $v_1$ of the object at the start of timing to its velocity $v_2$ at some later time $t$ and to the object's displacement $s$. They are:

$$v_2 = v_1 + at^2$$
$$s = (v_1 + v_2)t/2$$
$$s = v_1t + at^2/2$$
$$s = v_2t - at^2/2$$
$$v_2{}^2 = v_1{}^2 + 2as$$

**equator** On the Earth's surface, the circle formed by the plane cross-section that perpendicularly bisects the axis of rotation. The plane in which the circle lies is called the *equatorial plane*. A similar circle on any sphere with a defined axis is also called an equator, or equatorial circle.

**equatorial** *See* equator.

**equiangular** /ee-kwee-**ang**-gyŭ-ler, ek-wee-/ Having equal angles.

**equidistant** /ee-kwă-**dis**-tănt, ek-wă-/ At the same distance. For example, all points on the circumference of a circle are equidistant from the center.

**equilateral** /ee-kwă-**lat**-ĕ-răl, ek-wă-/ Having sides of equal length. For example, an equilateral triangle has three sides of equal length (and equal interior angles of 60°).

**equilibrant** /i-**kwil**-ă-brănt/ A single force that is able to balance a given set of forces and thus cause equilibrium. It is equal and opposite to the resultant of the given forces.

**equilibrium** /ee-kwă-**lib**-ree-ŭm, ek-wă-/ A state of constant momentum. An object is in equilibrium if: **1.** its linear momentum does not change (it moves in a straight line at constant speed and has constant mass, or is at rest); **2.** its angular momentum does not change (its rotation is zero or constant). For these conditions to be met: **1.** the resultant of all outside forces acting on the object must be zero (or there are no outside forces); **2.** there is no resultant turning-effect (moment).

An object is not in equilibrium if any of the following are true: **1.** its mass is changing;

2. its speed is changing;
3. its direction is changing;
4. its rotational speed is changing.
   *See also* stability.

**equivalence** /i-kwiv-ă-lĕns/ *See* biconditional.

**equivalence class** The set of elements [x] in a set S that is equivalent to x by an EQUIVALENCE RELATION on that set. The distinct equivalence classes of a set are said to be a *partition* of that set, with each element of the set belonging to only one of the equivalence classes.

**equivalence principle** *See* relativity; theory of.

**equivalence relation** A binary relation R defined on a set S is said to be an equivalence relation if it satisfies the following three properties: (1) $xRx$ for every x in S – this is the property of *reflexivity*, (2) if $xRy$ then $yRx$ – this is the property of *symmetry*, and (3) if $xRy$ and $yRz$ then $xRz$ – this is the property of *transitivity*. Such relations are especially important because they partition the set on which they are defined into disjoint classes, known as *equivalence classes*.

**equivalent fractions** /i-kwiv-ă-lĕnt/ Two or more fractions that represent the same rational number; they can be canceled until they are the same fraction (*see* cancellation). For example, 10/16, 15/24, 45/72 and 65/104 are equivalent fractions (equal to 5/8).

**erase head** The part of any magnetic recording device – tape or disk – that erases data (recorded signals) before the write head records new data. *See* write head.

**Eratosthenes, sieve of** /e-ră-**tos**-th'ĕ-neez/ A method of finding prime numbers. To find all the prime numbers less than a given number n one first goes through all the numbers from 2 to n removing all those that are multiples of 2. Then all those after 3 are examined and all the multiples of 3 are removed. One proceeds in this way

with all the numbers less than or equal to √n. Only prime numbers will remain. This method in number theory is named for the Greek astronomer Eratosthenes of Cyrene (*c.* 276 BC–*c.* 194 BC).

**erg** A former unit of energy used in the c.g.s. system. It is equal to $10^{-7}$ joule.

**Erlangen program** /**er**-lang-ĕn/ A view of geometry in which each type of geometry is considered to be characterized by a group of transformations and the invariants of such transformations. For example, the geometry of a crystal lattice is characterized by the SPACE GROUP of that lattice. The Erlangen program is so called because the eminent German mathematician Felix Klein put forward this idea when he became a member of the faculty of Erlangen University in 1872. Not all types of geometry can be incorporated into the Erlangen program. For example, RIEMANNIAN GEOMETRY cannot be incorporated directly, although there are some ways of generalizing the Erlangen program that can include Riemannian geometry.

**error** The uncertainty in a measurement or estimate of a quantity. For example, on a mercury thermometer, it is often possible to read temperature only to the nearest degree Celsius. A temperature of 20°C should then be written as $(20 \pm 0.5)$°C because it really means 'between 19.5°C and 20.5°C'. There are two basic types of error. *Random error* occurs in any direction, cannot be predicted, and cannot be compensated for. It includes the limitations in the accuracy of the measuring instrument and the limitations in reading it. *Systematic error* arises from faults or changes in conditions that can be corrected for. For example, if a 1 gram weight used on a balance is 2 milligrams underweight, every measurement taken with it will be 2 milligrams less than the correct value.

**escape speed** (**escape velocity**) The minimum initial speed (velocity) that an object must have in order to escape from the surface of a planet (or moon) against the gravitational attraction. The escape speed is equal to $\sqrt{(2GM/r)}$, where $G$ is the gravitational constant, $M$ is the mass of the planet, and $r$ is the radius of the planet. The concept also applies to the escape of the object from a distant orbit.

**estimate** A rough calculation, usually involving one or more approximations, made to give a preliminary answer to a problem.

**ether** /ee-th'er/ (**aether**) A hypothetical fluid, formerly thought to permeate all space and to be the medium through which electromagnetic waves were propagated. *See* relativity; theory of.

**Euclidean algorithm** /yoo-**klid**-ee-ăn/ A method of finding the highest common factor of two positive integers. The smaller number is divided into the larger. The remainder is then divided into the smaller number, obtaining a second remainder. This second remainder is then divided into the first remainder, to give a third remainder. This is divided into the second, and so on, until a zero remainder is obtained. The remainder preceding this is the highest common factor of the two numbers. For example, the numbers 54 and 930. Dividing 54 into 930 gives 17 with a remainder of 12. Dividing 12 into 54 gives 4 with a remainder of 6. Dividing 6 into 12 gives 2 with a remainder of 0. Thus 6 is the highest common factor of 54 and 930. The algorithm is named for the Greek mathematician Euclid (*c.* 330 BC–*c.* 260 BC).

**Euclidean geometry** A system of geometry described by the Greek mathematician Euclid in his book *Elements* (*c.* 300 BC). It is based on a number of definitions – point, line, etc. – together with a number of basic assumptions. These were axioms or 'common notions' – for example, that the whole is greater than the part – and postulates about geometric properties – for example, that a straight line is determined by two points. Using these basic ideas a large number of theorems were proved using formal deductive arguments. The basic assumptions of Euclid have been modified, but the

system is essentially that used today for 'pure' geometry.

One important postulate in Euclid's system is that concerned with parallel lines (the *parallel postulate*). Its modern form is that if a point lies outside a straight line only one straight line can be drawn through that point parallel to the other line. *See* non-Euclidean geometry.

**Euler characteristic** /**oi**-ler/ A topological property of a curve or surface. For a curve, the Euler characteristic is the number of vertices minus the number of closed continuous line segments between. For example, any polygon has an Euler characteristic of zero. For a surface, the Euler characteristic is equal to the number of vertices plus the number of faces minus the number of edges. For example, a cube has an Euler characteristic of 2, and a cylinder, a Möbius strip, and a Klein bottle have an Euler characteristic of zero. The Euler characteristic is named for the Swiss mathematician Leonhard Euler (1707–83).

**Euler line** A straight line on which the CENTROID G, the CIRCUMCENTER C and the ORTHOCENTER O of a triangle all lie. The ratio of CG to OG on this line is always 1/2.

**Euler's constant** Symbol: $\gamma$ A fundamental mathematical constant defined by the limit of:
$$1 + 1/2 + 1/3 + \dots + 1/n - \log n$$
as $n \to \infty$. To six figures its value is 0.577 216. It is not known whether Euler's constant is irrational or not.

**Euler's formula** 1. (for polyhedra) The formula that relates the number of vertices $v$, faces $f$, and edges $e$ in a polyhedron, that is:
$$v + f - e = 2$$
For example, a cube has eight vertices six faces and twelve edges:
$$8 + 6 - 12 = 2$$
Using the theorem it can be shown that there are only five regular polyhedrons.
2. The definition of the function $e^{i\theta}$ for any real value of $\theta$, where i is the square root of −1, is

$$e^{i\theta} = \cos\theta + i\sin\theta$$

Any complex number $z = x + iy$ can be written in this form. $x = r\cos\theta$ and $y = r\sin\theta$ are real, with $r$ and $\theta$ representing $z$ on an Argand diagram. Note that putting $\theta = \pi$ gives $e^{i\pi} = -1$ and $\theta = 2\pi$ gives $e^{2\pi i} = 1$.

**even** Divisible by two. The set of even numbers is $\{2,4,6,8,...\}$. *Compare* odd.

**even function** A function $f(x)$ of a variable $x$, for which $f(-x) = f(x)$. For example, $\cos x$ and $x^2$ are even functions of $x$. *Compare* odd function.

**event** In probability, an *event* is any subset of the possible outcomes of an experiment. The event is said to *occur* if the outcome is a member of the subset. For example, if two dice are thrown, an event is a subset of all ordered pairs $(m,n)$ where $m$ and $n$ are each one of the integers 1, 2, 3, 4, 5, 6. Thus $\{(1,3),(2,2),(3,1)\}$ is an event, which may also be described as 'obtaining a sum of four'. *See also* dependent.

**evolute** /ev-ŏ-loot/ The evolute of a given curve is the locus of the centers of curvature of all the points on the curve. The evolute of a surface is another surface formed by the locus of all the centers of curvature of the first surface.

**exact differential equation** A differential equation of the form $P(x,y)dx + Q(x,y)dy = 0$, which can be matched to the differential $d\phi$, where $d\phi = (d\phi/dx)dx + (d\phi/dy)dy$, where $\phi(x,y)$ is a function of $x$ and $y$. If this is the case, then $\phi(x,y) = C$, where $C$ is a constant and $d\phi = 0$. If this function $\phi(x,y)$ exists then: $P(x,y)dx + Q(x,y) = (d\phi/dx)dx + (d\phi/dy)dy$, and $P(x,y) = d\phi/dx$, $Q(x,y) = d\phi/dy$. The properties of an exact differential equation can be used to find solutions of that equation. For a differential equation to be exact it is necessary and sufficient that the second mixed partial derivatives of $\phi(x,y)$ do not depend on the order of differentiation, i.e. $\partial p(x,y)/\partial y = \partial^2\phi/(\partial y\partial x) = \partial^2\phi/(\partial x\partial y) = \partial Q(x,y)/\partial x$. For example, the equation $(x + 2y)dx + (2x + y)dy = 0$ is an exact differential equation.

**excluded middle, law of** *See* laws of thought.

**exclusive disjunction** (exclusive OR) *See* disjunction.

**exclusive OR gate** *See* logic gate.

**existence theorem** A theorem that proves that one or more mathematical entities of a certain kind exists; e.g. that a function has a zero or a fixed point. An existence proof may be indirect and show that a certain entity must exist without giving any information about it or how to find it.

**existential quantifier** In mathematical logic, a symbol meaning 'there is (are)'. It is usually written $\exists$. The quantifier is followed by a variable that it is said to *bind*. Thus $(\exists x)F(x)$ means 'There is something that has property F'.

**expansion** A quantity expressed as a sum of a series of terms. For instance, the expression:
$$(x + 1)(x + 2)$$
can be expanded to:
$$x^2 + 3x + 2$$
Often a function can be written as an infinite series that is convergent. The function can then be approximated to any required accuracy by taking the sum of a sufficient number of terms at the beginning of the series. There are general formulae for expanding some types of expression. For example, the expansion of $(1 + x)^n$ is $1 + nx + [n(n - 1)/2!]x^2 + [n(n - 1)(n - 2)/3!] x^3 + ...$ where $x$ is a variable between $-1$ and $+1$, and $n$ is an integer. *See* binomial expansion; determinant; Fourier series; Taylor series.

**expectation** *See* expected value.

**expected value** (expectation) The value of a variable quantity that is calculated to be most likely to occur. If $x$ can take any of the set of discrete values
$$\{x_1, x_2, ...x_n\}$$

which have corresponding probabilities $\{p_1, p_2, \ldots p_n\}$, then the expected value is
$$E(x) = x_1 p_1 + x_2 p_2 + \ldots + x_n p_n$$
If $x$ is a continuous variable with a probability density function $f(x)$, then

$$E(x) = \int_{-\infty}^{\infty} x f(x) \, dx$$

**explicit** Denoting a function that contains no dependent variables. *Compare* implicit.

**exponent** /eks-**poh**-nĕnt/ A number or symbol placed as a superscript after an expression to indicate the power to which it is raised. For example, $x$ is an exponent in $y^x$ and in $(ay + b)^x$.

The laws of exponents are used for combining exponents of numbers as follows:

Multiplication:
$$x^a x^b = x^{a+b}$$
Division:
$$x^a / x^b = x^{a-b}$$
Power of a power:
$$(x^a)^b = x^{ab}$$
Negative exponent:
$$x^{-a} = 1/x^a$$
Fractional exponent:
$$x^{a/b} = {}^b\sqrt{x^a}$$
A number raised to the power zero is equal to 1; i.e. $x^0 = 1$.

**exponential** /eks-pŏ-**nen**-shăl/ A function or quantity that varies as the power of another quantity. In $y = 4^x$, $y$ is said to vary exponentially with respect to $x$. The function $e^x$ (or exp$x$), where e is the base of natural logarithms, is the exponential of $x$.

The infinite series
$$1 + x + x^2/2! + x^3/3! + \ldots + x^n/n! + \ldots$$
is equal to $e^x$ and is known as the *exponential series*. The exponential form of a complex number is
$$re^{i\theta} = r(\cos\theta + i \sin\theta)$$
*See also* complex number; Euler's formula; Taylor series.

**exponential series** The infinite power series that is the expansion of the function $e^x$, namely:
$$1 + x + x^2/2! + x^3/3! + \ldots + x^n/n! + \ldots$$

This series is convergent for all real-number values of the variable $x$.

Replacing $x$ by $-x$ gives an alternating series for $e^{-x}$:
$$1 - x + x^2/2! - x^3/3! \ldots$$
Series for sinh$x$ and cosh$x$ can be obtained by combining series for $e^x$ and $e^{-x}$.

**expression** A combination of symbols (representing numbers of other mathematical entities) and operations; e.g. $3x^2$, $\sqrt{(x^2 + 2)}$, $e^x - 1$.

**exterior angle** The angle formed on the outside of a plane figure between the extension of one straight edge beyond a vertex, and the outer side of the other straight edge at that vertex. In a triangle, the exterior angle at one vertex equals the sum of the angles on the insides of the other two vertices, i.e. the sum of the interior opposite angles. *Compare* interior angle.

The exterior angle $\delta = 180° - \gamma = \alpha + \beta$.

**extraction** The process of finding a root of a number.

**extrapolation** The process of estimating a value outside a known range of values. For example, if the speed of an engine is controlled by a lever, and depressing the lever by two, four, and six centimeters gives speeds of 20, 30, and 40 revolutions per second respectively, then one can extrapolate from this information and assume that depressing it by a further two centimeters will increase the speed to 50 revolutions per second. Extrapolation can be carried out graphically; for example, a graph can be drawn over a known range of values and the resulting curve extended. The further from the known range, the greater will be the uncertainty in the extrapolation. The case in which the graph of the behavior is a straight line is a *linear extrapolation*. *Compare* interpolation.

**face** A flat surface on the outside of a solid figure. A cube has six identical faces.

**facet** A FACE, or flat side, of a many-sided object.

**factor** (**divisor**) A number by which another number is divided. *See also* common factor.

**factorial** The product of all the whole numbers less than or equal to a number. For example, factorial 7, written 7!, is equal to $7 \times 6 \times 5 \times 4 \times 3 \times 2 \times 1$. Factorial zero is defined as 1.

**factorization** The process of changing algebraic or numerical expressions from a sum of terms into a product. For example, the left side of the equation $4x^2 - 4x - 8 = 0$ can be factorized to $(2x + 2)(2x - 4)$ making it easy to solve for $x$. As the product of the two factors is 0 when either of the factors is 0, it follows that $(2x + 2) = 0$ and $(2x - 4) = 0$ will provide solutions, i.e. $x = -1$ and $x = 2$.

**factor theorem** The condition that $(x - a)$ is a factor of a polynomial $f(x)$ in a variable $x$ if and only if $f(a) = 0$. For example, if $f(x) = x^2 + x - 6$, $f(2) = 4 + 2 - 6 = 0$ and $f(-3) = 9 - 3 - 6 = 0$, so the factors of $f(x)$ are $(x - 2)$ and $(x + 3)$. The factor theorem is derived from the remainder theorem.

**Fahrenheit degree** /**fa**-rĕn-hÿt/ Symbol: °F A unit of temperature difference equal to one hundred and eightieth of the difference between the temperatures of freezing and boiling water. On the Fahrenheit scale water freezes at 32°F and boils at 212°F. To convert from a temperature on the Fahrenheit scale ($T_F$) to a temperature on the Celsius scale ($T_C$) the following formula is used: $T_F = 9T_C/5 + 32$. The scale is named for the German physicist (Gabriel) Daniel Fahrenheit (1686–1736).

**fallacy** *See* logic.

**family** A set of related curves or figures. For example, the equation $y = 3x + c$ represents a family of parallel straight lines.

**farad** /**fa**-răd, -rad/ Symbol: F The SI unit of capacitance. When the plates of a capacitor are charged by one coulomb and there is a potential difference of one volt between them, then the capacitor has a capacitance of one farad. $1\ F = 1\ CV^{-1}$, 1 farad = 1 coulomb per volt. The unit is named for the British physicist and chemist Michael Faraday (1791–1867).

**Farey sequence** /**fair**-ee/ The sequence of all fractions in lowest terms whose denominators are less than $n$, where $n$ is the order of the sequence, listed in increasing size. For example, the Farey sequence of order 5 is 1/5, 1/4, 1/3, 2/5, 1/2, 3/5, 2/3, 3/4, 4/5. The sequence is named for the English mathematician John Farey (1766–1826).

**fathom** A unit of length used to measure depth of water. It is equal to 6 feet (1.8288 m).

**F distribution** The statistical distribution followed by the ratio of variances, $s_1{}^2/s_2{}^2$, of pairs of random samples, size $n_1$ and $n_2$, taken from a normal distribution. It is used to compare different estimates of the same variance.

**feedback** *See* cybernetics.

**femto-** Symbol: f A prefix denoting $10^{-15}$. For example, 1 femtometer (fm) = $10^{-15}$ meter (m).

**Fermat's last theorem** /fair-mahz/ The theorem that the equation

$$x^n + y^n = z^n$$

where $n$ is an integer greater than 2, can have no solution for $x$, $y$, and $z$. Fermat wrote in the margin of a book on equations that he had discovered a 'truly wonderful' proof of the theorem but that the margin was too small to write it down. It is now generally thought that he was deluding himself; a proof was finally found in 1995 using advanced mathematics which did not exist at the time of Fermat (the seventeenth century). The theorem is named for the French mathematician and physicist Pierre de Fermat (1601–65).

**Fermat's principle** A fundamental principle of optics that states that the path a light ray takes is always the path that takes the least time. Fermat's principle can be used to derive the laws describing the reflection and refraction of light. It is an example of a VARIATIONAL PRINCIPLE.

**fermi** /fer-mee/ A unit of length equal to $10^{-15}$ meter. It was formerly used in atomic and nuclear physics. The unit is named for the Italian–American physicist Enrico Fermi (1901–54).

**Fibonacci numbers** /fee-bŏ-**nah**-chee/ The infinite sequence in which successive numbers are formed by adding the two previous numbers, that is:

$$1, 1, 2, 3, 5, 8, 13, 21, \ldots$$

The Fibonacci sequence of numbers is named for the Italian mathematician Leonardo Fibonacci (c. 1170–c. 1250).

**fictitious force** A force in a system that arises because of the frame of reference of the observer. Such 'forces' are said to be 'fictitious' because they do not actually exist; they can be removed by transfer to a different frame of refernce. Examples are centrifugal force and Coriolis force.

**field** 1. A set of entities with two operations, called addition and multiplication. The entities form a commutative group under addition with 0 as the identity element. If 0 is omitted, the entities form a commutative group under multiplication. Also the distribution law, $a(b + c) = ab + ac$, applies for all $a$, $b$, and $c$. An example of a field is the set of rational numbers. 2. A region in which a particle or body exerts a force on another particle or body through space. In a gravitational field a mass is supposed to affect the properties of the surrounding space so that another mass in this region experiences a force. The region is thus called a 'field of force'. Electric, magnetic, and electromagnetic fields can be similarly described. The concept of a field was introduced to explain action at a distance.

**figure** A shape formed by a combination of points, lines, curves, or surfaces. Circles, squares, and triangles are plane figures. Spheres, cubes, and pyramids are solid figures.

**finite decimal** *See* decimal.

**finite sequence** *See* sequence.

**finite series** *See* series.

**finite set** A set that has a fixed countable number of elements. For example, the set of 'months in the year' has 12 members and is therefore a finite set. *Compare* infinite set.

**first moment** For an area $A$ about an axis, the product of $A$ and the distance of C, the centroid of $A$, from the axis. Thus, if $A$ is in the $xy$ plane and the centroid C of the area has coordinates which are denoted by $(x,y)$ then the first moment of $A$ about the $x$-axis is $Ay$ and the first moment of $A$ about the $y$-axis is $Ax$. For a volume $V$ about the axis of rotation of a volume of revolution, the product of $V$ and the distance of C, the centroid of the volume, from the axis. Thus, if a volume $V$ which is generated by rotating an area about the $x$-axis has a centroid C which has coordinates denoted by $(x,0)$ then the first moment of $V$ about the $x$-axis is zero and the first moment of $V$ about the $y$-axis is $Vx$.

**first-order differential equation** A DIFFERENTIAL EQUATION in which the highest derivative of the dependent variable is a first derivative.

**fixed-point iteration** An approximate method for finding the root of an equation $f(x) = 0$. The first step in the procedure is to write the equation as $x = g(x)$. The next step is to take a value $x_0$ to be an approximation to a true root. Subsequent approximate values of the root are found by using the equation $x_i+1 = g(x_i)$. If the values of $x_i$ tend to a limiting value $a$ as $i$ increases then $a = g(a)$. This means that $a$ is a root of the equation $f(x) = 0$.

**fixed-point notation** *See* floating point.

**fixed-point theorem** A theorem that demonstrates that a function leaves one point in its domain unchanged, i.e. for which $f(x) = x$. One celebrated example is *Brouwer's fixed point theorem*, which states that any continuous transformation of a circular disk onto itself must have a fixed point.

**flip-flop** *See* bistable circuit.

**floating objects, law of** *See* flotation; law of.

**floating-point** A notation used to describe real numbers, particularly in computers. In this notation, a number is written in the form $a \times b^n$, where $a$ is a number between 0.1 and 1, $b$ is the number base being used (usually 10 or 2) and $n$ is an integer. The numbers $a$, $b$ and $n$ are called the *mantissa*, the *base*, and the *exponent* respectively. For example, the number 2538.9 can be written as $0.25389 \times 10^4$. Floating-point notation is contrasted with *fixed-point notation*, in which all numbers have both a fixed number of digits and a fixed number of digits after the decimal point. If the number in the example given is expressed by six digits, with

two of these digits being after the decimal point, it is written as 2538.90.

**floppy disk (diskette)** A device that can be used to store data in electronic form, consisting of a flexible plastic DISK with a magnetic coating on one or both sides. It is permanently encased in a stiff envelope inside which it can be made to rotate. A read–write head operates through a slot in the envelope.

**flotation, law of** An object floating in a fluid displaces its own weight of fluid. This follows from Archimedes' principle for the special case of floating objects. (A floating object is in equilibrium, its only support coming from the fluid. It may be totally or partly submerged.)

**flowchart** A diagram on which can be represented the major steps in a process used, say, in industry, or a problem to be investigated, or a task to be performed. A flowchart is built up from a number of boxes connected by arrowed lines. The boxes, of various shapes, have a label attached showing for example the operation or calculation to be done at each step. At a *decision box* a question is asked. The answer, either yes or no, determines which of two possible paths to take. Computer programs are often written by first drawing a flowchart of the problem or task in hand. *See also* program.

**fluctuation** A deviation from the average value of some quantity. If there are many particles in a system the fluctuations in the value of a quantity $Q$ can usually be described by the equation
$$<Q^2> - <Q>^2 = 1/<Q>$$
where $<Q>$ denotes the average value of a quantity. If it is proportional to the number of particles in the system then the fluctuations are usually very small. However, there are many problems in which fluctuations are important, a notable example being a system near to a phase transition.

**fluid ounce** *See* ounce.

**flywheel** A large heavy wheel (with a large moment of inertia) used in mechanical devices. Energy is used to make the wheel rotate at high speed; the inertia of the wheel keeps the device moving at constant speed, even though there may be fluctuations in the torque. A flywheel thus acts as an 'energy-storage' device.

**focal chord** A chord of a conic that passes through a focus.

**focal radius** A line from the focus of a conic to a point on the conic.

**focus** (*pl.* **focuses** or **foci**) A point associated with a CONIC. The distance between the focus and any point on the curve is in a fixed ratio (the eccentricity) to the distance between that point and a line (the directrix). An ellipse has two foci. The sum of the distances to each focus is the same for all points on the ellipse.

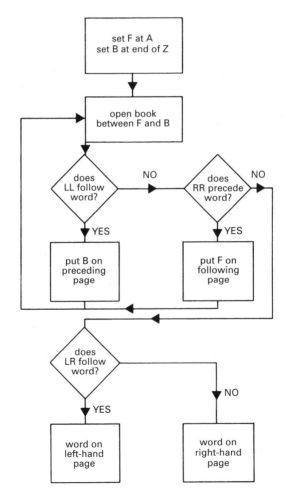

A flowchart for finding which page a word is on in this dictionary (assuming that it is in). F is a front marker, B a back marker, LL is the first word on a left-hand page, RR the last word on a right-hand page, and LR the first word on the right-hand page.

**foot** Symbol: ft The unit of length in the f.p.s. system (one third of a yard). It is equal to 0.304 8 meter.

**force** Symbol: $F$ That which tends to change an object's momentum. Force is a vector; the unit is the newton (N). In SI, this unit is so defined that:
$$F = \mathrm{d}(mv)/\mathrm{d}t$$
from Newton's second law.

**forced oscillation** (forced vibration) The oscillation of a system or object at a frequency other than its natural frequency. Forced oscillation must be induced by an external periodic force. *Compare* free oscillation. *See also* resonance.

**force ratio** (mechanical advantage) For a MACHINE, the ratio of the output force (load) to the input force (effort). There is no unit; the ratio is, however, sometimes given as a percentage. It is quite possible for force ratios far greater than one to be obtained. Indeed many machines are designed for this so that a small effort can overcome a large load. However the efficiency cannot be greater than one and a large force ratio implies a large distance ratio.

**forces, parallelogram (law) of** *See* parallelogram of vectors.

**forces, triangle (law) of** *See* triangle of vectors.

**formalism** A program for studying the foundations of mathematics in which the completeness and consistency of mathematical systems is examined. This program was dealt a fatal blow by the discovery of GÖDEL'S INCOMPLETENESS THEOREM.

**formal logic** *See* symbolic logic.

**format** The arrangement of information on a printed page, on a punched card, in a computer storage device, etc., that must or should be used to meet with certain requirements.

**formula** (*pl.* **formulas** or **formulae**) A general expression that can be applied to several different values of the quantities in question. For example, the formula for the area of a circle is $\pi r^2$, where $r$ is the radius.

**Foucault pendulum** /foo-**koh**/ A simple pendulum consisting of a heavy bob on a long string. The period is large and the plane of vibration rotates slowly over a period of time as a result of the rotation of the Earth below it. The apparent force causing this movement is the Coriolis force. The pendulum is named for the French physicist Jean Bernard Léon Foucault (1819–68).

**four-color problem** A problem in topology concerning the division of the surface of a sphere into regions. The name comes from the coloring of maps. It appears that in coloring a map it is not necessary to use more than four colors to distinguish regions from each other. Two regions with a common line boundary between them need different colors, but two regions meeting at a point do not. This was proved by Appel and Haken in 1976 with the extensive use of computers. On the surface of a torus only seven colors are necessary to distinguish regions.

**Fourier series** /**foo**-ree-ay, -er/ A method of expanding a function by expressing it as an infinite series of periodic functions (sines and cosines). The general mathematical form of the Fourier series is:
$$f(x) = a_0/2 + (a_1\cos x + b_1\sin x) + (a_2\cos 2x + b_2\sin 2x) + (a_3\cos 3x + b_3\sin 3x) + (a_n\cos nx + b_n\sin nx) + \dots$$
The constants $a_0$, $a_1$, $b_1$, etc., called *Fourier coefficients*, are obtained by the formulae:
$$a_0 = (1/\pi)\int_{-\pi}^{\pi} f(x)\mathrm{d}x$$
$$a_n = (1/\pi)\int_{-\pi}^{\pi} f(x)\cos nx\mathrm{d}x$$
$$b_n = (1/\pi)\int_{-\pi}^{\pi} f(x)\sin nx\mathrm{d}x$$
The series is named for the French mathematician Baron (Jean Baptiste) Joseph Fourier (1768–1830).

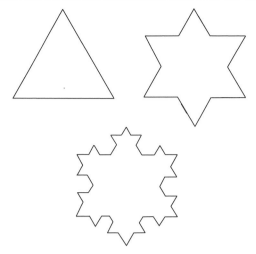

Fractal: generation of the snowflake curve.

**fourth root of unity** A complex number z which satisfies the relation $z^4 = 1$. This relation is satisfied by four values of z, namely 1, –1, i, and –i.

**f.p.s. system** A system of units that uses the foot, the pound, and the second as the base units. It has now been largely replaced by SI units for scientific and technical work although it is still used to some extent in the USA.

**fractal** /frak-tăl/ A curve or surface that has a fractional dimension and is formed by the limit of a series of successive operations. A typical example of a fractal curve is the *snowflake curve* (also known as the *Koch curve* and named for the Swedish mathematician Helge von Koch (1870–1924)). This is generated by starting with an equilateral triangle and dividing each side into three equal parts. The center part of each of these sides is then used as the base of three smaller equilateral triangles erected on the original sides. If the center parts are removed, the result is a star-shaped figure with 12 sides. The next stage is to divide each of the 12 sides into three and generate more triangles. The process is continued indefinitely with the resulting generation of a snowflake-shaped curve. A

curve of this type in the limit has a dimension that lies between 1 (a line) and 2 (a surface). The snowflake curve actually has a dimension of 1.26.

One important aspect of fractals is that they are generated by an iterative process and that a small part of the figure contains the information that could produce the whole figure. In this sense, fractals are said to be 'self-similar'. The study of fractals has applications in chaos theory and in certain scientific fields (e.g. the growth of crystals). It is also important in computer graphics, both as a method of generating striking abstract images and, because of the self-similarity property, as a method of compressing large graphics files.

*See also* Mandelbrot set.

**fraction** A number written as a quotient; i.e. as one number divided by another. For example, in the fraction ⅔, 2 is known as the *numerator* and 3 as the *denominator*. When both numerator and denominator are integers, the fraction is known as a *simple, common,* or *vulgar fraction*. A *complex fraction* has another fraction as numerator or denominator, for example (2/3)/(5/7) is a complex fraction. A *unit fraction* is a fraction that has 1 as the numerator. If the numerator of a fraction is less than the denominator, the fraction is known as a *proper fraction*. If not, it is an *improper fraction*. For example, 5/2 is an improper fraction and can be written as 2½. In this form it is called a *mixed number*.

In adding or subtracting fractions, the fractions are put in terms of their lowest common denominator. For example,

$$1/2 + 1/3 = 3/6 + 2/6 = 5/6$$

In multiplying fractions, the numerators are multiplied and the denominators multiplied. For example:

$$2/3 \times 5/7 = (2 \times 5)/(3 \times 7) = 10/21$$

In dividing fractions, one fraction is inverted thus:

$$2/3 \div 1/2 = 2/3 \times 2/1 = 4/3$$

*See also* ratio.

**frame of reference** A set of coordinate axes with which the position of any object may be specified as it changes with time. The origin of the axes and their spatial directions must be specified at every instant of time for the frame to be fully determined.

**framework** A collection of light rods that are joined together at their ends to give a rigid structure. In a framework the rods are said to be light if their weights are very much smaller than the loads they are bearing. When a framework has external forces acting on it each rod in the framework can either prevent the structure from collapsing or stop the joints connecting the rods from becoming separated. If a rod is stopping a collapse it exerts a push at both ends and is said to be in *compression* or in *thrust*. If a rod is stopping the structure from becoming separated it exerts a pull at both ends and is said to be in *tension*. If the whole of the framework is in equilibrium then the forces at each joint have to be in equilibrium. This means that in equilibrium the external forces on the framework are in equilibrium with the internal forces since the forces in the rods occur as equal and opposite forces.

**freedom, degrees of** The number of independent quantities that are necessary to determine an object or system. The number of degrees of freedom is reduced by constraints on the system since the number of independent quantities necessary to determine the system is also reduced. For example, a point in space has three degrees of freedom since three coordinates are needed to determine its postion. If the point is constrained to lie on a curve in space, it has then only one degree of freedom, since only one parameter is needed to specify its position on the curve.

**free oscillation** (free vibration) An oscillation at the natural frequency of the system or object. For example, a pendulum can be forced to swing at any frequency by applying a periodic external force, but it will swing freely at only one frequency, which depends on its length. *Compare* forced oscillation. *See also* resonance.

**free variable** In mathematical logic, a variable that is not within the scope of any quantifier. (If it is within the scope of a quantifier it is *bound*.) In the formula F($y$) → ($\exists x$)G($x$), $y$ is a free variable.

**French curve** A drawing instrument consisting of a rigid piece of plastic or metal sheeting with curved edges. The curvature of the edges varies from being almost straight to very tight curves, so that a part of the edge can be chosen to guide a pen or pencil along any desired curvature. Another instrument that serves the same purpose consists of a deformable strip of lead bar cut into short sections and surrounded by a thick layer of plastic. This is bent to form any required curvature.

French curve

**frequency** Symbol: $f$, $\nu$ The number of cycles per unit time of an oscillation (e.g. a pendulum, vibrating system, wave, alternating current, etc.). The unit is the hertz (Hz). The symbol $f$ is used for frequency, although $\nu$ is often employed for the frequency of light or other electromagnetic radiation.

Angular frequency ($\omega$) is related to frequency by $\omega = 2\pi f$.

The frequency of an event is the number of times that it has occurred, as recorded in a FREQUENCY TABLE.

**frequency curve** A smoothed FREQUENCY POLYGON for data that can take a continuous set of values. As the amount of data is increased and the size of class interval decreased, the frequency polygon more closely approximates a smooth curve. Relative frequency curves are smoothed relative frequency polygons. *See also* skewness.

**frequency function** 1. The function that gives the values of the frequency of each result or observation in an experiment. For a large sample that is representative of the whole population, the observed frequency function will be the same as the probability DISTRIBUTION FUNCTION $f(x)$ of a population variable $x$.
2. *See* random variable.

**frequency polygon** The graph obtained when the mid-points of the tops of the rectangles in a HISTOGRAM with equal class intervals are joined by line segments. The area under the polygon is equal to the total area of the rectangles.

**frequency table** A table showing how often each type (class) of result occurs in a sample or experiment. For example, the daily wages received by 100 employes in a company could be shown as the number in each range from $50.00 to $74.99, $75.00 to $99.99, and so on. In this case the representative value of each class (the *class mark*) is $(50 + 74.99)/2$, etc. *See also* histogram.

**Fresnel integrals** /fray-**nel**/ Two integrals $C(x)$ and $S(x)$ defined by:
$$C(x) = \int x_0 \cos(\pi u^2/2)\mathrm{d}u$$
$$S(x) = \int x_0 \sin(\pi u^2/2)\mathrm{d}u.$$

These integrals occur in the theory of diffraction and have been extensively tabulated. The Fresnel integrals can be combined to give:
$$C(x) - \mathrm{i}\,S(x) = \int x_0 \exp[-\mathrm{i}(\pi u^2/2)]\,\mathrm{d}u,$$
$$C(x) + \mathrm{i}\,S(x) = \int x_0 \exp[\mathrm{i}(\pi u^2/2)]\,\mathrm{d}u.$$

**friction** A force opposing the relative motion of two surfaces in contact. In fact, each surface applies a force on the other in the opposite direction to the relative motion; the forces are parallel to the line of contact. The exact causes of friction are still not fully understood. It probably results from minute surface roughness, even on apparently 'smooth' surfaces. Frictional forces do not depend on the area of contact. Presumably lubricants act by separating the surfaces. For friction between two solid surfaces, *sliding friction* (or *kinetic friction*) opposes friction between two moving surfaces. It is less than the force of *static* (or *limiting*) *friction*, which opposes slip between surfaces that are at rest. *Rolling friction* occurs when a body is rolling on a surface: here the surface in contact is constantly changing. Frictional force ($F$) is proportional to the force holding the bodies together (the 'normal reaction' $R$). The constants of proportionality (for different cases) are called *coefficients of friction* (symbol: $\mu$):
$$\mu = F/R$$
Two *laws of friction* are sometimes stated:
1. The frictional force is independent of the area of contact (for the same force holding the surfaces together).
2. The frictional force is proportional to the force holding the surfaces together. In sliding friction it is independent of the relative velocities of the surfaces.

**frustum** /**frus**-tŭm/ (*pl.* **frustums** or **frusta**) A geometric solid produced by two parallel planes cutting a solid, or by one plane parallel to the base.

**fulcrum** /**fŭl**-krŭm/ (*pl.* **fulcrums** or **fulcra**) The point about which a lever turns.

**function** (**mapping**) Any defined procedure that relates one number, quantity, etc., to another or others. In algebra, a function of a variable $x$ is often written as $f(x)$. If two variable quantities, $x$ and $y$, are related by the equation $y = x^2 + 2$, for example, then $y$ is a function of $x$ or $y = f(x)$ $= x^2 + 2$. The function here means 'square the number and add two'. $x$ is the *independent variable* and $y$ is the *dependent variable*. The *inverse function* – the one that expresses $x$ in terms of $y$ in this case – would be $x = \pm\sqrt{(y-2)}$, which might be written as $x = g(y)$.

A function can be regarded as a relationship between the elements of one set (the *range*) and those of another set (the *domain*). For each element of the first set there is a corresponding element of the second set into which it is 'mapped' by the function. For example, the set of numbers $\{1,2,3,4\}$ is mapped into the set $\{1,8,27,64\}$ by taking the cube of each element. A function may also map elements of a set into others in the same set. Within the set {all women}, there are two subsets {mothers} and {daughters}. The mapping between them is 'is the mother of' and the inverse is 'is the daughter of'.

**functional** A function in which both the domain and range can be sets of functions. Roughly speaking, a functional can be considered to be a function of a function. Functionals are used extensively in analysis and physics, particularly for problems involving many degrees of freedom. A functional F of a function f is denoted by F[f]. *See also* functional analysis.

**functional analysis** A branch of analysis that deals with mappings between classes of functions and the OPERATORS that bring about such mappings. In functional analysis a function can be regarded as a point in an abstract space. Functional analysis was extensively investigated in the twentieth century, with applications to differential equations, integral equations, and quantum mechanics. These enabled the foundations of these subjects to be laid on a sound axiomatic basis.

**fundamental** The simplest way (mode) in which an object can vibrate. The fundamental frequency is the frequency of this vibration. The less simple modes of vibration are the higher *harmonics*; their frequencies are higher than that of the fundamental.

**fundamental theorem of algebra** Every polynomial equation of the form:
$$a_0 z^n + a_1 z^{n-1} + a_2 z^{n-2} + \dots$$
$$a_{n-1} z + a_n = 0$$
in which $a_0$, $a_1$, $a_2$, etc., are complex numbers, has at least one complex root. *See also* polynomial.

**fundamental theorem of calculus** The theorem used in calculating the value of a DEFINITE INTEGRAL. If $f(x)$ is a continuous function of $x$ in the interval $a \leq x \leq b$, and if $g(x)$ is any INDEFINITE INTEGRAL of $f(x)$, then:
$$\int_a^b f(x)\mathrm{d}x = [g(x)_a^b] = g(b) - g(a)$$

**fundamental units** The units of length, mass, and time that form the basis of most systems of units. In SI, the fundamental units are the meter, the kilogram, and the second. *See also* base unit.

**furlong** A unit of length equal to one eighth of a mile. It is equivalent to 201.168 m.

**gallon** A unit of capacity usually used to measure volumes of liquids. In the USA it is defined as 231 cubic inches and is equal to $3.785\ 4 \times 10^{-3}$ m$^3$. In the UK it is defined as the space occupied by 10 pounds of pure water and is equal to $4.546\ 1 \times 10^{-3}$ m$^3$. 1 UK gallon is equal to 1.2 US gallons.

**game theory** A mathematical theory of the optimal behavior in competitive situations in which the outcomes depend not only on the participants' choices but also on chance and the choices of others. A *game* may be defined as a set of rules describing a competitive situation involving a number of competing individuals or groups of individuals. These rules give their permissible actions at each stage of the game, the amount of information available, the probabilities associated with the chance events that might occur, the circumstances under which the competition ends, and a pay-off scheme specifying the amount each player pays or receives at such a conclusion. It is assumed that the players are rational in the sense that they prefer better rather than worse outcomes and are able to place the possible outcomes in order of merit. Game theory has applications in military science, economics, politics, and many other fields.

**gamma function** The integral function

$$\Gamma(x) = \int_0^\infty t^{x-1}e^{-t}\mathrm{d}t$$

If $x$ is a positive integer $n$, then $\Gamma(n) = n!$ If $x$ is an integral multiple of ½, the function is a multiple of $\sqrt{\pi}$.

$$\Gamma(½) = \sqrt{\pi}$$
$$\Gamma(3/2) = (½)\sqrt{\pi}$$
etc.

**gate** *See* logic gate.

**gauss** /gows/ Symbol: G The unit of magnetic flux density in the c.g.s. system. It is equal to $10^{-4}$ tesla.

**Gaussian distribution** /gow-see-ăn/ *See* normal distribution.

**Gaussian elimination** A technique used in solving a set of linear equations for several unknown quantities. The set of equations is expressed in terms of a matrix that is formed from the coefficients and constants of the equations then converted into ECHELON FORM by elementary row operations, i.e. by multiplying a row by a number, adding a row which has been multiplied by a number to another row or swapping two rows. The set of solutions corresponding to an equation that has been transformed so as to give values of the unknown quantities directly is the same set of solutions for the original untransformed equations.

**Gauss's theorem** *See* divergence theorem.

**general conic** *See* conic.

**general form** (of an equation) A formula that defines a type of relationship between variables but does not specify values for constants. For example, the general form of a polynomial equation in $x$ is

$$ax^n + bx^{n-1} + cx^{n-2} + \ldots = 0$$

$a$, $b$, $c$, etc., are constants and $n$ is the highest integer power of $x$, called the *degree* of the polynomial. Similarly, the general form of a quadratic equation is

$$ax^2 + bx + c = 0$$

*See also* equation; polynomial.

**general theory** *See* relativity; theory of.

**generator** A line that generates a surface; for example, in a cone, cylinder, or solid of revolution.

**geodesic** /jee-ŏ-**dess**-ik/ A line on a surface between two points that is the shortest distance between the points. On a plane a geodesic is a straight line. On a spherical surface it is part of a great circle of the sphere.

**geometric distribution** The distribution of the number of independent Bernoulli trials before a successful result is obtained; for example, the distribution of the number of times a coin has to be tossed before a head comes up. The probability that the number of trials $(x)$ is $k$ is
$$P(x=k) = q^{k-1}p$$
The mean and variance are $1/p$ and $q/p^2$ respectively. The moment generating function is $e^t p/(1 - qe^t)$.

**geometric mean** *See* mean.

**geometric progression** *See* geometric sequence.

**geometric sequence** (**geometric progression**) A SEQUENCE in which the ratio of each term to the one after it is constant, for example, 1, 3, 9, 27, ... The general formula for the $n$th term of a geometric sequence is $u_n = ar^n$. The ratio is called the *common ratio*. In the example the first term, $a$, is 1, the common ratio, $r$, is 3, and so $u_n$ equals $3^n$. If a geometric sequence is convergent, $r$ lies between 1 and $-1$ (exclusive) and the limit of the sequence is 0. That is, $u_n$ approaches zero as $n$ becomes infinitely large. *Compare* arithmetic sequence. *See also* convergent sequence; divergent sequence; geometric series.

**geometric series** A SERIES in which the ratio of each term to the one after it is constant, for example, $1 + 2 + 4 + 8 + 16 + ...$ . The general formula for a geometric series is
$$S_n = a + ar + ar^2 + ... + ar^n = a(r^n - 1)/(r - 1)$$

In the example, the first term, $a$, is 1, the *common ratio*, $r$, is 2, and so the $n$th term $ar^n$ equals $2^n$. If $r$ is greater than 1, the series will not be convergent. If $-1 < r < 1$ and the sum of all the terms after the $n$th term can be made as small as required by making $n$ large enough, then the series is convergent. This means that there is a finite sum even when $n$ is infinitely large. The *sum to infinity* of a convergent geometric series is $a/(1 - r)$. *Compare* arithmetic series. *See also* convergent series; divergent series; geometric sequence.

**geometry** The branch of mathematics concerned with points, lines, curves, and surfaces – their measurement, relationships, and properties that are invariant under a given group of transformations. For example, geometry deals with the measurement or calculation of angles between straight lines, the basic properties of circles, and the relationship between lines and points on a surface. *See* analytical geometry; Euclidean geometry; non-Euclidean geometry; topology.

**giga-** Symbol: G A prefix denoting $10^9$. For example, 1 gigahertz (GHz) = $10^9$ hertz (Hz).

**gill** /gil/ A unit of capacity equal to one quarter of a pint. A US gill is equivalent to $1.182\ 9 \times 10^{-4}$ m$^3$ and a UK gill is equivalent to $1.420 \times 10^{-4}$ m$^3$. *See* pint.

**glide** A symmetry that can occur in crystals. It consists of the combination of a reflection and a translation. *See also* space group.

**Gödel's incompleteness theorem** /goh-dĕlz/ A fundamental result of mathematical logic showing that any formal system powerful enough to express the truths of arithmetic must be incomplete; that is that it will contain statements that are *true* but cannot be proved using the system itself. The incompleteness theorem is named for the Austrian–American mathematician Kurt Gödel (1906–78).

**Goldbach conjecture** /**golt**-bahkh/ The conjecture that every even number other than 2 is the sum of two prime numbers; so far, unproved. The conjecture is named for the Prussian mathematician and historian Christian Goldbach (1690–1764).

**golden rectangle** A rectangle in which the adjacent sides are in the ratio (1 + √5)/2.

**golden section** The division of a line of length $l$ into two lengths $a$ and $b$ so that $l/a = a/b$, that is, $a/b = (1 + √5)/2$. Proportions based on the golden section are particularly pleasing to the eye and occur in many paintings, buildings, designs, etc.

**goniometer** /gon-ee-**om**-ĕ-ter/ An instrument for measuring the angles between adjoining flat surfaces, such as the faces of a crystal.

**governor** A mechanical device to control the speed of a machine. One type of simple governor consists of two loads attached to a shaft so that as the speed of rotation of the shaft increases, the loads move farther outward from the center of rotation, while still remaining attached to the shaft. As they move outward they operate a control that reduces the rate of fuel or energy input to the machine. As they reduce speed and move inward they increase the fuel or energy input. Thus, on the principle of negative feedback, the speed of the machine is kept fairly constant under varying conditions of load.

**grad** (**gradient**) Symbol: ∇ A vector operator that, for any scalar function f($x,y,z$), has components in the $x$, $y$, and $z$ directions equal to the PARTIAL DERIVATIVES with respect to $x$, $y$, and $z$ in that order. It is defined as:

grad f = ∇f = $i$∂f/∂$x$ + $j$∂f/∂$y$ + $k$∂f/∂$z$

where $i$, $j$, and $k$ are the unit vectors in the $x$, $y$, and $z$ directions. In physics, ∇F is often used to describe the spatial variation in the magnitude of a force $F$ in, for example, a magnetic or gravitational field. It is a vector with the direction in which the rate of change of $F$ is a maximum, if such a

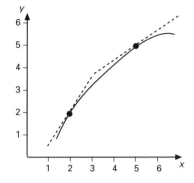

The gradient of the curve at the point (2,2) is 2, and at the point (5,5) is 1/2.

maximum exists. In the Earth's gravitational field this would be radially toward the center of the Earth (downward). In a magnetic field, ∇F would point along the lines of force.

**grade** Symbol: g A unit of plane angle equal to one hundredth of a right angle. $1^g$ is equal to 0.9°.

**gradient** 1. (slope) In rectangular Cartesian coordinates, the rate at which the $y$-coordinate of a curve or a straight line changes with respect to the $x$-coordinate. The straight line $y = 2x + 4$ has a gradient of +2; $y$ increases by two for every unit increase in $x$. The general equation of a straight line is $y = mx + c$, where $m$ is the gradient and $c$ is a constant. (0,$c$) is the point at which the line cuts the $y$-axis, i.e. the intercept. If $m$ is negative, $y$ decreases as $x$ increases.

For a curve, the gradient changes continuously; the gradient at a point is the gradient of the straight line that is a tangent to the curve at that point. For the curve $y = $ f($x$), the gradient is the DERIVATIVE d$y$/d$x$. For example, the curve $y = x^2$ has a gradient given by d$y$/d$x = 2x$, at any particular value of $x$.
2. See grad.

**gram** Symbol: g A unit of mass defined as $10^{-3}$ kilogram.

**gram-atom** *See* mole.

**gramme** An alternative spelling of *gram*.

**gram-molecule** *See* mole.

**graph** 1. A drawing that shows the relationship between numbers or quantities. Graphs are usually drawn with coordinate axes at right angles. For example, the heights of children of different ages can be shown by making the distance along a horizontal line represent the age in years and the distance up a vertical line represent the height in meters. A point marked on the graph ten units along and 1.5 units up represents a ten-year-old who is 1.5 meters tall. Similarly, graphs are used to give a geometric representation of equations. The graph of $y = x^2$ is a parabola. The graph of $y = 3x + 10$ is a straight line. Simultaneous equations can be solved by drawing the graphs of the equations, and finding the points where they cross. For the two equations above, the graphs cross at two points: $x = -2$, $y = 4$ and $x = 5$, $y = 25$.

There are various types of graph. Some, such as the HISTOGRAM and the PIE CHART, are used to display numerical information in a form that is simple and quickly understood. Some, such as CONVERSION GRAPHS, are used as part of a calculation. Others, such as SCATTER DIAGRAMS, may be used in analysing the results of a scientific experiment. *See also* bar chart.
2. (topology) A network of lines and vertices. *See* Königsberg bridge problem.

**graphics display** (graphical display unit) *See* visual display unit.

**gravitation** The concept originated by Isaac Newton around 1666 to account for the apparent motion of the Moon around the Earth, the essence being a force of attraction, called gravity, between the Moon and the Earth. Newton used this theory of gravitation to give the first satisfactory explanations of many facts, such as Kepler's laws of planetary motion, the ocean tides, and the precession of the equinoxes. *See also* Newton's law of universal gravitation.

**gravitational constant** Symbol: $G$ The constant of proportionality in the equation that expresses NEWTON'S LAW OF UNIVERSAL GRAVITATION:
$$F = Gm_1m_2/r^2$$
where $F$ is the gravitational attraction between two point masses $m_1$ and $m_2$ separated by a distance $r$. The value of $G$ is 6.67 $\times 10^{-11}$ N m$^2$ kg$^{-2}$. It is regarded as a universal constant, although it has been suggested that the value of $G$ may be changing slowly owing to the expansion of the Universe. (There is no current evidence that $G$ has changed with time.)

**gravitational field** The region of space in which one body attracts other bodies as a result of their mass. To escape from this field a body has to be projected outward with a certain speed (the *escape speed*). The strength of the gravitational field at a point is given by the ratio force/mass, which is equivalent to the acceleration of free fall, $g$. This may be defined as $GM/r^2$, where $G$ is the gravitational constant, $M$ the mass of the object at the center of the field, and $r$ the distance between the object and the point in question. The standard value of the acceleration of free fall at the Earth's surface is 9.8 m s$^{-2}$, but it varies with altitude (i.e. with $r^2$).

**gravitational mass** The MASS of a body as measured by the force of attraction between masses. The value is given by Newton's law of universal gravitation. Inertial and gravitational masses appear to be equal in a uniform gravitational field. *See also* inertial mass.

**gravity** The gravitational pull of the Earth (or other celestial body) on an object. The force of gravity on an object causes its weight.

**gravity, center of** *See* center of mass.

**great circle** A circle on the surface of a sphere that has the same radius as the sphere. A great circle is formed by a cross-section by any plane that passes through the center of the sphere. *Compare* small circle.

**greatest upper bound**  *See* bound.

**Green's function**  A solution for certain types of partial differential equation. The method of Green's functions is used extensively in many branches of theoretical physics, particularly mechanics, electrodynamics, acoustics, the many-body problem in quantum mechanics and quantum field theory. In these physical applications the Green's function can be considered to be a response to an impulse, with the Green's function method for finding the solution to a partial differential equation being the sum or integral of the responses to impulses. An example of a Green's function is the electric potential due to a unit point source of charge.

**Green's theorem**  A result in VECTOR CALCULUS that is a corollary of the DIVERGENCE THEOREM (Gauss theorem). If $u$ and $v$ are scalar functions, $S$ indicates a surface integral and $V$ a volume integral, Green's theorem states that
$$\int_V (U\nabla.\nabla v - v\nabla.\nabla u)dV =$$
$$\int S(U\nabla v - v\nabla u).dS.$$
An alternative form of Green's theorem is
$$\int_S u\nabla v.dS = \int_V u\nabla.\nabla v + \int_V \nabla u.\nabla v\, dV.$$
Green's theorem is used extensively in physical applications of vector calculus such as problems in electrodynamics.

**gross**  **1.** Denoting a weight of goods including the weight of the container or packing.
**2.** Denoting a profit calculated before deducting overhead costs, expenses, and (usually) taxes.
*Compare* net.

**group**  A set having certain additional properties: **1.** In a group there is a binary operation for which the elements of the set can be related in pairs, giving results that are also members of the group (the property of *closure*). For example, the set of all positive and negative numbers and zero form a group under the operation of addition. Adding any member to any other gives an element that is also a member of the group; e.g. 3 + (−2) = 1, etc.

**2.** There is an identity element for the operation – i.e. an element that, combined with another, leaves it unchanged. In the example, the identity element is zero: adding zero to any member leaves it unchanged; 3 + 0 = 3, etc.
**3.** For each element of the group there is another element – its *inverse*. Combining an element with its inverse leads to the identity element. In the example, the number +3 has an inverse −3 (and vice versa); thus +3 + (−3) = 0.
**4.** The associative law holds for the members of the group. In the example:
$$2 + (3 + 5) = (2 + 3) + 5$$
Any set of elements obeying the above rules forms a group. Note that the binary operation need not be addition. *Group theory* is important in many branches of mathematics – for instance, in the theory of roots of equations. It is also very useful in diverse branches of science. In chemistry, group theory is used in describing symmetries of atoms and molecules to determine their energy levels and explain their spectra. In physics, certain elementary particles can be classified into mathematical groups on the basis of their quantum numbers (this led to the discovery of the omega-minus particle as a missing member of a group). Group theory has also been applied to other subjects, such as linguistics.
*See also* Abelian group; cyclic group.

**group speed**  If a wave motion has a phase speed that depends on wavelength, the disturbance of a progressive wave travels with a different speed than the phase speed. This is called the *group speed*. It is the speed with which the group of waves travels, and is given by:
$$U = c - \lambda dc/d\lambda$$
where $c$ is the phase speed. The group speed is the one that is usually obtained by measurement. If there is no dispersion of the wave motion, as for electromagnetic radiation in free space, the group and phase speeds are equal.

**Guldinus theorem**  *See* Pappus' theorems.

**gyroscope** /jÿ-rŏ-skohp/  A rotating object

that tends to maintain a fixed orientation in space. For example, the axis of the rotating Earth always points in the same direction toward the Pole Star (except for a small PRECESSION). A spinning top or a cyclist are stable when moving at speed because of the gyroscopic effect. Practical applications are the navigational gyrocompass and automatic stabilizers in ships and aircraft.

**half-angle formulae** *See* addition formulae.

**half-plane** A set of points that lie on one side of a straight line. If the line $l$ is denoted by $ax + by + c = 0$, where $a$, $b$, and $c$ are constants, then the *open half-plane* on one side of the line is given by the set of points $(x,y)$ that satisfy the inequality $ax + by + c > 0$. The open half-plane on the other side of the plane is given by the set of points $(x,y)$ that satisfy the inequality $ax + by + c < 0$. The inequalities $ax + by + c > 0$ and $ax + by + c \leq 0$ define *closed half planes* on opposite sides of the plane. Half planes (both open and closed) are used extensively in linear programming.

**Hamiltonian** /ham-ăl-**toh**-nee-ăn/ In classical mechanics, a function of the coordinates $q_i$, $i = 1, 2, \dots n$, and momenta, $p_i$, $i = 1, 2, \dots n$, generally denoted by $H$ and defined by
$$H = \Sigma p_i q_i - L$$
where $q_i$ denotes the derivative with respect to time and $L$ is the *Lagrangian function* of the system expressed as a function of the coordinates momenta and time. If the Lagrangian function does not depend explicitly on time, the system is said to be *conservative* and $H$ is the total energy of the system. The Hamiltonian function is named for the Irish mathematician Sir William Rowan Hamilton (1805–65).

**Hamiltonian graph** A graph that has a *Hamiltonian cycle*, i.e. a cycle contains each vertex, where a cycle of a graph is a sequence of alternating vertices and edges which can be written as $v_0, l_1, v_1, l_2, v_2 \dots, l_l, v_l$, where $l_i$ is an edge which connects the vertices $v_{i-1}$ and $v_i$. In this definition all the edges are different and all the vertices are different, with the important exception that $v_0 = v_l$. The Hamiltonian graph is of interest in graph theory when the problem of going around the graph along edges so as to visit each vertex only once is considered.

**ham-sandwich theorem 1.** A ham sandwich can be cut with one stroke of a knife so that the ham and each slice of bread are exactly cut in half. More formally, if $A$, $B$, and $C$ are bounded connected sets in space, then there is a plane that cuts each set into two sets with equal volume.
**2.** If $f(x) \leq g(x) \leq h(x)$ for all $x$ and the functions f and h have the same limit, then g also has this limit.

**hard copy** In computer science, a document that can be read, such as a computer printout in plain language. *See* printer.

**hard disk** A rigid magnetic disk that stores programs and data in a computer. They are usually fixed in the machine and cannot be removed. *See also* disk; floppy disk.

**hardware** The physical embodiment of a computer system, i.e. its electronic circuitry, disk and magnetic tape units, line printers, cabinets, etc. *Compare* software.

**harmonic analysis** The use of trigonometric series to study mathematical functions. *See* Fourier series.

**harmonic mean** *See* mean.

**harmonic motion** A regularly repeated sequence that can be expressed as the sum of a set of sine waves. Each component sine wave represents a possible simple harmonic motion. The complex vibration of sound sources (with fundamental and overtones), for instance, is a harmonic motion, as is the sound wave produced. *See also* simple harmonic motion.

**harmonic progression** *See* harmonic sequence.

**harmonic sequence** (**harmonic progression**) An ordered set of numbers, the reciprocals of which have a constant difference between them; for example, {1, ½, ⅓, ¼, … 1/$n$}. In this example {1, 2, 3, 4, … $n$} have a constant difference – i.e. they form an ARITHMETIC SEQUENCE. The reciprocals of the terms in a harmonic sequence form an arithmetic sequence, and vice versa.

**harmonic series** The sum of the terms in a harmonic sequence; for example: $1 + \frac{1}{2} + \frac{1}{3} + \frac{1}{4} + \ldots$ The harmonic series is a divergent series.

**HCF** Highest common factor. *See* common factor.

**head** An input/output device in a computer that can read, write, or erase signals onto or from magnetic tape or disk. *See* erase head; write head.

**heat equation** An equation which describes the flow of heat. In three spatial dimensions it states that the rate of change of the absolute temperature $T$ with respect to time $t$ is proportional to the LAPLACIAN $\nabla^2$

of $T$: $\partial T / \partial t = c \nabla^2 T$, where $c$ is a constant. In the case of one space dimension with coordinate $x$ this equation becomes $\partial T / \partial t = c \partial^2 T / \partial x^2$. The heat equation is different from the wave equation since the heat equation has a first derivative with respect to time whereas the wave equation has a second derivative with respect to time. Physically, this corresponds to heat conduction being an irreversible process, like friction, whereas wave motion is reversible. The heat equation has the same form as the general equation for diffusion.

**hectare** /**hek**-tair/ Symbol ha. A unit of area equal to 10 000 square metres. tHE hectare is most often used as a convenient unit of land.

**hecto-** Symbol: h A prefix denoting $10^2$. For example, 1 hectometer (hm) = $10^2$ meters (m).

**height** A vertical distance, usually upward, from a base line or plane. For example, the perpendicular distance from the base of a triangle to the vertex opposite, and the distance between the uppermost and the base planes of a cuboid, are both known as the height of the figure.

**helix** /**hee**-liks/ A spiral-shaped space curve. A *cylindrical helix* lies on a cylinder. A *conical helix* lies on a cone. For example, the shape of the thread on a screw is a helix. In a straight screw, it is a cylindrical helix and in a conically tapered screw it is a conical helix.

**Helmholtz's theorem** /**helm**-holts/ A theorem in VECTOR CALCULUS that states that if a vector $V$ satisfies certain general mathematical conditions then $V$ can be written as the sum of an IRROTATIONAL VECTOR and a SOLENOIDAL VECTOR. This theorem is important in electrodynamics. It is named for the German physicist Hermann von Helmholtz (1821–94).

**hemisphere** The surface bounded by half of a SPHERE and a plane through the center of the sphere.

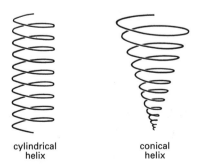

cylindrical
helix

conical
helix

Types of helix

**henry** /**hen**-ree/ Symbol: H The SI unit of inductance, equal to the inductance of a closed circuit that has a magnetic flux of one weber per ampere of a current in the circuit. 1 H = 1 Wb A$^{-1}$. The unit is named for the American physicist Joseph Henry (1797–1878).

**heptagon** /**hep**-tă-gon/ A plane figure with seven straight sides. A *regular heptagon* has seven equal sides and seven equal angles.

**Hermitian matrix** /her-**mee**-shăn, -**mish**-ăn, air-**mee**-shăn/ The *Hermitian conjugate* of a matrix is the transpose of the complex conjugate of the matrix, where the complex conjugate of a matrix is the matrix whose elements are the complex conjugates of the corresponding elements of the given matrix (*see* conjugate complex numbers). A *Hermitian matrix* is a matrix that is its own Hermitian conjugate; i.e. a square matrix such that $a_{ij}$ is the complex conjugate of $a_{ij}$ for all $i$ and $j$ where $a_{ij}$ is the element in the $i$th row and $j$th column. The matrix is named for the French mathematician Charles Hermite (1822–1901).

**Hero's formula** A formula for the area of a triangle with sides $a$, $b$, and $c$:
$$A = \sqrt{[s(s - a)(s - b)(s - c)]}$$
where $s$ is half the perimeter; i.e. $\frac{1}{2}(a + b + c)$. The formula is named for the Greek mathematician and inventor Hero of Alexandria (*fl.* AD 62).

**hertz** /herts/ Symbol: Hz The SI unit of frequency, defined as one cycle per second (s$^{-1}$). Note that the hertz is used for regularly repeated processes, such as vibration or wave motion. An irregular process, such as radioactive decay, would have units expressed as s$^{-1}$ (per second). The unit is named for the German physicist Heinrich Rudolf Hertz (1857–94).

**heuristic** /hyû-**ris**-tik/ Based on trial and error, as for example some techniques in iterative calculations. *See also* iteration.

**hexadecimal** /heks-ă-**dess**-ă-măl/ Denoting or based on the number sixteen. A hexadecimal number is made up with sixteen different digits instead of the ten in the decimal system. Normally these are 0, 1, 2, 3, 4, 5, 6, 7, 8, 9, A, B, C, D, E, F. For example, 16 is written as 10, 21 is written as 15 (16 + 5), 59 is written as 3B [(3 × 16) + 11]. Hexadecimal numbers are sometimes used in computer systems because they are much shorter than the long strings of binary digits that the machine normally uses. Binary numbers are easily converted into hexadecimal numbers by grouping the digits in fours. *Compare* binary; decimal; duodecimal; octal.

**hexagon** /**heks**-ă-gon/ A plane figure with six straight sides. A *regular hexagon* is one with all six sides and all six angles equal, the angles all being 120°. Congruent regular hexagons can be fitted together to cover completely a plane surface. Apart from squares and equilateral triangles, they are the only regular polygons with this property.

**hexahedron** /heks-ă-**hee**-drŏn/ A POLYHEDRON that has six faces. For example, the cube, the cuboid, and the rhombohedron are all hexahedrons. The cube is a *regular hexahedron*; all six faces are congruent squares.

**higher derivative** The $n$-th derivative of a function f of $x$, where $n$ is greater than or equal to two. If one writes $y = f(x)$ the higher derivatives are denoted by d$^n$y/dx$^n$ or $y^{(n)}$. The *second derivative* of $y$ with respect to $x$ is denoted by d$^2$y/dx$^2$ or $y''$. It is found by differentiating dy/dx with respect to $x$. Similarly, the *third derivative* of $y$ with respect to $x$ is denoted by d$^3$y/dx$^3$ or $y'''$. For example, if $y = x^4 + x^3 + 1$, dy/dx = $4x^3 + 3x^2$, d$^2$y/dx$^2$ = $12x^2 + 6x$, d$^3$y/dx$^3$ = $24x + 6$.

**highest common factor** *See* common factor.

**high-level language** *See* program.

**Hilbert's problems** /**hil**-berts/ A set of 23 important mathematical problems posed by the German mathematician David

Hilbert (1862–1941) at the International Congress of Mathematics in 1900. This set of problems has stimulated a great deal of important work in mathematics since 1900. Some of Hilbert's problems have been solved but, at the time of writing, some of them remain unsolved.

**histogram** /**his**-tŏ-gram/ A statistical graph that represents, by the length of a rectangular column, the number of times that each class of result occurs in a sample or experiment. *See also* frequency polygon.

**holomorphic** /hol-ŏ-**mor**-fik, hoh-lŏ-/ *See* analytic.

**homeomorphism** /hoh-mee-ŏ-**mor**-fiz-ăm, hom-ee-/ A one-one transformation between two topological spaces that is continuous in both directions. What this means is that if two figures are homeomorphic one can be continuously deformed into the other without tearing. For example, any two spheres of any size are homeomorphic. But a sphere and a torus are not.

**homogeneous** /hoh-mŏ-**jee**-nee-ŭs, hom-ŏ-/ 1. (in a function) Having all the terms to the same degree in the variables. For a homogeneous function $f(x,y,z,...)$ of degree $n$

$$F(kx,ky,kz,...) = k^n f(x,y,z)$$

for all values of $k$. For example, $x^2 + xy + y^2$ is a homogeneous function of degree 2 and

$$(kx)^2 + kx.ky + (ky)^2 = k^2(x^2 + xy + y^2)$$

2. Describing a substance or object in which the properties do not vary with position; in particular, the density is constant throughout.

**homomorphism** /hoh-mŏ-**mor**-fiz-ăm, hom-ŏ-/ If $S$ and $T$ are sets on which binary relations * and • are defined respectively, a mapping h from $S$ and $T$ is a *homomorphism* if it satisfies the condition $h(x*y) = h(x)•h(y)$ for all $x$ and $y$ in $S$, i.e. it preserves structure. If the mapping is one-to-one it is called an *isomorphism* and the sets $S$ and $T$ are *isomorphic*.

**Hooke's law** For an elastic material below its elastic limit, the extension resulting from the application of a load (force) is proportional to the load. The law is named for the English physicist Robert Hooke (1635–1703).

**horizontal** Describing a line that is parallel with the horizon. It is at right angles to the vertical.

**horsepower** Symbol: HP A unit of power equal to 550 foot-pounds per second. It is equivalent to 746 W.

**hour** A unit of time equal to 60 minutes or 1/24 of a day.

**hundredweight** Symbol: cwt In the UK, a unit of mass equal to 112 pounds. It is equivalent to 50.802 3 kg. In the USA a hundredweight is equal to 100 pounds, but this unit is rarely used.

**Huygens' principle** /**hy**-gĕnz/ (also known as HUYGENS' CONSTRUCTION) A result in the theory of waves which states that each point of a wave-front which is propagating serves as the source of secondary waves, with the secondary waves having the same frequency and speed as the original wave. Huygens' principle is used extensively in optics to analyze light waves. The modification of Huygens' principle to incorporate diffraction and interference is sometimes called the *Huygens–Fresnel principle*. Both Huygens' principle and the Huygens–Fresnel principle can be derived as mathematical consequence of the WAVE EQUATION. The principle is named for the Dutch astronomer and physicist Christiaan Huygens (1629–95).

**hybrid computer** A computer system containing both analog and digital devices so that the properties or each can be used to the greatest advantage. For instance, a digital and an analog computer can be interconnected so that data can be transferred between them. This is achieved by means of a *hybrid interface*. Hybrid computers are designed for specific tasks and have a variety of uses, mainly in scientific

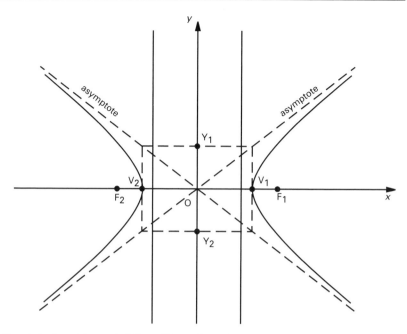

Hyperbola: $F_1$ and $F_2$ aere foci. $V_1$ and $V_2$ are the vertices with coordinates $(a,0)$ and $(-a,0)$ respectively. $Y_1$ and $Y_2$ have coordinates $(b,0)$ and $(-b,0)$.

and technical fields. *See also* analog computer; computer.

**hydraulic press** A MACHINE in which forces are transferred by way of pressure in a fluid. In a hydraulic press the effort $F_1$ is applied over a small area $A_1$ and the load $F_2$ exerted over a larger area $A_2$. Since the pressure is the same, $F_1/A_1 = F_2/A_2$. The force ratio for the machine, $F_2/F_1$, is $A_1/A_2$. Thus, in this case (and in the related hydraulic braking system and hydraulic jack) the force exerted by the user is less than the force applied; the force ratio is greater than 1. If the distance moved by the effort is $s_1$ and that moved by the load is $s_2$ then, since the same volume is transmitted through the system, $s_1A_1 = s_2A_2$; i.e. the distance ratio is $A_2/A_1$. In practical terms, the device is not very efficient since frictional effects are large.

**hydrostatics** /hȳ-drŏ-**stat**-iks/ The study of fluids (liquids and gases) in equilibrium.

**hyperbola** /hȳ-**per**-bŏ-lă/ (*pl.* **hyperbolas** or **hyperbolae**) A CONIC with an eccentricity greater than 1. The hyperbola has two branches and two axes of symmetry. An axis through the foci cuts the hyperbola at two *vertices*. The line segment joining these vertices is the *transverse axis* of the hyperbola. The *conjugate axis* is a line at right angles to the transverse axis through the center of the hyperbola. A chord through a focus perpendicular to the transverse axis is a *latus rectum*.

In Cartesian coordinates the equation:
$$x^2/a^2 - y^2/b^2 = 1$$
represents a hyperbola with its center at the origin and the transverse axis along the *x*-axis. $2a$ is the length of the transverse axis. $2b$ is the length of the conjugate axis. This is the distance between the vertices of a different hyperbola (the *conjugate hyperbola*) with the same asymptotes as the given one. The foci of the hyperbola are at the points $(ae,0)$ and $(-ae,0)$, where $e$ is the eccentricity. The asymptotes have the equations:
$$x/a - y/b = 0$$

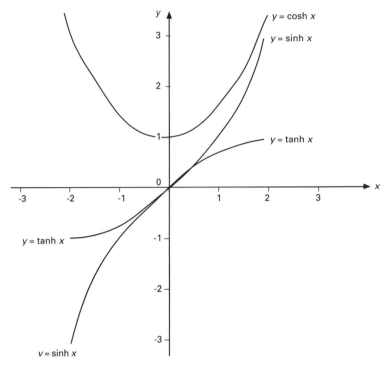

The graphs of the hyperbolic functions cosh x, sinh x, and tanh x.

$$x/a + y/b = 0$$

The equation of the conjugate hyperbola is

$$x^2/a^2 - y^2/b^2 = -1$$

The length of the latus rectum is $2b^2/ae$. A hyperbola for which $a$ and $b$ are equal is a rectangular hyperbola:

$$x^2 - y^2 = a^2$$

If a rectangular hyperbola is rotated so that the x- and y-axes are asymptotes, then its equation is

$$xy = k$$

where $k$ is a constant.

**hyperbolic functions** /hȳ-per-**bol**-ik/ A set of functions that have properties similar in some ways to the trigonometric functions, called the hyperbolic sine, hyperbolic cosine, etc. They are related to the hyperbola in the way that the trigonometric functions (circular functions) are related to the circle.

The *hyperbolic sine* (sinh) of an angle $a$ is defined as:

$$\sinh a = \tfrac{1}{2}(e^a - e^{-a})$$

The *hyperbolic cosine* (cosh) of an angle $a$ is defined as:

$$\cosh a = \tfrac{1}{2}(e^a + e^{-a})$$

The *hyperbolic tangent* (tanh) of an angle $a$ is defined as:

$$\tanh a = \sinh a/\cosh a =$$
$$(e^a - e^{-a})/(e^a + e^{-a})$$

*Hyperbolic secant* (sech), *hyperbolic cosecant* (cosech), and *hyperbolic cotangent* (coth) are defined as the reciprocals of cosh, sinh, and tanh respectively. Some of the fundamental relationships between hyperbolic functions are:

$$\sinh(-a) = -\sinh a$$
$$\cosh(-a) = +\cosh a$$
$$\cosh^2 a - \sinh^2 a = 1$$
$$\mathrm{sech}^2 a + \tanh^2 a = 1$$
$$\coth^2 a - \mathrm{cosech}^2 a = 1$$

**hyperbolic paraboloid** A CONICOID that is described by the equation $x^2/a^2 - y^2/b^2 = 2z/c$, where $a$, $b$, and $c$ are constants and the coordinate system is such that the origin is a SADDLE POINT. Cross-sections through the $z$-axis cut this surface in parabolas, with the origin being the vertex of all these parabolas. Cross-sections of the surface formed by planes parallel to the $xy$-planes are hyperbolas. The cross-section with the $xy$-plane is a pair of straight lines. Cross-sections with planes which are parallel to the $xz$- and $yz$-planes are parabolas. The $xz$- and $yz$-planes are symmetry planes.

**hyperboloid** /hȳ-**per**-bŏ-loid/ A surface generated by rotating a hyperbola about one of its axes of symmetry. Rotation about the conjugate axis gives a *hyperboloid of one sheet*. Rotation about the transverse axis gives a *hyperboloid of two sheets*

**hypertext** /**hȳ**-per-text/ A method of coding and displaying text on a computer screen in such a way that key words or phrases in the document can act as direct links to other documents or to other parts of the document. It is extensively used on the World Wide Web. An extension of hypertext, known as *hypermedia*, allows linkage to sounds, images, and video clips.

**hypocycloid** /hȳ-pŏ-**sȳ**-kloid/ A cusped curve that is the locus of a point on the circumference of a circle that rolls around the inside of a larger fixed circle. *See also* cusp; epicycloid.

**hypotenuse** /hȳ-**pot**-ĕ-news/ The side opposite the right angle in a right-angled triangle. The ratios of the hypotenuse length to the lengths of the other sides are used in trigonometry to define the sine and cosine functions of angles.

**hypothesis** /hȳ-**poth**-ĕ-sis/ (*pl.* hypotheses) A statement, theory, or formula that has yet to be proved but is assumed to be true for the purposes of the argument.

**hypothesis test** (**significance test**) A rule for deciding whether an assumption (hypothesis) about the distribution of a random variable should be accepted or rejected, using a sample from the distribution. The assumption is called the *null hypothesis*, written $H_0$, and it is tested against some *alternative hypothesis*, $H_1$. For example, when a coin is tossed $H_0$ can be P(heads) = ½ and $H_1$ that P(heads) > ½. A statistic is computed from the sample data. If it falls in the critical region, where the value of the statistic is significantly different from that expected under $H_0$, $H_0$ is rejected in favor of $H_1$. Otherwise $H_0$ is accepted. A type I error occurs if $H_0$ is rejected when it should be accepted. A type II error occurs if it is accepted when it should be rejected. The significance level of the test, $\alpha$, is the maximum probability with which a type I error can be risked. For example, $\alpha = 1\%$ means $H_0$ is wrongly rejected in one case out of 100.

**icosahedron** /ȳ-kos-ă- **hee**-drŏn/ (*pl.* **icosahedrons** or **icosahedra**) A POLYHEDRON that has twenty triangular faces. A *regular icosahedron* has twenty congruent faces, each one an equilateral triangle.

**identity, law of** *See* laws of thought.

**identity element** An element of a set that, combined with another element, leaves it unchanged. *See* group.

**identity matrix** *See* unit matrix.

**identity set** A set consisting of the same elements as another. For example, the set of natural numbers greater than 2 and the set of integers greater than 2 are identity sets.

**if and only if** (iff) *See* biconditional.

**if...then...** *See* implication.

**image** The result of a geometrical transformation or a mapping. For example, in geometry, when a set of points are transformed into another set by reflection in a line, the reflected figure is called the image. Similarly, the result of a rotation or a projection is called the image. The algebraic equivalent of this occurs when a function f(x) acts on a set A of values of $x$ to produce an image set B. *See also* domain; range; transformation.

**imaginary axis** The axis on the complex plane that purely imaginary numbers lie on. This axis is usually drawn as the $y$-axis.

**imaginary number** A multiple of i, the square root of minus one. The use of imaginary numbers is needed to solve equations

such as $x^2 + 2 = 0$, for which the solutions are $x = +i\sqrt{2}$ and $x = -i\sqrt{2}$. *See* complex number.

**imaginary part** Symbol Im$z$. The part i$y$ of a complex number $z$ that can be written as $z = x + iy$, where $x$ and $y$ are both real numbers.

**impact** A collision between two bodies. If the initial masses and velocities of the bodies are known then the velocities following the impact can be calculated in terms of the initial velocities and the COEFFICIENT OF RESTITUTION by using the principle of the conservation of momentum. The instance of a direct (head-on) impact is easiest to analyze theoretically but the motions of bodies which are not initially moving along the same line can also be analyzed by a modification of the method used for bodies that are initially moving along the same line.

**Imperial units** The system of measurement based on the yard and the pound. The f.p.s. system was a scientific system based on Imperial units.

**implication 1. (material implication; conditional)** Symbol: → or ⊃ In logic, the relationship *if...then...* between two propositions or statements. Strictly, implication reflects its ordinary language inter-

| P Q | P → Q |
|-----|-------|
| T T | T |
| T F | F |
| F T | T |
| F F | T |

Implication

pretation (*if...then...*) much less than conjunction, disjunction, and negation do theirs. Formally, $P \rightarrow Q$ is equivalent to 'either not $P$ or $Q$' ($\sim P \vee Q$), hence $P \rightarrow Q$ is false *only* when $P$ is true and $Q$ is false. Thus, logically speaking, if 'pigs can fly' is substituted for $P$ and 'grass is green' for $Q$ then 'if pigs can fly then grass is green' is true. The truth-table definition of implication is given in the illustration.

2. In algebra, the symbol $\Rightarrow$ is used between two equations when the first implies the second. For example:

$$x = y \Rightarrow x^2 = y^2$$

*See also* condition; truth table.

**implicit** Denoting a function that contains two or more variables that are not independent of each other. An *implicit function* of $x$ and $y$ is one of the form $f(x,y) = 0$, for example,

$$x^2 + y^2 - 4 = 0$$

Sometimes an *explicit function*, that is, one expressed in terms of an independent variable, can be derived from an implicit function. For example,

$$y + x^2 - 1 = 0$$

can be written as

$$y = 1 - x^2$$

where $y$ is an explicit function of $x$.

**improper fraction** *See* fraction.

**improper integral** An integral in which either the interval of integration is infinite or the value of the function $f(x)$ which is being integrated becomes infinite at some point in the interval of integration.

An improper integral of the type $\int_a^\infty f(x)dx$ is said to exist if the value of the integral of the function $f(x)$ with respect to $x$ in which the interval of integration runs from $a$ to $b$ tends to a finite limit $l$ as $b \rightarrow \infty$, i.e. $\int_a^\infty f(x)dx$ exists if $\lim \int_a^b f(x)dx = l$. An example of an improper integral of this type is given by $\int_1^\infty (1/x^2)dx = 1$.

It is also possible to define an improper integral if the interval of integration is from $-\infty$ to $a$. This means that if the integral $\int_{-\infty}^\infty f(x)dx$ is written as the sum $\int_{-\infty}^a f(x)dx$ and $\int_a^\infty f(x)dx$ and if these two integrals exist and have values $l_1$ and $l_2$ respectively

then the improper integral $\int_{-\infty}^\infty f(x)dx$ exists and has the value $l_1 + l_2$.

If the function $f(x)$ which is being integrated becomes infinite at some point over the interval of integration it is possible to examine whether the improper integral exists. This is the case whether the point at which the function becomes infinite is one of the limits of integration or is within the interval of integration. The case when the point is one of the limits is easier to analyze. If the point is the lower limit $a$ then one can consider the integral with the lower limit $a+\delta$, where $\delta$ is a small number, perform the integration and take the limit as $\delta \rightarrow 0$. If such a limit exists then the improper integral exists and has a value given by this limit. The case when the function becomes infinite at the upper limit $b$ can be investigated in a similar way.

If the function becomes infinite at some point $c$ which is between $a$ and $b$ the integral is split into two integrals, with $c$ being the upper limit of one integral and the lower limit of the other integral. If both these improper integrals exist then the original improper integral exists, with its value being equal to the sum of the values of the two integrals into which it was split.

**impulse** (**impulsive force**) A force acting for a very short time, as in a collision. If the force ($F$) is constant the impulse is $F\delta t$, $\delta t$ being the time period. If the force is variable the impulse is the integral of this over the short time period. An impulse is equal to the change of momentum that it produces.

**impulsive force** *See* impulse.

**incenter** /**in**-sen-ter/ *See* incircle.

**inch** Symbol: in or $''$ A unit of length equal to one twelfth of a foot. It is equivalent to 0.025 4 m.

**incircle** /**in**-ser-kăl/ (**inscribed circle**) A circle drawn inside a figure, touching all the sides. The center of the circle is the *incenter* of the figure. *Compare* circumcircle.

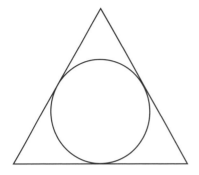

Incircle: an incircle touches all the sides of its surrounding figure.

**inclined plane** A type of MACHINE. Effectively a plane at an angle, it can be used to raise a weight vertically by movement up an incline. Both distance ratio and force ratio depend on the angle of inclination ($\theta$) and equal $1/\sin\theta$. The efficiency can be fairly high if friction is low. The screw and the wedge are both examples of inclined planes.

**inclusion** *See* subset.

**inclusive disjunction** (**inclusive or**) *See* disjunction.

**inconsistent** Describing a set of equations for which the solution set is empty. Geometrically, a set of equations is inconsistent if there is no point common to all the lines or curves represented by the equations. In the case of two linear equations in two variables $x$ and $y$ the two equations can be represented by straight lines $L_1$ and $L_2$. The two equations are inconsistent if $L_1$ and $L_2$ are parallel. In the case of three linear equations in three variables $x$, $y$, and $z$ the three equations can be represented by planes $P_1$, $P_2$, and $P_3$. There are three ways in which the equations can be inconsistent. The first way is that the three planes are all parallel and distinct. The second way is that two of the planes are parallel and distinct. The third way is that one plane is parallel to the line of intersection of the other two planes.

**increasing function** A real function f of $x$ for which $f(x_1) \le f(x_2)$ for any $x_1$ and $x_2$ in an interval I for which $x_1 < x_2$. If $f(x_1) < f(x_2)$ when $x_1 < x_2$ then f is said to be a *strictly increasing function*.

**increment** A small difference in a variable. For example, $x$ might change by an increment $\Delta x$ from the value $x_1$ to the value $x_2$; $\Delta x = x_2 - x_1$. In calculus, infinitely small increments are used. *See also* differentiation; integration.

**indefinite integral** The general integration of a function $f(x)$, of a single variable, $x$, without specifying the interval of $x$ to which it applies. For example, if $f(x) = x^2$, the indefinite integral
$$\int f(x)dx = \int x^2 dx = (x^3/3) + C$$
where $C$ is an unknown constant (the constant of integration) that depends on the interval. *Compare* definite integral. *See also* integration.

**independence** *See* probability.

**independent** *See* dependent.

**independent variable** *See* variable.

**indeterminate equation** An equation that has an infinite number of solutions. For example,
$$x + 2y = 3$$
is indeterminate because an infinite number of values of $x$ and $y$ will satisfy it. An indeterminate equation in which the variables can take only integer values is called a *Diophantine equation* and it has an infinite but denumerable set of solutions. Diophantine equations are named for the Greek mathematician Diophantus of Alexandria (*fl.* AD 250).

**indeterminate form** An expression that can have no quantitative meaning; for example 0/0.

**index** (*pl.* **indexes** or **indices**) A number that indicates a characteristic or function in a mathematical expression. For example, in $y^4$, the exponent, 4, is also known as the index. Similarly in $^3\sqrt{27}$ and $\log_{10}x$, the numbers 3 and 10 respectively are called indices (or indexes).

**indirect proof** (reductio ad absurdum) A logical argument in which a proposition or statement is proved by showing that its negation or denial leads to a CONTRADICTION. *Compare* direct proof.

**induction** /in-**duk**-shŏn/ **1.** (mathematical induction) A method of proving mathematical theorems, used particularly for series sums. For instance, it is possible to show that the series $1 + 2 + 3 + 4 + \dots$ has a sum to $n$ terms of $n(n + 1)/2$. First we show that if it is true for $n$ terms it must also be true for $(n + 1)$ terms. According to the formula
$$S_n = n(n + 1)/2$$
if the formula is correct, the sum to $(n + 1)$ terms is obtained by adding $(n + 1)$ to this
$$S_{n+1} = n(n + 1)/2 + (n + 1)$$
$$S_{n+1} = (n + 1)(n + 2)/2$$
This agrees with the result obtained by replacing $n$ in the general formula by $(n + 1)$, i.e.:
$$S_{n+1} = (n + 1)(n + 1 + 1)/2$$
$$S_{n+1} = (n + 1)(n + 2)/2$$
Thus, the formula is true for $(n + 1)$ terms if it is true for $n$ terms. Therefore, if it is true for the sum to one term ($n = 1$), it must be true for the sum to two terms ($n + 1$). Similarly, if true for two terms, it must be true for three terms, and so on through all values of $n$. It is easy to show that it is true for one term:
$$S_n = 1(1 + 1)/2$$
$$S_n = 1$$
which is the first term in the series. Hence the theorem is true for all integer values of $n$.
**2.** In logic, a form of reasoning from individual cases to general ones, or from observed instances to unobserved ones. Inductive arguments can be of the form: $F_1$ is $A$, $F_2$ is $A$ … $F_n$ is $A$, therefore all $F$s are $A$ ('this swan has wings, that swan has wings … therefore all swans have wings'); or: all $F$s observed so far are $A$, therefore all $F$s are $A$ ('all swans observed so far are white, therefore all swans are white'). Unlike deduction, asserting the premises

while denying the conclusion in an induction does not lead to a CONTRADICTION. The conclusion is *not* guaranteed to be true if the premises are. *Compare* deduction.

**inelastic collision** /in-i-**las**-tik/ A collision for which the restitution coefficient is less than one. In effect, the relative velocity after the collision is less than that before; the kinetic energy of the bodies is not conserved in the collision, even though the system may be closed. Some of the kinetic energy is converted into internal energy. *See also* restitution, coefficient of.

**inequality** A relationship between two expressions that are not equal, often written in the form of an equation but with the symbols > or < meaning 'is greater than' and 'is less than'. For example, if $x < 4$ then $x^2 < 16$. If $y^2 > 25$, then $y > 5$ or $y < -5$. If the end values are included, the symbols ≥ (is greater than or equal to) and ≤ (is less than or equal to) are used. When one quantity is very much smaller or greater than another, it is shown by << or >>. For example, if $x$ is a large number $x >> 1/x$ or $1/x << x$. *See also* equality.

**inequation** /in-i-**kway**-zhŏnz/ Another word for inequality.

**inertia** /i-**ner**-shă/ An inherent property of matter implied by Newton's first law of motion: inertia is the tendency of a body to resist change in its motion. *See also* inertial mass; Newton's laws of motion.

**inertial mass** /i-**ner**-shăl/ The mass of an object as measured by the property of inertia. It is equal to the ratio force/acceleration when the object is accelerated by a constant force. In a uniform gravitational field, it appears to be equal to GRAVITATIONAL MASS – all objects have the same gravitational acceleration at the same place.

**inertial system** A frame of reference in which an observer sees an object that is free of all external forces to be moving at constant velocity. The observer is called an *inertial observer*. Any FRAME OF REFERENCE that moves with constant velocity and without rotation relative to an inertial frame is also an inertial frame. NEWTON'S LAWS OF MOTION are valid in any inertial frame (but not in an accelerated frame), and the laws are therefore independent of the velocity of an inertial observer.

**inf** *See* infimum.

**inference** /in-fĕ-rĕns/ **1.** The process of reaching a conclusion from a set of premisses in a logical argument. An inference may be deductive or inductive. *See also* deduction; induction. **2.** *See* sampling.

**infimum** (inf) The greatest lower BOUND of a set.

**infinite number** The smallest infinite number is $\aleph_0$ (aleph zero). This is the number of members in the set of integers. A whole hierarchy of increasingly large infinite numbers can be defined on this basis. $\aleph_1$, the next largest, is the number of subsets of the set of integers. *See also* aleph; continuum; countable.

**infinite sequence** *See* sequence.

**infinite series** *See* series.

**infinite set** A set in which the number of elements is infinite. For example, the set of 'positive integers', z = {1, 2, 3, 4, …}, is infinite but the set of 'positive integers less than 20' is a *finite set*. Another example of an infinite set is the number of circles in a particular plane. *Compare* finite set.

**infinitesimal** /in-fi-nă-**tess**-ă-măl/ Infinitely small, but not equal to zero. Infinitesimal changes or differences are made use of in CALCULUS (infinitesimal calculus).

**infinity** Symbol: ∞ The value of a quantity that increases without limit. For example, if $y = 1/x$, then $y$ becomes infinitely large, or approaches infinity, as $x$ approaches 0. An infinitely large negative quantity is denoted by $-\infty$ and an infinitely large positive

quantity by $+\infty$. If $x$ is positive, $y = -1/x$ tends to $-\infty$ as $x$ tends to 0.

**inflection** *See* point of inflection.

**information theory** The branch of probability theory that deals with uncertainty, accuracy, and information content in the transmission of messages. It can be applied to any system of communication, including electrical signals and human speech. Random signals (noise) are often added to a message during the transmission process, altering the signal received from that sent. Information theory is used to work out the probability that a particular signal received is the same as the signal sent. Redundancy, for example simply repeating a message, is needed to overcome the limitations of the system. Redundancy can also take the form of a more complex checking process. In transmitting a sequence of numbers, their sum might also be transmitted so that the receiver will know that there is an error when the sum does not correspond to the rest of the message. The sum itself gives no extra information since, if the other numbers are correctly received, the sum can easily be calculated. The statistics of choosing a message out of all possible messages (letters in the alphabet or binary digits for example) determines the amount of information contained in it. Information is measured in bits (binary digits). If one out of two possible signals are sent then the information content is one bit. A choice of one out of four possible signals contains more information, although the signal itself might be the same.

**inner product** Consider a vector space $V$ over a scalar field $F$. An *inner product* on $V$ is a mapping of ordered pairs of vectors in $V$ into $F$; i.e. with every pair of vectors $x$ and $y$ there is associated a scalar, which is written $\langle x,y \rangle$ and called the inner product of $x$ and $y$, such that for all vectors $x$, $y$, $z$ and scalars $\alpha$
(i) $\langle x+y,z \rangle = \langle x,z \rangle + \langle y,z \rangle$
(ii) $\langle x,y \rangle = \alpha \langle x,z \rangle$
(iii) $\langle x,y \rangle = \langle \overline{y,x} \rangle$, where $\langle \overline{a,b} \rangle$ is the complex conjugate of $\langle a,b \rangle$
(iv) $\langle x,x \rangle \geq 0$, $\langle x,x \rangle = 0$ if and only if $x = 0$.

An inner product on $V$ defines a *norm* on $V$ given by $\|x\| = \sqrt{\langle x,x \rangle}$. *See* norm.

**input** 1. The signal or other form of information that is applied (fed in) to an electrical device, machine, etc. The input to a computer is the data and programmed instructions that a user communicates to the machine. An *input device* accepts computer input in some appropriate form and converts the information into a code of electrical pulses. The pulses are then transmitted to the central processor of the computer.
2. The process or means by which input is applied.
3. To feed information into an electrical device or machine.
     *See also* input/output; output.

**input/output** (I/O) The equipment and operations used to communicate with a computer, and the information passed in or out during the communication. Input/output devices include those used only for INPUT or for OUTPUT of information and those, such as visual display units, used for both input and output.

**inscribed** Describing a geometric figure that is drawn inside another geometrical figure. *Compare* circumscribed.

**inscribed circle** *See* incircle.

**instantaneous value** /in-stăn-**tay**-nee-ŭs/ The value of a varying quantity (e.g. velocity, acceleration, force, etc.) at a particular instant in time.

**integer** /**in**-tĕ-jer/ Symbol: $z$ Any of the set of whole numbers, $\{\dots, -2, -1, 0, 1, 2, \dots\}$ used for counting. The integers include zero and the negative whole numbers.

**integer part** Symbol $[x]$. An integer $n$ which is related to a real number $x$ by $n \le$ $x < n + 1$. The integer $n$ is said to be the *integer part* of x. For example, $[13/3] = 4$ and $[e] = 2$. Care needs to be taken in the case of negative numbers. For example, $[-13/3] = -5$. The difference between positive numbers can be significant if one has a computer program that converts a real number into an integer by truncation. In the case of positive numbers and zero truncation and taking the integer part of the real number would give the same integer but this is not the case for negative numbers.

**integer variable** *See* variable.

**integral** /**in**-tĕ-grăl, in-**teg**-răl/ The result of integrating a function. *See* integration.

**integral equation** An equation in which an unknown function appears under an integral sign. Many problems in physical science and engineering can be expressed as integral equations. It is sometimes easier to solve problems which can be expressed as ordinary or partial differential equations by expressing them in terms of integral equations. In particular, solving initial value problems and boundary-value problems of differential equations is frequently performed by converting the differential equation into an integral equation. Just as there is no general technique for performing integration so there is no general technique for solving all integral equations.

**integral transform** A transform in which a function $f(x)$ of a variable $x$ is transformed into a function $g(y)$ of a variable $y$ by the integral equation:
$$g(y) = \int K(x, y)f(x)\mathrm{d}x,$$
where $K(x,y)$ is a part of the integral equation known as its *kernel*. For example, integral transforms can map a function of time into another function in frequency when wave phenomena are being analyzed. *See* Laplace transform.

Integers: a number line showing positive and negative integers.

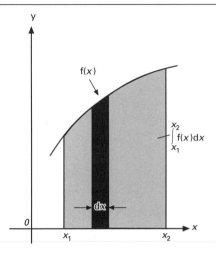

The integration of a function $y = f(x)$ as the area between the curve and the $x$-axis.

**integrand** /in-tĕ-grand/ A function that is to be integrated. For example, in the integral of $f(x).dx$, $f(x)$ is the integrand. *See also* integration.

**integrating factor** A multiplier used to simplify and solve DIFFERENTIAL EQUATIONS, usually given the symbol $\xi$. For example, $x dy - y dx = 2x^3 dx$ may be multiplied by $\xi(x) = 1/x^2$ to give the standard form:
$$d(y/x) = (x.dy - y.dx)/x^2 = 2x dx$$
which has the solution $y/x = x^2 + C$, where $C$ is a constant.

**integration** The continuous summing of change in a function, $f(x)$, over an interval of the variable $x$. It is the inverse process of DIFFERENTIATION in calculus, and its result is known as the *integral* of $f(x)$ with respect to $x$. An integral:
$$\int_{x_1}^{x_2} v dt$$
can be regarded as the area between the curve and the $x$-axis, between the values $x_1$ and $x_2$. It can be considered as the sum of a number of column areas of width $\Delta x$ and heights given by $f(x)$. As $\Delta x$ approaches zero, the number of columns increases infinitely and the sum of the column areas approaches the area under the curve. The

integral of velocity is distance. For example, a car traveling with a velocity $V$ in a time interval $t_1$ to $t_2$ goes a distance $s$ given by
$$\int_{t_1}^{t_2} v dt$$
Integrals of this type, between definite *limits*, are known as *definite integrals*. An *indefinite integral* is one without limits. The result of an indefinite integral contains a constant – the *constant of integration*. For example,
$$\int x dx = x^2/2 + C$$
where $C$ is the constant of integration. A table of integrals is given in the Appendix.

**integration by parts** A method of integrating a function of a variable by expressing it in terms of two parts, both of which are differentiable functions of the same variable. A function $f(x)$ is written as the product of $u(x)$ and the derivative $dv/dx$. The formula for the differential of a product is:
$$d(u.v)/dx = u.dv/dx + v.du/dx$$
Integrating both sides over $x$ and rearranging the equation gives
$$\int u.(dv/dx)dx = uv - \int v(du/dx)dx$$
which can be used to evaluate the integral of a product. For example, to integrate $x\sin x$, $dv/dx$ is taken to be $\sin x$, so $v = -\cos x + C$ ($C$ is a constant of integration).

$u$ is taken to be $x$, so $du/dx = 1$. The integral is then given by

$$\int x\sin x\,dx =$$
$$x(-\cos x + C) - \int(-\cos x + C) =$$
$$-x\cos x + \sin x + k$$

where $k$ is a constant of integration. Usually a trigonometric or exponential function is chosen for $dv/dx$.

**integration by substitution** A method of integrating a function of one variable by expressing it as a simpler or more easily integrated function of another variable. For example, to integrate $\sqrt{(1 - x^2)}$ with respect to $x$, we can make $x = \sin u$, so that $\sqrt{(1 - x^2)} = \sqrt{(1 - \sin^2 u)} = \sqrt{(\cos^2 u)} = \cos u$, and $dx = (dx/du).du = \cos u\,du$. Therefore:

$$\int_a^b \sqrt{(1 - x^2)}dx = \int_c^d \cos^2 u.du$$
$$= [u/2 - \tfrac{1}{2}\sin u\cos u]_c^d$$

Note that for a definite integral the limits must also be changed from values of $x$ to corresponding values of $u$.

**integro-differential equation** /in-**teg**-roh-dif-ĕ-**ren**-shăl/ A differential equation that also has an integral as part of it. Such equations occur in the quantitative description of transport processes. General methods for the solution of integro-differential equations do not exist and approximation methods have to be used.

**interaction** Any mutual action between particles, systems, etc. Examples of interactions include the mutual forces of attraction between masses (gravitational interaction) and the attractive or repulsive forces between electric charges (electromagnetic interaction).

**intercept** A part of a line or plane cut off by another line or plane.

**interest** The amount of money paid each year at a stated rate on a borrowed capital, or the amount received each year at a stated rate on loaned or invested capital. The interest rate is usually expressed as a percentage per annum. *See* compound interest; simple interest.

**interface** A shared boundary. It is the area(s) or place(s) at which two devices or two systems meet and interact. For example, there is a simple interface between an electric plug and socket. A far more complicated interface of electronic circuits provides the connection between the central processor of a computer and each of its peripheral units. Human-machine interface refers to the interaction between people and machines, including computers. For good, i.e. efficient, interaction, devices such as visual display units and easily understandable programming languages have been introduced.

**interior angle** An angle formed on the inside of a plane figure by two of its straight sides. For example, there are three interior angles in a triangle, which add up to 180°. *Compare* exterior angle.

**intermediate value theorem** A theorem concerning continuous functions which states that if a real function f is continuous on an interval I bounded by $a$ and $b$ for which $f(a) \neq f(b)$ then if $n$ is some real number between $f(a)$ and $f(b)$ there must be some number $c$ in the interval, i.e. $a < c < b$ for which $f(c) = n$.

The intermediate value theorem can be used to find the solutions of an equation. For example, if $f(a) > 0$ and $f(b) < 0$ then $f(x) = 0$ between $a$ and $b$.

**internal store** *See* central processor; store.

**interpolation** /in-terp-ŏ-**lay**-shŏn/ The process of estimating the value of a function from known values on either side of it. For example, if the speed of an engine, controlled by a lever, increases from 40 to 50 revolutions per second when the lever is pulled down by four centimeters, one can interpolate from this information and assume that moving it two centimeters will give 45 revolutions per second. This is the simplest method of interpolation, called *linear interpolation*. If known values of one variable $y$ are plotted against the other variable $x$, an estimate of an unknown value of $y$ can be made by drawing a

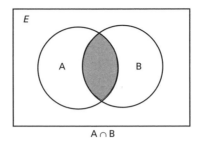

$A \cap B$

The shaded area in the Venn diagram is the intersection of set A and set B.

straight line between the two nearest known values.

The mathematical formula for linear interpolation is:

$$y_3 = y_1 + (x_3 - x_1)(y_2 - y_1)/(x_2 - x_1)$$

$y_3$ is the unknown value of $y$ (at $x_3$) and $y_2$ and $y_1$ (at $x_2$ and $x_1$) are the nearest known values, between which the interpolation is made. If the graph of $y$ against $x$ is a smooth curve, and the interval between $y_1$ and $y_2$ is small, linear interpolation can give a good approximation to the true value, but if $(y_2 - y_1)$ is large, it is less likely that $y$ will fit sufficiently well to a straight line between $y_1$ and $y_2$. A possible source of error occurs when $y$ is known at regular intervals, but oscillates with a period shorter than this interval.

*Compare* extrapolation.

**interquartile range** /in-ter-**kwor**-tȳl/ A measure of dispersion given by $(P_{75} - P_{25})$ where $P_{75}$ is the upper QUARTILE and $P_{25}$ the lower quartile. The semi-interquartile range is $\frac{1}{2}(P_{75} - P_{25})$.

**intersection 1.** The point at which two or more lines cross each other, or a set of points that two or more geometrical figures have in common.
**2.** In set theory, the set formed by the elements common to two or more sets. For example, if set A is {black four-legged animals} and set B is {sheep} then the intersection of A and B, written A∩B, is {black sheep}. This can be represented on a VENN DIAGRAM by the intersection of two circles, one representing A and the other B.

**interval** A set of numbers, or points in a coordinate system, defined as all the values between two end points. *See also* closed interval; open interval.

**into** A mapping from one set to another is said to be *into* if the range of the mapping is a proper subset of the second set; i.e. if there are members of the second set which are not the image of any element of the first set under the mapping. *Compare* onto.

**intransitive** /in-**tran**-să-tiv/ Describing a relation that is not transitive; i.e. when $x$R$y$ and $y$R$z$ then it is not true that $x$R$z$. An example of an intransitive relation is being the square of. If $x = y^2$ and $y = z^2$ it does not follow that $x = z^2$.

**intuitionism** /in-too-**ish**-ŏ-niz-ăm/ An approach to the foundations of mathematics that was developed in the late nineteenth century and the early decades of the twentieth century. Intuitionism tried to build up mathematics based on intuitive concepts rather than from a rigid series of axioms. Using this approach the intuitionists were able to build up various parts of algebra and geometry and to formulate calculus, albeit in a complicated way. As with other attempts such as FORMALISM to provide foundations for mathematics, intuitionism was not accepted universally as providing a foundation for mathematics.

The main difficulties for all attempts to provide foundations for mathematics, including formalism and intuitionism involve dealing with infinite sets and infinite processes.

**invariant** /in-**vair**-ee-ănt/ Describing a property of an equation, function, or geometrical figure that is unaltered after the application of any member of some given family of transformations. For example, the area of a polygon is invariant under the group of rotations.

**inverse element** An element of a set that, combined with another element, gives the identity element. *See* group.

**inverse function** The reverse mapping of the set B into the set A when the FUNCTION that maps A into B has already been defined.

**inverse hyperbolic functions** The inverse functions of hyperbolic sine, hyperbolic cosine, hyperbolic tangent, etc., defined in an analogous way to the inverse trigonometric functions. For instance, the inverse hyperbolic sine of a variable $x$, written arc sinh$x$ or $\sinh^{-1}x$, is the angle (or number) of which $x$ is the hyperbolic sine. Similarly, the other inverse hyperbolic functions are:
*inverse hyperbolic cosine* of $x$ (written arc cosh$x$ or $\cosh^{-1}x$)
*inverse hyperbolic tangent* of $x$ (written arc tanh$x$ or $\tanh^{-1}x$)
*inverse hyperbolic cotangent* of $x$ (written arc coth$x$ or $\coth^{-1}x$)
*inverse hyperbolic secant* of $x$ (written arc sech$x$ or $\text{sech}^{-1}x$)
*inverse hyperbolic cosecant* of $x$ (written arc cosech$x$ or $\text{cosech}^{-1}x$).

**inverse mapping** A mapping, denoted $f^{-1}$, from one set $B$ to another set $A$, where f is a one-to-one and onto mapping from $A$ to $B$ which satisfies the definition in the following sentence. The inverse mapping $f^{-1}$ from $B$ to $A$ exists if, for an element $b$ of $B$, there is a unique element $a$ of $A$ given by $f^{-1}(b) = a$ which satisfies $f(a) = b$. Like the mapping f, the inverse mapping $f^{-1}$ is a one-to-one and onto mapping.

**inverse matrix** The unique $n \times n$ matrix, denoted $A^{-1}$ corresponding to the $n \times n$ matrix $A$ that satisfies $AA^{-1} = A^{-1}A = I$, where $I$ is the UNIT MATRIX of order $n$. If a matrix is not a square matrix it cannot have an inverse matrix. It is not necessarily the case that a square matrix has an inverse. If a matrix does have an inverse then the matrix is *invertible* or *non-singular*.

There are several techniques for finding inverse matrices. In general, the inverse of $A$ is given by $(1/\det A)$ adj $A$, where adj$A$ IS the ADJOINT of $A$, provided that $\det A \neq 0$. If $\det A = 0$ then the matrix is said to be SINGULAR and an inverse does not exist. In the case of a $2 \times 2$ matrix $A$ of the form

$$\mathbf{A} = \begin{pmatrix} a & b \\ c & d \end{pmatrix}$$

The inverse $A^{-1}$ exists if $ad - bc \neq 0$ and is given by

$$\mathbf{A}^{-1} = \frac{1}{ad - bc}\begin{pmatrix} d & -b \\ -c & a \end{pmatrix}$$

In the case of large matrices it is not convenient to find the inverse of a matrix

$$x + 3y = 5 \quad \Longleftrightarrow \quad \begin{pmatrix} 1 & 3 \\ 2 & 4 \end{pmatrix} \times \begin{pmatrix} x \\ y \end{pmatrix} = \begin{pmatrix} 5 \\ 6 \end{pmatrix}$$
$$2x + 4y = 6$$

$$\begin{pmatrix} -2 & \frac{3}{2} \\ 1 & -\frac{1}{2} \end{pmatrix} \times \begin{pmatrix} 1 & 3 \\ 2 & 4 \end{pmatrix} = \begin{pmatrix} 1 & 0 \\ 0 & 1 \end{pmatrix}$$

$$\Rightarrow \begin{pmatrix} x \\ y \end{pmatrix} = \begin{pmatrix} -2 & \frac{3}{2} \\ 1 & -\frac{1}{2} \end{pmatrix} \times \begin{pmatrix} 5 \\ 6 \end{pmatrix} = \begin{pmatrix} -1 \\ 2 \end{pmatrix}$$

The solution of simultaneous equations by finding the inverse of a matrix. The equations are written as the equivalent matrix equation, both sides of which are then multiplied by the inverse of the coefficient matrix.

by using the general formula involving the determinant. It is more convenient to use a technique similar to GAUSSIAN ELIMINATION, which is used to solve sets of linear equations.

**inverse of a complex number** A complex number, denoted by $1/z$ or $z^{-1}$, which is the inverse of the complex number $z = x + i\,y$, i.e. it is the number $z^{-1}$ for which $zz^{-1} = 1$. It is given by $z = x/(x^2 + y^2) - i\,y/(x^2 + y^2)$. If $z$ is written in terms of polar coordinates $z = r\,(\cos\theta + i\sin\theta) = r\exp(i\theta)$ then $z^{-1}$ is given by $(1/r)\exp(-i\theta) = (1/r)(\cos\theta - i\sin\theta)$, which exists if $r \neq 0$.

**inverse ratio** A reciprocal ratio. For example, the inverse ratio of $x$ to $y$ is the ratio of $1/x$ to $1/y$.

**inverse square law** A physical law in which an effect varies inversely as the square of the distance from the source producing the effect. An example is Newton's law of universal gravitation.

**inverse trigonometric functions** The inverse functions of sine, cosine, tangent, etc. For example, the *inverse sine* of a variable is called the arc sine of $x$; it is written arc sin$x$ or $\sin^{-1}x$ and is the angle (or number) of which the sine is $x$. Similarly, the other inverse trigonometric functions are:
*inverse cosine* of $x$ (arc cosine, written arc cos$x$ or $\cos^{-1}x$)
*inverse tangent* of $x$ (arc tangent, written arc tan$x$ or $\tan^{-1}x$)
*inverse cotangent* of $x$ (arc cotangent, written arc cot$x$ or $\cot^{-1}x$)
*inverse cosecant* of $x$ (arc cosecant, written arc cosec$x$ or $\text{cosec}^{-1}x$)
*inverse secant* of $x$ (arc secant, written arc sec$x$ or $\sec^{-1}x$).

**inverter gate** *See* logic gate.

**involute** /**in**-vŏ-loot/ The involute of a curve is a second curve that would be obtained by unwinding a taut string wrapped around the first curve. The involute is the curve traced out by the end of the string.

**I/O** *See* input/output.

**irrational number** /i-**rash**-ŏ-năl/ A number that cannot be expressed as a ratio of two integers. The irrational numbers are precisely those infinite decimals that are not repeating. Irrational numbers are of two types:
(i) *Algebraic irrational numbers* are roots of polynomial equations with rational numbers as coefficients. For example, $\sqrt{3} = 1.732\,050\,8\ldots$ is a root of the equation $x^2 = 3$. This equation does not have a rational solution since such a solution could be written $x = m/n$ with $m^2 = 3n^2$, but this is impossible since 3 divides the left-hand side an even number of times and the right-hand side an odd number of times.
(ii) *Transcendental numbers* are irrational numbers that are not algebraic, e.g. e and π.

**irrational term** A term in which at least one of the indices is an irrational number. For example, $xy^{\sqrt{2}}$ and $2x^\pi$ are irrational terms.

**irrotational vector** A vector $V$ for which the curl of $V$ vanishes, i.e. $\nabla \times V = 0$. Examples of irrotational vectors include the gravitational force described by Newton's law of gravity and the electrostatic force governed by Coulomb's law. If $u$ is the displacement of a plane wave in an elastic medium (or a spherical wave a long way from the source of the wave) then the condition that $u$ is an irrotational vector means that the wave is a longitudinal wave. When this result is combined with the concept of a SOLENOIDAL VECTOR and HELMHOLTZ'S THEOREM it is useful in analyzing waves in seismology.

If a vector is an irrotational vector then it can be written as $-1$ times the gradient of a scalar function. In physical applications the scalar function is usually called the *scalar potential*.

**isolated point** A point that satisfies the equation of a curve but is not on the main arc of the curve. For example, the equation $y^2(x^2 - 4) = x^4$ has a solution at $x = 0$ and $y = 0$, but there is no real solution at any point near the origin, so the origin is an isolated point. *See also* double point; multiple point.

**isolated system** *See* closed system.

**isometric paper** /ȳ-sŏ-**met**-rik/ Paper on which there is a grid of equilaterial triangles printed. This type of paper is useful for drawing shapes such as cubes and cuboids, with each face being a parallelogram in the drawing. Using this type of paper it is possible to draw all the dimensions of the figures correctly and to measure them by using the regular grid pattern.

**isometry** /ȳ-**som**-ĕ-tree/ A transformation in which the distances between the points remain constant.

**isomorphic** /ȳ-sŏ-**mor**-fik/ *See* homomorphism.

**isomorphism** /ȳ-sŏ-**mor**-fiz-ăm/ *See* homomorphism.

**isosceles** /ȳ-**soss**-ĕ-leez/ Having two equal sides. *See* triangle.

**issue price** *See* nominal value.

**iterated integral** /it-ĕ-ray-tid/ (**multiple integral**) A succession of integrations performed on the same function. For example, a DOUBLE INTEGRAL or a TRIPLE INTEGRAL.

**iteration** /it-ĕ-**ray**-shŏn/ A method of solving a problem by successive approximations, each using the result of the preceding approximation as a starting point to obtain a more accurate estimate. For example, the square root of 3 can be calculated by writing the equation $x^2 = 3$ in the form $2x^2 = x^2 + 3$, or $x = \frac{1}{2}(x + 3/x)$. To obtain a solution for $x$ by iteration, we might start with a first estimate, $x_1 = 1.5$. Substituting this in the equation gives the second estimate, $x_2 = \frac{1}{2}(1.5 + 2) = 1.750\ 00$. Continuing in this way, we obtain:

$$x_3 = \tfrac{1}{2}(1.75 + 3/1.75) = 1.732\ 14$$
$$x_4 = \tfrac{1}{2}(1.732\ 14 + 3/1.732\ 14) = 1.732\ 05$$

and so on, to any required accuracy. The difficulty in solving equations by iteration is in finding a formula for iteration (algorithm) that gives convergent results. In this case, for example, the algorithm $x_{n+1} = 3/x_n$ does not give convergent results. There are several standard techniques, such as NEWTON'S METHOD, for obtaining convergent algorithms. Iterative calculations, although often tedious for manual computation, are widely used in computers.

**j** An alternative to i for the square root of –1. The use of j rather than i in complex numbers is particularly common among electrical engineers.

**job** A unit of work submitted to a computer. It usually includes several programs. The information necessary to run a job is input in the form of a short program written in the *job-control language* (JCL) of the computer. The JCL is interpreted by the OPERATING SYSTEM and is used to identify the job and describe its requirements to the operating system. *See also* batch processing.

**joule** /jool/ Symbol: J The SI unit of energy and work, equal to the work done where the point of application of a force of one newton moves one meter in the direction of action of the force. 1 J = 1 N m. The joule is the unit of all forms of energy. The unit is named for the British physicist James Prescott Joule (1818–89).

**Julia set** /**joo**-lee-ă/ A set of points in the complex plane defined by iteration of a complex number. One takes the expression $z^2 + c$, where $z$ and $c$ are complex numbers, and calculates it for a given value of $z$ and takes the result as a new starting value of $z$. This process can be repeated indefinitely and three possibilities occur, depending on the initial value for $z$. One is that the value tends to zero with successive iterations. Another is that the value diverges to infinity. There is, however, a set of initial values of $z$ for which successive iterations give values that stay in the set. This set of values is a Julia set and it can be represented by points in the complex plane. The Julia set is the boundary between values that have an attractor at zero and values that have an ATTRACTOR at infinity. The actual form of the Julia set depends on the value of the constant complex number $c$. Thus, if $c = 0$, the iteration is $z \rightarrow z^2$. In this case the Julia set is a circle with radius 1. There is an infinite number of Julia sets showing a wide range of complex patterns. The set is named for the French mathematician Gaston Julia (1893–1978). *See also* Mandelbrot set.

# K

**kelvin** Symbol: K The SI base unit of thermodynamic temperature. It is defined as the fraction 1/273.16 of the thermodynamic temperature of the triple point of water. Zero kelvin (0 K) is absolute zero. One kelvin is the same as one degree on the Celsius scale of temperature. The unit is named for the British theoretical and experimental physicist Baron William Thomson Kelvin (1824–1907).

**Kendall's method** A method of measuring the degree of association between two different ways of ranking $n$ objects, using two variables ($x$ and $y$), which give data $(x_1,y_1),...,(x_n,y_n)$. The objects are ranked using first the $x$s and then the $y$s. For each of the $2n(n-1)/2$ pairs of objects a score is assigned. If the RANK of the $j$th object is greater (or less) than that of the $k$th, regardless of whether the $x$s or $y$s are used, the score is plus one. If the rank of the $j$th is less than that of the $k$th using one variable but greater using the other, the score is minus one. Kendall's coefficient of rank correlation $\tau = $ (sum of scores)/$\frac{1}{2}n(n-1)$. The closer $\tau$ is to one, the greater the degree of association between the rankings. The method is named for the British statistician Maurice Kendall (1907–83). *See also* Spearman's method.

**Kepler's laws** /**kep**-lerz/ Laws of planetary motion deduced in about 1610 by the German astronomer Johannes Kepler (1571–1630) using astronomical observations made by the Danish astronomer Tycho Brahe (1546–1601):
(1) Each planet moves in an elliptical orbit with the Sun at one focus of the ellipse.
(2) The line between the planet and the Sun sweeps out equal areas in equal times.

(3) The square of the period of each planet is proportional to the cube of the semi-major axis of the ellipse.

Application of the third law to the orbit of the Moon about the Earth gave support to Newton's theory of gravitation.

**keyboard** A computer input device that a human user can operate to type in data in the form of alphanumeric characters. It has a standard QWERTY key layout with some additional characters and function keys. *See* alphanumeric; input device.

**kilo-** Symbol: k A prefix denoting $10^3$. For example, 1 kilometer (km) = $10^3$ meters (m).

**kilogram** /**kil**-ŏ-gram/ (**kilogramme**) Symbol: kg The SI base unit of mass, equal to the mass of the international prototype of the kilogram, which is a piece of platinum–iridium kept at Sèvres in France.

**kilogramme** /**kil**-ŏ-gram/ An alternative spelling of *kilogram*.

**kilometer** /**kil**-ŏ-mee-ter, kă-**lom**-ĕ-ter/ (km) A unit of distance equal to 1000 metres, approximating to 0.62 mile.

**kilowatt-hour** /**kil**-ŏ-wot/ Symbol: kwh A unit of energy, usually electrical, equal to the energy transferred by one kilowatt of power in one hour. It is the same as the Board of Trade unit and has a value of 3.6 $\times 10^6$ joules.

**kinematics** /kin-ĕ-**mat**-iks/ The study of the motion of objects without consideration of its cause. *See also* mechanics.

**kinetic energy** /ki-**net**-ik/ Symbol: $T$ The

work that an object can do because of its motion. For an object of mass $m$ moving with velocity $v$, the kinetic energy is $mv^2/2$. This gives the work the object would do in coming to rest. The rotational kinetic energy of an object of moment of inertia $I$ and angular velocity $\omega$ is given by $I\omega^2/2$. *See also* energy.

**kinetic friction** *See* friction.

**kite** A plane figure with four sides, and two pairs of adjacent sides equal. Two of the angles in a kite are opposite and equal. Its diagonals cross perpendicularly, one of them (the shorter one) being bisected by the other. The area of a kite is equal to half the product of its diagonal lengths. In the special case in which the two diagonals have equal lengths, the kite is a rhombus.

**Klein bottle** /klÿn/ A curved surface with the unique topological property that it has only one surface, no edges, and no inside and outside. It can be thought of as formed by taking a length of flexible stretchy tubing, cutting a hole in the side through which one end can fit exactly, passing an end of the tube through this hole, and then joining it to the other end from the inside. Starting at any point on the surface a continuous line can be drawn along it to any other point without crossing an edge. The bottle is named for the German mathematician (Christian) Felix Klein (1849–1925). *See also* topology.

**knot 1.** A curve formed by looping and interlacing a string and then joining the ends. The mathematical theory of knots is a branch of TOPOLOGY.
**2.** A unit of velocity equal to one nautical mile per hour. It is equal to $0.414$ m s$^{-1}$.

**Koch curve** /kokh/ *See* fractal.

**Königsberg bridge problem** /**koh**-nigz-berg/ A classical problem in topology. The river in the Prussian city of Königsberg divided into two branches and was crossed by seven bridges in a certain arrangement. The problem was to show that it is impossible to walk in a continuous path across all the bridges and cross each one only once. The problem was solved by Euler in the eighteenth century, by replacing the arrangement by an equivalent one of lines and vertices. He showed that a network like this (called a *graph*) can be traversed in a single path if and only if there are fewer than three vertices at which an odd number of line segments meet. In this case there are four.

**Kronecker delta** /**kroh**-nek-er/ Symbol $\delta_{ij}$. A quantity defined by $\delta_{ij} = 1$, if $i = j$, and $\delta_{ij} = 0$ if $i \neq j$. It is named for the German mathematician Leopold Kronecker (1823–91).

**Kuratowski–Zorn lemma** *See* Zorn lemma.

map         graph

Königsberg bridge problem

**lambda calculus** /lam-dă/ A branch of mathematical logic that is used to express the idea of COMPUTABILITY. Any mathematical process that can be performed by using an ALGORITHM can be achieved using the lambda calculus. The ideas of the lambda calculus are very closely related to the foundations of the theory of how computers operate, particularly the concept of a TURING MACHINE.

**lamina** /lam-ă-nă/ (*pl.* **laminae**) An idealization of a thin object that is taken to have uniform density and has the dimensions of area but zero thickness. A sheet of metal or paper or a card can be represented by a lamina.

**Lami's theorem** /lam-ee, lah-mee/ A relation between three forces *A*, *B*, and *C* that are in equilibrium and the angles α, β, and γ between the forces shown in the figure. The theorem states that $A/\sin\alpha = B/\sin\beta = C/\sin\gamma$. Lami's theorem is derived by applying the SINE RULE to the triangle of forces associated with *A*, *B*, and *C* and using the result that $\sin(180° - \theta) - \sin\theta$. It is used to solve problems involving three forces.

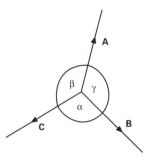

Lami's theorem

**Langlands program** /lang-lănds/ An attempt to unify different, apparently unconnected, branches of mathematics that was initiated by the American mathematician Robert Langlands in the 1960s. This program started off as a series of conjectures relating different branches of mathematics. A major boost to the Langlands program was supplied in 1995 when FERMAT'S LAST THEOREM was proved by showing that a conjecture relating two different branches of mathematics is true. If the other conjectures in the Langlands program are true it would mean that problems in one branch of mathematics could be related to problems in another branch of mathematics, with the hope that problems that are very difficult to solve in one branch of mathematics might be easier to solve in another branch. In addition, the program would, if realized, give a great deal of unity to mathematics.

**Laplace's equation** /lah-plahs/ *See* partial differential equation. The equation is named for the French mathematician Pierre Simon, Marquis de Laplace (1749–1827).

**Laplace transform** An integral transform f(*x*) of a function F(*t*) defined by:

$$f(x) = \int_0^\infty \exp(-xt)F(t)dt$$

It is possible for this integral not to exist but it is possible to specify under what conditions it does exist.

The Laplace transform of a function F(*t*) is sometimes denoted by L{F(*t*)}. For example, if F(*t*) = 1, L{F(*t*)} = 1/*x*. If F(*t*) = sin *at*, L{F(*t*)} = $a/(x^2 + a^2)$. If F(*t*) = exp(−*at*), L{F(*t*)} = 1/(*x* + *a*). Tables of Laplace transforms are available and they can be used in the solution of ordinary and partial differential equations.

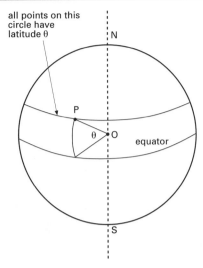

all points on this circle have latitude θ

N

P

θ O

equator

S

The latitude θ of a point P on the Earth's surface.

**Laplacian** /lă-**plass**-ee-ăn/ (Laplace operator) Symbol $\nabla^2$. The operator defined in three dimensions by
$$\nabla^2 = \partial^2/\partial x^2 + \partial^2/\partial y^2 + \partial^2/\partial z^2.$$
The Laplacian occurs in many PARTIAL DIF-FERENTIAL EQUATIONS, including equations of physical interest such as the LAPLACE EQUATION and the POISSON EQUATION.

**laptop** /**lap**-top/ A type of small lightweight portable computer with a flip-up screen and powered by rechargeable batteries.

**laser printer** A type of computer printer in which the image is formed by scanning a charged plate with a laser. Powdered ink adheres to the charged areas and is transferred to paper as in a photocopier. Laser printers can produce text almost as good as typesetting machines.

**lateral** /**lat**-ĕ-răl/ Denoting the side of a solid geometrical figure, as opposed to the base. For instance, a lateral edge of a pyramid is one of the edges from the vertex (apex). A lateral face of a pyramid or prism is a face that is not a base. The lateral surface (or area) of a cylinder or cone is the curved surface (or area), excluding the plane base.

**latin square** An $n \times n$ square array of $n$ different symbols with the property that each symbol appears once and only once in each row and each column. Such an arrangement is possible for every $n$. For example, for $n = 4$ and the letters A, B, C, D:
$$\begin{array}{cccc} A & B & C & D \\ C & D & A & B \\ D & C & B & A \\ B & A & D & C \end{array}$$
Such arrays are used in statistics to analyze experiments with three factors influencing the outcome; for example the experimenter, the method, and the material under test. The significance of each factor in the experiment may be tested by using a latin square. For example, denoting the four methods by A, B, C, D we may use the latin square

material

experimenter
$$\begin{array}{cccc} A & B & C & D \\ C & D & A & B \\ D & C & B & A \\ B & A & D & C \end{array}$$
The latin-square array is used since, in studying the effects of one factor, the influences of the other factors occur to the same extent.

**latitude** The distance of a point on the Earth's surface from the equator, measured as the angle in degrees between a plane through the equator (the equatorial plane) and the line from the point to the center of the Earth. A point on the equator has a latitude of 0° and the North Pole has a latitude of 90° N. *See also* longitude.

**lattice** A partially ordered set such that each pair of elements *a* and *b* has both:
**1.** A greatest lower bound *c*; i.e. an element *c* such that $c \leq a$ and $c \leq b$ and if $c' \leq a$ and $c' \leq b$ then $c' \leq c$.
**2.** A least upper bound *d*; i.e. an element *d* such that $d \geq a$ and $d \geq b$ and if $d' \geq a$ and $d' \geq b$ then $d' \geq d$.

The elements *c* and *d* are called the *meet* and *join* respectively of *a* and *b* and are denoted by $c = a \cap b$ and $d = a \cup b$. An example of a lattice is the set of all subsets of a given set, where $A \leq B$ means that each element of *A* is also an element of *B*. In this example, $A \cap B$ is the intersection of the sets *A* and *B* and $A \cup B$ is their union.

**latus rectum** /**lay**-tŭs **rek**-tŭm/ (*pl.* **latera recta**) *See* ellipse; hyperbola; parabola.

**law of flotation** *See* flotation; law of.

**law of large numbers** A theorem in probability stating that if an event *E* has probability *p*, and if N(*E*) represents the number of occurrences of the event in *n* trials, then N(*E*)/*n* is very close to *p* if *n* is a

large number and N(*E*)/*n* converges to *p* as *n* tends to infinity.

**law of moments** *See* moment.

**law of the mean** (**mean value theorem**) The rule in differential calculus that, if f(*x*) is continuous in the interval $a \leq x \leq b$, and the derivative f'(*x*) exists everywhere in this interval, then there is at least one value of *x* ($x_0$) between *a* and *b* for which:
$$[f(b) - f(a)]/(b - a) = f'(x_0)$$
Geometrically this means that if a straight line is drawn between two points, (*a*,f(*a*)) and (*b*,f(*b*)), on a continuous curve, then there is at least one point between these where the tangent to the curve is parallel to the line. This law is derived from ROLLE'S THEOREM.

**laws of conservation** *See* conservation law.

**laws of friction** *See* friction.

**laws of thought** Three laws of logic that are traditionally considered – as are other logic rules – to exemplify something fundamental about the way we think; that is, it is not arbitrary that we say certain forms of reasoning are correct. On the contrary, it would be impossible to think otherwise.
**1.** *The law of contradiction* (law of noncontradiction). It is not the case that some-

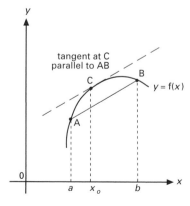

The law of the mean for a function f(*x*) that is continuous between *x* = *a* and *x* = *b*.

thing can both be true and not be true: symbolically

$$\sim(p \wedge \sim p)$$

**2.** *The law of excluded middle.* Something must either be true or not be true: symbolically

$$p \vee \sim p$$

**3.** *The law of identity.* If something is true, then it is true: symbolically

$$p \to p$$

**LCD** Lowest common denominator. *See* common denominator.

**LCM** Lowest common multiple. *See* common multiple.

**leading diagonal** *See* square matrix.

**least common denominator** *See* common denominator.

**least common multiple** *See* common multiple.

**least squares method** A method of fitting a REGRESSION LINE to a set of data. If the data are points $(x_1,y_1),...(x_n,y_n)$, the corresponding points $(x_1,y_1'),...(x_n,y_n')$ are found using the linear equation $y = ax + b$. The least squares line minimizes

$$[(y_1 - y_1')^2 + (y_2 - y_2')^2 + ... + (y_n - y_n')^2]$$

It is found by solving the normal equations

$$\Sigma y = an + b\Sigma x$$
$$\Sigma xy = a\Sigma x + b\Sigma x^2$$

for $a$ and $b$. The technique is extended for regression quadratics, cubics, etc.

**least upper bound** *See* bound.

**left-handed system** *See* right-handed system.

**Legendre polynomial** /lǎ-**zhahn**-drě/ A series of functions that arises as solutions to Laplace's equation in spherical polar coordinates. They form an infinite series. The polynomial is named for the French mathematician Adrien-Marie Legendre (1752–1833). *See also* partial differential equation.

**Leibniz theorem** /**lȳb**-nits/ A formula for finding the *n*th derivative of a product of two functions. The *n*th derivative with respect to $x$ of a function $f(x) = u(x)v(x)$, written $D^n(uv) = d^n(uv)/dx^n$, is equal to the series:

$$uD^nv + {}_nC_1DuD^{n-1}v + {}_nC_2D^2uD^{n-2}v + ... + {}_nC_{n-1}D^{n-1}uDv + vD^nu$$

where ${}_nC_r = n!/[(n - r)!r!]$.

The formula holds for all positive integer values of $n$.

For $n = 1$, $D(uv) = uDv + vDu$
For $n = 2$, $D^2(uv) = uD^2v + 2DuDv + vD^2u$
For $n = 3$, $D^3(uv) = uD^3v + 3DuD^2v + 3D^2uDv + D^3u$

Note the similarity between the differential coefficients and the binomial expansion coefficients. The theorem is named for the German mathematician, philosopher, historian, and physicist Gottfried Wilhelm Leibniz (1646–1716).

**lemma** /**lem**-ǎ/ (*pl.* **lemmas** or **lemmata**) A THEOREM proved for use in the proof of another theorem.

**length** The distance along a line, plane figure, or solid. In a rectangle, the greater of the two dimensions is usually called the length and the smaller the breadth.

**lever** A class of MACHINE; a rigid object able to turn around some point (pivot or fulcrum). The force ratio and the distance ratio depend on the relative positions of the fulcrum, the point where the user exerts the effort, and the point where the lever applies force to the load. There are three types (orders) of lever.

*First order*, in which the fulcrum is between the load and the effort. An example is a crowbar.

*Second order*, in which the load is between the effort and the fulcrum. An example is a wheelbarrow.

*Third order*, in which the effort is between the load and the fulcrum. An example is a pair of sugar tongs.

Levers can have high efficiency; the main energy losses are by friction at the pivot, and by bending of the lever itself.

**library programs** Collections of computer programs that have been purchased, contributed by users, or supplied by the computer manufacturers for the use of the computing community.

**light pen** A computer input device that enables a user to write or draw on the screen of a VISUAL DISPLAY UNIT (VDU).

**light-year** Symbol: ly A unit of distance used in astronomy, defined as the distance that light travels through space in one year. It is approximately equal to $9.460\ 5 \times 10^{15}$ meters.

**limit** In general, the value approached by a function as the independent variable approaches some value. The idea of a limit is the basis of the branch of mathematics known as *analysis*. There are several examples of the use of limits.
1. The limit of a function is the value it approaches as the independent variable tends to some value or to infinity. For instance, the function $x/(x + 3)$ for positive values of $x$ is less than 1. As $x$ increases it approaches 1 – the value approached as $x$ becomes infinite. This is written
$$\mathop{\mathrm{Lim}}_{x \to \infty} x/(x + 3) = 1$$
stated as 'the limit of $x/(x + 3)$ as $x$ approaches (or tends to) infinity is 1'. 1 is the *limiting value* of the function.
2. The limit of a CONVERGENT SEQUENCE is the limit of the $n$th term as $n$ approaches infinity.
3. The limit of a CONVERGENT SERIES is the limit of the sum of $n$ terms as $n$ approaches infinity.
4. A DERIVATIVE of a function f($x$) is the limit of [f($x$ + $\delta x$) – f($x$)]/$\delta x$ as $\delta x$ approaches zero.
5. A definite INTEGRAL is the limit of a finite sum of terms $y\delta x$ as $\delta x$ approaches zero.

**limit cycle** A closed curve in phase space to which a system evolves. A limit cycle is a type of ATTRACTOR characteristic of oscillating systems.

**limiting friction** *See* friction.

**line** A join between two points in space or on a surface. A line has length but no breadth, that is, it has only one dimension. A straight line is the shortest distance between two points on a flat surface.

**linear** Relating to a straight line. Two variables $x$ and $y$ have a linear relationship if it can be represented graphically by a straight line, i.e. by an equation of the form $y = mx + c$ (where $m$ is the slope of the line when plotted and $c$ is a constant). The equation is known as a LINEAR EQUATION.

**linear dependence** *See* dependent.

**linear equation** An EQUATION in which the highest power of an unknown variable is one. The general form of a linear equation is
$$mx + c = 0$$
where $m$ and $c$ are constants. On a Cartesian coordinate graph
$$y = mx + c$$
is a straight line that has a gradient $m$ and crosses the $y$-axis at $y = c$. The equation
$$x + 4y^2 = 4$$
is linear in $x$ but not in $y$. *See also* equation.

**linear extrapolation** *See* extrapolation.

**linear independence** *See* dependent.

**linear interpolation** *See* interpolation.

**linearly ordered** A *linearly ordered set* is a partially ordered set that satisfies the *trichotomy principle*: for any two elements $x$ and $y$ exactly one of $x > y$, $x < y$, is true. For example, the set of positive integers with their natural order is a linearly ordered set.

**linear momentum** *See* momentum.

**linear momentum, conservation of** *See* constant linear momentum; law of.

**linear programming** The process of finding maximum or minimum values of a linear function under limiting conditions or constraints. For example, the function $x – 3y$ might be minimized subject to the constraints that $x + y \leq 10$, $x \leq y – 2$, $8 \geq x \geq 0$

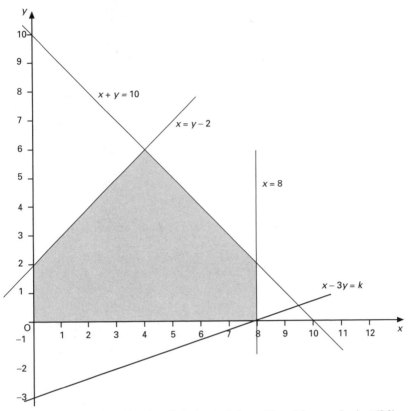

Linear programing: possible values lie in the shaded area. The minimum value is at (8,0).

and $y \geq 0$. The constraints can be shown as the area on a Cartesian coordinate graph bounded by the lines $x + y = 10$; $x = y - 2$, $x = 8$, and $y = 0$. The minimum value for $x - 3y$ is chosen from points within this area. A series of parallel lines $x - 3y = k$ are drawn for different values of $k$. The line $k = -9$ just reaches the constraint area at the point $(10,0)$. Lower values are outside it, and so $x = 10$, $y = 0$ gives the minimum value of $x - 3y$ within the constraints. Linear programming is used to find the best possible combination of two or more variable quantities that determine the value of another quantity. In most applications, for example, finding the best combination of quantities of each product from a factory to give the maximum profit, there are many variables and constraints. Linear functions with large numbers of variables

and constraints are maximized or minimized by computer techniques that are similar in principle to this graphical technique for two variables.

**linear transformation 1.** In one dimension, the general linear transformation is given by:
$$x' = (ax + b)/(dx + c)$$
where $a$, $b$, $c$, and $d$ are constants. In two dimensions, the general linear transformation is given by:
$$x' = (a_1x + b_1y + c_1)/(d_1x + e_1y + f_2)$$
and
$$y' = (a_2x + b_2y + c_2)/(d_2x + e_2y + f_2)$$
General linear transformations in more than two dimensions are defined similarly.
**2.** In an $n$-dimensional vector space, a linear transformation has the form $y = Ax$, where $x$ and $y$ are column vectors and $A$ is

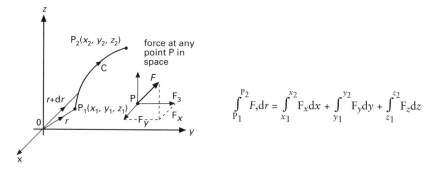

$$\int_{P_1}^{P_2} F.dr = \int_{x_1}^{x_2} F_x dx + \int_{y_1}^{y_2} F_y dy + \int_{z_1}^{z_2} F_z dz$$

The line integral of a force vector $F$ along a path C from a point $P_1$ to a point $P_2$.

a matrix. A linear transformation takes $ax + by$ into $ax' + by'$ for all $a$ and $b$ if it takes $x$ and $y$ into $x'$ and $y'$.

**line integral** (**contour integral; curvilinear integral**) The integration of a function along a particular path, C, which may be a segment of straight line, a portion of space curve, or connected segments of several curves. The function is integrated with respect to the position vector $r = ix + jy + kz$, which denotes the position of each point $P(x,y,z)$ on a curve C.

For example, the direction and magnitude of a force vector $F$ acting on a particle may depend on the particle's position in a gravitational field or a magnetic field. The work done by the force in moving the particle over a distance $dr$ is $F.dr$. The total work done in moving the particle along a particular path from point $P_1$ to point $P_2$ is the line integral shown in the diagram.

**line printer** An output device of a computer system that prints characters (letters, numbers, punctuation marks, etc.) on paper a complete line at a time and can therefore operate very rapidly.

**Lissajous figures** /lee-sa-zhoo/ Patterns obtained by combining two simple harmonic motions in different directions. For example, an object moving in a plane so that two components of the motion at right angles are simple harmonic motions, traces out a Lissajous figure. If the components have the same frequency and amplitude and are in phase the motion is a straight line. If they are out of phase by 90°, it is a circle. Other phase differences produce ellipses. If the frequencies of the components differ, more complex patterns are formed. Lissajous figures can be demonstrated with an oscilloscope by deflecting the spot with one oscillating signal along one axis and with another signal along the other axis. The patterns are named for the French physicist Jules Antoine Lissajous (1822–80).

**liter** /lee-ter/ Symbol: l A unit of volume defined as $10^{-3}$ meter$^3$. The name is not recommended for precise measurements. Formerly, the liter was defined as the volume of one kilogram of pure water at 4°C and standard pressure. On this definition, $1\ l = 1000.028\ cm^3$.

**literal equation** An equation in which the numbers are, like the unknowns, represented by letters. For example, $ax + b = c$ is a literal equation in $x$. To solve it, it is necessary to find $x$ in terms of $a$, $b$, and $c$, i.e. $x = (c - b)/a$. Finding the solution of a literal equation can be regarded as changing the subject of a formula.

**load** The force generated by a machine. *See* machine.

**local maximum** (**relative maximum**) A value of a function f($x$) that is greater than for the adjacent values of $x$, but is not the greatest of all values of $x$. *See* maximum point.

**local meridian** *See* longitude.

**local minimum** (relative minimum) A value of a function $f(x)$ that is less than for the adjacent values of $x$ but is not the lowest of all values of $x$. *See* minimum point.

**location** *See* store.

**locus of points** A set of points, often defined by an equation relating coordinates. For example, in rectangular Cartesian coordinates, the locus of points on a line through the origin at $45°$ to the $x$-axis and the $y$-axis is defined by the equation $x = y$; the line is said to be the locus of the equation. A circle is the locus of all points that lie a fixed distance from a given point.

**logarithm** /lôg-ă-rith-m/ A number expressed as the exponent of another number. Any number $x$ can be written in the form $x = a^y$. $y$ is then the logarithm to the base $a$ of $x$. For example, the logarithm to the base ten of 100 ($\log_{10}100$) is two, since $100 = 10^2$. Logarithms that have a base of ten are known as *common logarithms* (or *Briggsian logarithms*), named for the English mathematician Henry Briggs (1561–1630). They are used to carry out multiplication and division calculations because numbers can be multiplied by adding their logarithms. In general $p \times q$ can be written as $a^c \times a^d = a^{(c+d)}$, $p = a^c$ and $q = a^d$. Both logarithms and antilogarithms (the inverse function) are available in the form of printed tables, one used for calculations. For example, $4.91 \times 5.12$ would be calculated as follows: $\log_{10}4.91$ is 0.6911 (from tables) and $\log_{10}5.12$ is 0.7093 (from tables). Therefore, $4.91 \times 5.12$ is given by antilog (0.6911 + 0.7093), being 25.14 (from antilog tables). Similarly, division can be carried out by subtraction of logarithms and the $n$th root of a number $(x)$ is the antilogarithm of $(\log x)/n$.

For numbers between 0 and 1 the common logarithm is negative. For example, $\log_{10}0.01 = -2$. The common logarithm of any positive real number can be written in the form $n + \log_{10}x$, where $x$ is between 1 and 10 and $n$ is an integer. For example,
$\log_{10}15 = \log_{10}(10 \times 1.5) = \log_{10}10 + \log_{10}1.5 = 1 + 0.1761$
$\log_{10}150 = \log_{10}(100 \times 1.5) = 2.1761$
$\log_{10}0.15 = \log_{10}(0.1 \times 1.5) = -1 + 0.1761$.
This is written in the notation $\bar{1}.1761$.

The integer part of the logarithm is called the *characteristic* and the decimal fraction is the *mantissa*. *Natural logarithms* (*Napierian logarithms*) use the base $e = 2.718\ 28...$, $\log_e x$ is often written as $\ln x$.

**logarithmic function** /lôg-ă-rith-mik/ The function $\log_a x$, where $a$ is a constant. It is defined for positive values of $x$.

**logarithmic scale 1.** A line in which the distance, $x$, from a reference point is proportional to the logarithm of a number. For example, one unit of length along the line might represent 10, two units 100, three units 1000, and so on. In this case the distance $x$ along the logarithmic scale is given by the equation $x = \log_{10}a$. Logarithmic scales form the basis of the slide rule since two numbers may be multiplied by adding lengths on a logarithmic scale $\log(a \times b) = \log a + \log b$.

The graph of the curve $y = x^n$, when plotted on graph paper with logarithmic coordinate scales on both axes (log-log graph paper), is a straight line since $\log y = n\log x$. This method can be used to establish the equation of a non-linear curve. Known values of $x$ and $y$ are plotted on log-log graph paper and the gradient $n$ of the resulting line is measured, enabling the equation to be found. *See also* log-linear graph.

**2.** Any scale of measurement that varies logarithmically with the quantity measured. For instance, pH in chemistry is a measure of acidity or alkalinity – i.e. of hydrogen-ion concentration. It is defined as $\log_{10}(1/[H^+])$. An increase in pH from 5 to 6 represents a decrease in $[H^+]$ from $10^{-5}$ to $10^{-6}$, i.e. a factor of 10. An example of a logarithmic scale in physics is the decibel scale used for noise level.

Logarithmic scale

**logarithmic series** The infinite power series that is the expansion of the function $\log_e(1+x)$, namely:

$$x - x^2/2 + x^3/3 - x^4/4 + \ldots$$

This series is convergent for all values of $x$ greater than –1 and less than or equal to 1.

**log graph** *See* logarithmic scale.

**logic** The study of the methods and principles used in distinguishing correct or valid arguments and reasoning from incorrect or invalid arguments and reasoning. The main concern in logic is not whether a conclusion is in fact accurate, but whether the process by which it is derived from a set of initial assumptions (*premisses*) is correct. Thus, for example, the following form of argument is valid:

> all *A* is *B*
> all *B* is *C*
> therefore all *A* is *C*,

and thus the conclusion

> all fish have wings

can be derived, correctly, from the premisses

> all fish are mammals

and

> all mammals have wings

even though the premisses and the conclusion are untrue. Similarly, true premisses and true conclusions are no guarantee of a valid argument. Therefore, the true conclusion

> all cats are mammals

does *not* follow, logically, from the true premisses:

> all cats are warm-blooded

and

> all mammals are warm-blooded

because it is an example of the *invalid* argument form

> all *A* is *B*
> all *C* is *B*
> therefore, all *A* is *C*.

The incorrectness of the argument shows up clearly when, after making reasonable substitutions for *A*, *B*, and *C*, we get *true* premisses but a *false* conclusion:

> all dogs are mammals
> all cats are mammals
> therefore all dogs are cats.

Such an argument is called a *fallacy*.

Logic puts forward and examines rules that will insure that – given true premisses – a true conclusion can automatically be reached. It is not concerned with examining or assessing the truth of the premisses; it is concerned with the form and structure of arguments, not their content. *See* deduction; induction; symbolic logic; truth-value; validity.

**logic circuit** An electronic switching circuit that performs a logical operation, such as 'and' and 'implies' on its input signals. There are two possible levels for the input and output signals, high and low, sometimes indicated by the binary digits 1 and 0, which can be combined like the values 'true' and 'false' in a truth table. For example, a circuit with two inputs and one output might have a high output only when the inputs are different. The output therefore is the logical function 'either…or…' of the two inputs (the exclusive disjunction). *See* truth table.

**logic gate** An electronic circuit that carries out logical operations. Examples of such operations are 'and', 'either – or',

| input 1 | input 2 | output |
|---------|---------|--------|
| high | high | low |
| high | low | high |
| low | high | high |
| low | low | low |

Table for a logic circuit

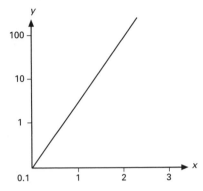

In this log-linear graph, the function $y = 4.9\,e^{1.5x}$ is shown as a straight line with a gradient of 1.5.

'not', 'neither – nor', etc. Logic gates operate on high or low input and output voltages. Binary logic circuits, those that switch between two voltage levels (high and low), are widely used in digital computers. The *inverter gate* or *NOT gate* simply changes a high input to a low output and vise versa. In its simplest form, the *AND gate* has two inputs and one output. The output is high if and only if both inputs are high. The *NAND gate* (not and) is similar, but has the opposite effect; that is, a low output if and only if both inputs are high. The *OR gate* has a high output if one or more of the inputs are high. The *exclusive OR gate* has a high input only if one of the inputs, but not more than one, is high. The *NOR gate* has a high output only if all the inputs are low. Logic gates are constructed using transistors, but in a circuit diagram they are often shown by symbols that denote only their logical functions. These functions are, in effect, those relationships that can hold between propositions in symbolic logic, with combinations that can be represented in a TRUTH TABLE. *See also* conjunction; disjunction; negation.

**log-linear graph (semilogarithmic graph)** A graph on which one axis has a logarithmic scale and the other has a linear scale. On a log-linear graph, an exponential function (one of the form $y = ke^{ax}$ where $k$ and $a$ are constants) is a straight line. Values of $x$ are plotted on the linear scale and values of $y$ on the LOGARITHMIC SCALE.

**log–log graph** A graph on which both axes have LOGARITHMIC SCALES.

**long division** A method used to divide one number by another number or one algebraic expression by another algebraic expression. In both cases it is customary to work from the left. In the case of numbers, the first figure of the number to be divided is divided by the dividing number. This gives an integer (which may be zero) and a remainder. The remainder is added to the next figure of the number to be divided and then divided by the dividing number. This process is repeated until all the figures in the number to be divided have been divided by the dividing number, with the answer being an integer and a remainder. The calculation of numbers using long division has largely been replaced by the use of electronic calculators. Nevertheless, the same sort of procedure is used to divide one algebraic expression by another algebraic expression.

**longitude** The east-west position of a point on the Earth's surface measured as the angle in degrees from a standard meridian (taken as the Greenwich meridian). A *meridian* is a great circle that passes through the North and South poles. The *local meridian* of a point is a great circle

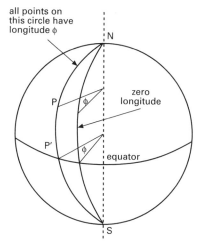

all points on this circle have longitude φ

N

zero longitude

P φ

P' φ

equator

S

The longitude φ of a point P on the Earth's surface.

passing through that point and the two poles. *See also* latitude.

**longitudinal wave** A wave motion in which the vibrations in the medium are in the same direction as the direction of energy transfer. Sound waves, transmitted by alternate compression and rarefaction of the medium, are an example of longitudinal waves. *Compare* transverse wave.

**long multiplication** A technique for the multiplication of two numbers or algebraic expressions. It is customary to write the two numbers to be multiplied above each other. The top number is multiplied by the figures of the bottom number separately, with a separate line for each product in the appropriate place underneath the initial numbers, with a final line at the bottom giving the final product. The calculation of numbers by long multiplication has largely been replaced by the use of electronic calculators. Long multiplication of algebraic expressions can be performed by a similar procedure.

**loop** A sequence of instructions in a computer program that is performed either a specified number of times or is performed repeatedly until some condition is satisfied. *See also* branch.

**Lorentz–Fitzgerald contraction** /lo-rents fits-je-răld/ A reduction in the length of a body moving with a speed $v$ relative to an observer, as compared with the length of an identical object at rest relative to the observer. The object is supposed to contract by a factor $\sqrt{(1 - v^2/c^2)}$, $c$ being the speed of light in free space. The contraction was postulated to account for the negative result of the Michelson–Morley experiment using the ideas of classical physics. The idea behind it was that the electromagnetic forces holding atoms together were modified by motion through the ether. The idea was made superfluous (along with the concept of the ether) by the theory of relativity, which supplied an alternative explanation of the Michelson–Morley experiment. The contraction was named for the Dutch theoretical physicist Hendrik Antoon Lorentz (1853–1928) and the Irish physicist George Francis Fitzgerald (1851–1901), who arrived at the solution independently.

**Lorentz force** /lo-**rents**/ The force $F$ exerted on an electric charge $q$ that is moving in a magnetic field of field strength $B$ with a velocity $v$. This force is given by: $F = q v \times b$. The force is perpendicular to both the direction of the magnetic field and the velocity. This means that the Lorentz force deflects the electric charge but does not change its speed. There are many physical applications of the Lorentz force.

**Lorentz transformation** A set of equations that correlate space and time coordinates in two frames of reference. *See also* relativity, theory of.

**lower bound** *See* bound.

**lowest common denominator** *See* common denominator.

**lowest common multiple** *See* common multiple.

**low-level language** *See* program.

**lumen** /loo-měn/ Symbol: lm (*pl.* **lumens** or **lumina**) The SI unit of luminous flux, equal to the luminous flux emitted by a point source of one candela in a solid angle of one steradian. 1 lm = 1 cd sr.

**lune** /loon/ A portion of the area of a sphere bounded by two great semicircles that have common end points.

**lux** /luks/ Symbol: lx (*pl.* **lux**) The SI unit of illumination, equal to the illumination produced by a luminous flux of one lumen falling on a surface of one square meter. 1 lx = 1 lm m$^{-2}$.

**machine** A device for transmitting force or energy between one place and another. The user applies a force (the effort) to the machine; the machine applies a force (the load) to something. These two forces need not be the same; in fact the purpose of the machine is often to overcome a large load with a small effort. For any machine this relationship is measured by the *force ratio* (or mechanical advantage) – the force applied by the machine (load, $F_2$) divided by the force applied by the user (effort, $F_1$).

The work done by the machine cannot exceed the work done to the machine. Therefore, for a 100% efficient machine:
if $F_2 > F_1$ then $s_2 < s_1$
and if $F_2 < F_1$ then $s_2 > s_1$.
Here $s_2$ and $s_1$ are the distances moved by $F_2$ and $F_1$ in a given time.

The relationship between $s_1$ and $s_2$ in a given case is measured by the *distance ratio* (or velocity ratio) – the distance moved by the effort ($s_1$) divided by the distance moved by the load ($s_2$).

Neither distance ratio nor force ratio has a unit; neither has a standard symbol. *See also* hydraulic press; inclined plane; lever; pulley; screw; wheel and axle.

**machine code** *See* program.

**machine language** *See* program.

**Maclaurin series** /mă-**klor**-in/ *See* Taylor series.

**magic square** A square array of numbers whose columns, rows and diagonals add to the same total. For the magic square:

```
6 7 2
1 5 9
8 3 4
```

the total is 15.

**magnetic disk** *See* disk.

**magnetic tape** A long strip of flexible plastic with a magnetic coating on which information can be stored. Its use in the recording and reproduction of sound is well-known. It is also widely used in com-

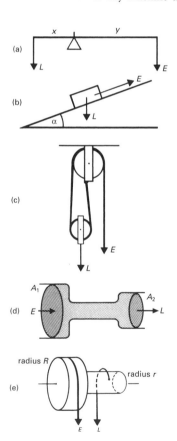

Machine: (a) lever; (b) inclined plane; (c) pulley; (d) hydraulic press; (e) wheel and axle.

puting to store information. The data is stored on the tape in the form of small closely packed magnetic spots arranged in rows across the tape. The spots are magnetized in one of two directions so that the data is in binary form. The magnetization pattern of a row of spots represents a letter, digit (0–9), or some other character.from the central processor. It is widely used as a backing store.

**magnificent seven problems** A set of seven mathematical problems put forward by a committee of mathematicians in 2000 as the most outstanding problems at that time. The POINCARÉ CONJECTURE was one of the seven problems. It is hoped that attempts to solve these problems will stimulate progress in mathematics in the twenty-first century just as attempts to solve HILBERT'S PROBLEMS stimulated progress in mathematics in the twentieth century. A reward of one million dollars is being offered for the solution of each of these problems. In order to claim this reward the solution has to be published in a mathematics journal and be scrutinized for two years. At the time of writing, no rewards have yet been given.

**magnitude 1.** The absolute value of a number (without regard to sign).
**2.** The non-directional part of a VECTOR, corresponding to the length of the line representing it.

**mainframe** *See* computer.

**main store** *See* store; central processor.

**major arc** *See* arc.

**major axis** *See* ellipse.

**major sector** *See* sector.

**major segment** The segment of a CIRCLE divided by a chord that contains the larger part of the circle.

**Mandelbrot set** /man-dĕl-brot/ A set of points in the complex plane generated by considering the iterations of the form $z \rightarrow$ $z^2 + c$, where $z$ and $c$ are complex numbers. This process can produce an infinite number of patterns depending on the value chosen for the constant complex number $c$ (*see* Julia set). These patterns are of two types: they are either connected, so that there is a single area within a boundary, or they are disconnected and broken into distinct parts. The Mandelbrot set is the set of all Julia sets that are connected. It can be represented by points on the complex plane and has a characteristic shape.

It is not necessary to generate Julia sets for the Mandelbrot set to be produced. It can be shown that a given value of $c$ is in the Mandelbrot set if the starting value $z = 0$ is bounded for the iteration. Thus, if the sequence

$$c, c^2 + c, (c^2 + c)^2 + c, \dots$$

remains bounded, then the point $c$ is in the Mandelbrot set.

The set has been the subject of much interest. The figure is a FRACTAL and can be examined on the computer screen in high 'magnification' (i.e. by calculating the set over a short range of values of $c$). It shows an amazingly complex self-similar structure characterized by the presence of smaller and smaller copies of the set at increasing levels of detail. The set is named for the American mathematician Benoit Mandelbrot (1924–    ).

**manifold** A space that is locally Euclidean but is not necessarily Euclidean globally. Examples of manifolds that are not Euclidean globally include the circle and the surface of a sphere. It is convenient to discuss many aspects of geometry and topology in terms of manifolds.

**mantissa** /man-**tiss**-ă/ *See* logarithm.

**many-valued function** A function for which one value of $x$ gives more than one value of $y$. If one value of $x$ gives $n$ values of $y$ then the function is said to be an $n$-valued function of $x$. If $n = 2$ then the function is called a two-valued function of $x$. For example, $y^2 = x$ is a two-valued function of $x$ since every real value of $x$ gives two values of $y$.

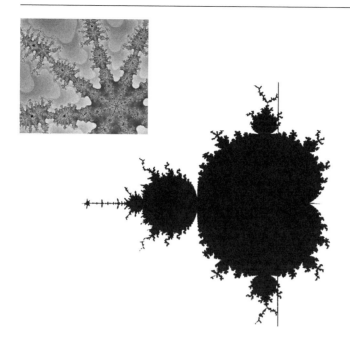

Mandelbrot set: the shape of the set with a detail showing the fine structure (top left).

**map** *See* function.

**mapping** *See* function.

**market price** *See* nominal value; yield.

**Markov chain** /**mar**-koff/ A sequence of discrete random events or variables that have probabilities depending on previous events in the chain. The sequence is named for the Russian mathematician Andrey Andreyevich Markov (1856–1922).

**mass** Symbol: $m$ A measure of the quantity of matter in an object. The SI unit of mass is the kilogram. Mass is determined in two ways: the *inertial mass* of a body determines its tendency to resist change in motion; the GRAVITATIONAL MASS determines its gravitational attraction for other masses. *See also* inertial mass; weight.

**mass, center of** *See* center of mass.

**mass-energy equation** The equation $E = mc^2$, where $E$ is the total energy (rest mass energy + kinetic energy + potential energy) of a mass $m$, $c$ being the speed of light in free space. The equation is a consequence of Einstein's special theory of relativity. It is a quantitative expression of the idea that mass is a form of energy and energy also has mass. Conversion of rest-mass energy into kinetic energy is the source of power in radioactive substances and the basis of nuclear-power generation.

**material implication** *See* implication.

**mathematical induction** *See* induction.

**mathematical logic** *See* symbolic logic.

**matrix** /**may**-triks, **mat**-riks/ (*pl.* **matrices** or **matrixes**) A set of quantities arranged in rows and columns to form a rectangular array. The common notation is to enclose these in parentheses. Matrices do not have a numerical value, like DETERMINANTS. They are used to represent relations between the quantities. For example, a plane vector can be represented by a single col-

$$\begin{pmatrix} 1 & 2 & 3 \\ 4 & 5 & 6 \end{pmatrix} + \begin{pmatrix} 2 & 6 & 10 \\ 4 & 8 & 12 \end{pmatrix} = \begin{pmatrix} 3 & 8 & 13 \\ 8 & 13 & 18 \end{pmatrix}$$

Matrix addition.

$$3 \times \begin{pmatrix} 1 & 2 & 3 \\ 4 & 5 & 6 \end{pmatrix} = \begin{pmatrix} 3 & 6 & 9 \\ 12 & 15 & 18 \end{pmatrix}$$

Multiplication of a matrix by a number.

$$\begin{pmatrix} 1 & 2 & 3 \\ 4 & 5 & 6 \end{pmatrix} \times \begin{pmatrix} 6 & 7 \\ 8 & 9 \\ 10 & 11 \end{pmatrix} = \begin{pmatrix} (6 + 16 + 30) & (7 + 18 + 33) \\ (24 + 40 + 60) & (28 + 45 + 66) \end{pmatrix}$$

Matrix multiplication

Matrix: addition and multiplication of matrices.

umn matrix with two numbers, a $2 \times 1$ matrix, in which the upper number represents its component parallel to the $x$-axis and the lower number represents the component parallel to the $y$-axis. Matrices can also be used to represent, and solve, simultaneous equations. In general, an $m \times n$ matrix – one with $m$ rows and $n$ columns – is written with the first row:

$$a_{11}a_{12} \cdots a_{1n}$$

The second row is:

$$a_{21}a_{22} \cdots a_{2n}$$

and so on, the $m$th row being:

$$a_{m1}a_{m2} \cdots a_{mn}$$

The individual quantities $a_{11}$, $a_{21}$, etc., are called *elements* of the matrix. The number of rows and columns, $m \times n$, is the *order* or *dimensions* of the matrix. Two matrices are equal only if they are of the same order and if all their corresponding elements are equal. Matrices, like numbers, can be added, subtracted, multiplied, and treated algebraically according to certain rules. However, the commutative, associative, and distributive laws of ordinary arithmetic do not apply. *Matrix addition* consists of adding corresponding elements together to obtain another matrix of the same order. Only matrices of the same order can be added. Similarly, the result of subtracting two matrices is the matrix formed by the differences between corresponding elements.

*Matrix multiplication* also has certain rules. In multiplication of an $m \times n$ matrix by a number or constant $k$, the result is another $m \times n$ matrix. If the element in the $i$th row and $j$th column is $a_{ij}$ then the corresponding element in the product is $ka_{ij}$. This operation is distributive over matrix addition and subtraction, that is, for two matrices $A$ and $B$,

$$k(A + B) = kA + kB$$

Also, $kA = Ak$, as for multiplication of numbers. In the multiplication of two matrices, the matrices $A$ and $B$ can only be multiplied together to form the product $AB$ if the number of columns in $A$ is the same as the number of rows in $B$. In this case they are called *conformable matrices*. If $A$ is an $m \times p$ matrix with elements $a_{ij}$ and $B$ is a $p \times n$ matrix with elements $b_{ij}$, then their product $AB = C$ is an $m \times n$ matrix with elements $c_{ij}$, such that $c_{ij}$ is the sum of the products

$$a_{i1}b_{1j} + a_{i2}b_{2j} + a_{i3}b_{3j} + \cdots + a_{ip}b_{pj}$$

Matrix multiplication is not commutative, that is, $AB \neq BA$.

*See also* square matrix.

**matrix of coefficients** An $m \times n$ matrix, with general entry $a_{ij}$ associated with the coefficients in the set of $m$ linear equations for $n$ unknown variables $x_1, x_2, \ldots x_n$:

$$a_{11}x_1 + a_{12}x_2 + \cdots a_{1n}x_n = b_1,$$
$$a_{21}x_1 + a_{22}x_2 + \cdots a_{2n}x_n = b_2,$$
$$a_{m1}x_1 + a_{m2}x_2 + \cdots a_{mn}x_n = b_m,$$

Expressing a set of linear equations in terms of the matrix of coefficients can be a step towards finding the set of solutions of

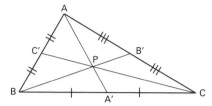

Median: the three medians intersect at P, the centroid of the triangle.

the equations. The matrix formed by including the constants $b_1$, $b_2$, etc., is the *augmented matrix*.

**maximum likelihood** A method of estimating the most likely value of a parameter. If a series of observations $x_1, x_2, ..., x_n$ are made, the likelihood function, $L(x)$, is the joint probability of observing these values. The likelihood function is maximized when $[d\log L(x)]/dp = 0$. In many cases, an intuitive estimate, such as the mean, is also the maximum likelihood estimate.

**maximum point** A point on the graph of a function at which it has the highest value within an interval. If the function is a smooth continuous curve, the maximum is a TURNING POINT, that is, the slope of the tangent to the curve changes continuously from positive to negative by passing through zero. If there is a higher value of the function outside the immediate neighborhood of the maximum, it is a *local maximum* (or *relative maximum*). If it is higher than all other values of the function it is an *absolute maximum*. See also stationary point.

**mean** A representative or expected value for a set of numbers. The *arithmetic mean* or *average* (called *mean*) of $x_1, x_2, ..., x_n$ is given by:
$$(x_1 + x_2 + x_3 + ... + x_n)/n$$
If $x_1, x_2, ..., x_k$ occur with frequencies $f_1, f_2, ..., f_k$ then the arithmetic mean is
$$(f_1 x_1 + f_2 x_2 + ... + f_k x_k)/(f_1 + f_2 + ... + f_k)$$

When data is classified, as for example in a frequency table, $x_1$ is replaced by the class mark.

The *weighted mean* W =
$$(w_1 x_1 + w_2 x_2 + ... + w_n x_n)/(w_1 + w_2 + ... + w_n)$$
where weight $w_i$ is associated with $x_i$.

The *harmonic mean* H =
$$n/[(1/x_1) + (1/x_2) + ... + (1/x_n)]$$

The *geometric mean* G =
$$(x_1.x_2 ... x_n)^{1/n}$$

The mean of a random variable is its expected value.

**mean center** *See* centroid.

**mean deviation** A measure of the dispersion of a set of numbers. It is equal to the average, for a set of numbers, of the differences between each number and the set's mean value. If $x$ is a random variable with mean value $\mu$, then the mean deviation is the average, or expected value, of $|x - \mu|$, written
$$\sum_i |x_i - \mu|/n$$

**mean value theorem** *See* law of the mean.

**measurements of central tendency** The general name given for the types of average used in statistics, i.e. the MEAN, the MEDIAN, and the MODE.

**measure theory** A branch of mathematics concerned with constructing the theory of integration in a mathematically rigorous way, starting from set theory.

**mechanical advantage** *See* force ratio; machine.

**mechanics** The study of forces and their effect on objects. If the forces on an object or in a system cause no change of momentum the object or system is in equilibrium. The study of such cases is *statics*. If the forces acting do change momentum the study is of *dynamics*. The ideas of dynamics relate the forces to the momentum changes produced. *Kinematics* is the study of motion without consideration of its cause.

**median** /mee-dee-ăn/ 1. The middle number of a set of numbers arranged in order. When there is an even number of numbers the median is the average of the middle two. For example, the median of 1,3,5,11,11 is 5, and of 1,3,5,11,11,14 is $(5 + 11)/2 = 8$. The median of a large population is the 50th percentile ($P_{50}$). *Compare* mean. *See also* percentile; quartile.
2. In geometry, a straight line joining the vertex of a triangle to the mid-point of the opposite side. The medians of a triangle intersect at a single point, which is the centroid of the triangle.

**mediator** A line that bisects another line at right angles.

**mega-** Symbol: M A prefix denoting $10^6$. For example, 1 megahertz (MHz) = $10^6$ hertz (Hz).

**member** *See* element.

**memory** *See* store.

**mensuration** /men-shŭ-ray-shŏn/ The study of measurements, especially of the dimensions of geometric figures in order to calculate their areas and volumes.

**Mercator's projection** /mer-kay-terz/ A method of mapping points from the surface of a sphere onto a plane surface. Mercator's projection is used to make maps of the World. The lines of longitude on the sphere become straight vertical lines on the plane. The lines of latitude on the sphere become straight horizontal lines. Areas further from the equator are more stretched out in the horizontal direction. For a particular point on the surface of the sphere at angle θ latitude and angle φ longitude, the corresponding Cartesian coordinates on the map are:

$$x = k\theta$$
$$y = k \log \tan(\phi/2)$$

Mercator's projection is an example of a conformal mapping, in which the angles between lines are preserved (except at the poles). This method of mapping is named for the Dutch cartographer and geographer Gerardus Mercator (1512–94). *See also* projection.

**meridian** /mĕ-rid-ee-ăn/ *See* longitude.

**Mersenne prime** /mair-sen/ a prime number that can be written as $2^p-1$, where $p$ is a prime number. Mersenne primes are discovered using computers, with the number being discovered increasing as the power of computers increases. For example, the Mersenne prime $2^{19937}-1$ was discovered in this way. Mersenne primes are related to even PERFECT NUMBERS since the only even perfect numbers are numbers of the form $2^{(p-1)}(2^p-1)$. The prime is named for the French mathematician Marin Mersenne (1588–1648).

**meter** /mee-ter/ Symbol: m The SI base unit of length, defined as the distance traveled by light in a vacuum in 1/299 792 458 of a second.

**metric** /met-rik/ An expression for the distance between two points in some geometrical space that is not necessarily Euclidean. The expressions for the metric in non-Euclidean geometrics are generalizations of the distance between two points in Euclidean geometry which is given by applying PYTHAGORAS' THEOREM to Cartesian coordinates. The concept of the metric is very important in the geometries that describe flat space–time in special relativity theory and curved space–time in general relativity theory.

**metric prefix** Any of various numerical prefixes used in the metric system.

**metric space** Any set of points, such as a plane or a volume in geometrical space, in which a pair of points $a$ and $b$ with a distance $d(a,b)$ between them, satisfy the conditions that $d(a,b) \geq 0$ and $d(a,b) = 0$ if and only if $a$ and $b$ are the same point. Another property of a metric space is that $d(a,b) + d(b,c) \geq d(a,c)$. The set of all the functions of $x$ that are continuous in the interval $x = a$ to $x = b$ is also a metric space. If $f(x)$ and $g(x)$ are in the space,

$$\int_a^b [f(x) - g(x)]dx$$

is defined for all values of $x$ between $a$ and $b$, and the integral is zero if and only if $f(x) = g(x)$ for all values of $x$ between $a$ and $b$.

**metric system** A system of units based on the meter and the kilogram and using multiples and submultiples of 10. SI units, c.g.s. units, and m.k.s. units are all scientific metric systems of units.

**metric ton** *See* tonne.

**metrology** /mĕ-**trol**-ŏ-jee/ The study of units of measurements and the methods of making precise measurements. Every physical quantity that can be quantified is expressed by a relationship of the type $q = nu$, where $q$ is the physical quantity, $n$ is a number, and $u$ is a unit of measurement. One of the prime concerns of metrologists is to select and define units for all physical quantities.

**m.g.f.** *See* moment generating function.

**Michelson–Morley experiment** /**mӯ**-kĕl-sŏn **mor**-lee/ A famous experiment conducted in 1887 in an attempt to detect the ether, the medium that was supposed to be necessary for the transmission of electromagnetic waves in free space. In the experiment, two light beams were combined to produce interference patterns after traveling for short equal distances perpendicular to each other. The apparatus was then turned through 90° and the two interference patterns were compared to see if there had been a shift of the fringes. If light has a velocity relative to the ether and there is an ether 'wind' as the Earth moves through space, then the times of travel of the two beams would change, resulting in a fringe shift. No such shift was detected, not even when the experiment was repeated six months later when the ether wind would have reversed direction. The experiment is named for its instigators, the American physicist Albert Abraham Michelson (1852–1931) and the American chemist and physicist Edward William Morley (1838–1923). *See also* relativity; theory of.

**micro-** Symbol: μ A prefix denoting $10^{-6}$. For example, 1 micrometer (μm) = $10^{-6}$ meter (m).

**microcomputer** /mӯ-kroh-kŏm-**pyoo**-ter/ *See* computer.

**micron** /**mӯ**-kron/ (**micrometer**) Symbol: μm A unit of length equal to $10^{-6}$ meter.

| METRIC PREFIXES | | | | | |
|---|---|---|---|---|---|
| Prefix | Symbol | Multiple | Prefix | Symbol | Multiple |
| atto- | a | $\times 10^{-18}$ | deca- | da | $\times 10$ |
| femto- | f | $\times 10^{-15}$ | hecto- | h | $\times 10^2$ |
| pico- | p | $\times 10^{-12}$ | kilo- | k | $\times 10^3$ |
| nano- | n | $\times 10^{-9}$ | mega- | M | $\times 10^6$ |
| micro- | μ | $\times 10^{-6}$ | giga- | G | $\times 10^9$ |
| milli- | m | $\times 10^{-3}$ | tera- | T | $\times 10^{12}$ |
| centi- | c | $\times 10^{-2}$ | peta- | P | $\times 10^{15}$ |
| deci- | d | $\times 10^{-1}$ | exa- | E | $\times 10^{18}$ |

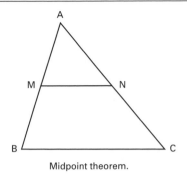

Midpoint theorem.

**microprocessor** /mȳ-kroh-**pross**-ess-er/ *See* central processor.

**mid-point theorem** A result in geometry that states that the line that joins the midpoints of two sides of a triangle has half the length of the third side of the triangle and is parallel to the third side. This result can be proved very easily using vectors in the following way.
Denote AB by *b* and AC by *c*. The directed line BC is given by BC = BA + AC = –*b* + *c* = *c*–*b*. One also has AM = (1/2) AB = (1/2) *b* and AN = (1/2)AC = (1/2) *c*. Thus, MN = MA + AN = –(1/2) *b* + (1/2) *c* = (1/2) *c* –(1/2) *b* = (1/2) (*c* – *b*). The last result can be written as MN = (1/2)BC. This means that MN has half the magnitude of BC and is parallel to it.

**mil** 1. A unit of length equal to one thousandth of an inch. It is commonly called a 'thou' and is equivalent to $2.54 \times 10^{-5}$ m.
2. A unit of area, usually called a *circular mil*, equal to the area of a circle having a diameter of 1 mil.

**mile** A unit of length equal to 1760 yards. It is equivalent to 1.6093 km.

**milli-** Symbol: m A prefix denoting $10^{-3}$. For example, 1 millimeter (mm) = $10^{-3}$ meter (m).

**million** A number equal to 1 000 000 ($10^6$).

**minicomputer** /min-ee-kŏm-**pyoo**-ter/ *See* computer.

**minimum point** A point on the graph of a function at which it has the lowest value within an interval. If the function is a smooth continuous curve, the minimum is a turning point, that is, the slope of the tangent to the curve changes continuously from negative to positive by passing through zero. If there is a lower value of the function outside the immediate neighborhood of the minimum, it is a *local minimum* (or *relative minimum*). If it is lower than all other values of the function it is an *absolute minimum*. *See also* stationary point; turning point.

**minor arc** *See* arc.

**minor axis** *See* ellipse.

**minor sector** *See* sector.

**minor segment** The segment of a CIRCLE divided by a chord that contains the smaller part of the circle.

**minuend** /**min**-yoo-end/ The term from which another term is subtracted in a difference. In 5 – 4 = 1, 5 is the minuend (4 is the subtrahend).

**minute** (of arc) A unit of plane angle equal to one sixtieth of a degree.

**mirror line** A line that is the axis of symmetry for reflection symmetry. If part of an object is in contact with the mirror line then the image of the object is also in contact with the mirror line. If part of an object is a certain distance from the mirror line then the image of that part of the object is the same distance from the mirror line. If the mirror line passes through an object then it also passes through the image of the object.

**mixed number** *See* fraction.

**m.k.s. system** A system of units based on the meter, the kilogram, and the second. It formed the basis for SI units.

**mmHg (millimeter of mercury)** A former unit of pressure defined as the pressure that

Möbius strip, which has one surface and one edge.

will support a column of mercury one millimeter high under specified conditions. It is equal to 133.322 4 Pa. It is also called the *torr*.

**Möbius strip** /**moh**-bee-ŭs/ (Möbius band) A continuous flat loop with one twist. It is formed by taking a flat rectangular strip, twisting it in the middle so that each end turns through 180° with respect to the other, and then joining the ends together. Because of the twist, a continuous line can be traced along the surface between any two points, without crossing an edge. The unique topological property of the Möbius strip is that it has one surface and one edge. If a Möbius strip is cut along a line parallel to the edge it is transformed into a doubly twisted band that has two edges and two sides. It is named for the German mathematician August Ferdinand Möbius (1790–1868). *See also* topology.

**modal class** /**moh**-d'l/ The CLASS that occurs with the greatest frequency, for example in a frequency table. *See also* mode.

**mode** The number that occurs most frequently in a set of numbers. For example, the mode (modal value) of {5, 6, 2, 3, 2, 1, 2} is 2. If a continuous random variable has probability density function f($x$), the mode is the value of $x$ for which f($x$) is a maximum. If such a variable has a frequency curve that is approximately symmetrical and has only one mode, then

(mean – mode) = 3(mean – median).

**modem** /**moh**-dem/ A device for sending computer data over long distances using telephone lines. It is short for modulator/demodulator.

**modulation** /moj-ŭ-**lay**-shŏn/ A process in which the characteristics of one wave are altered by some other wave. Modulation is used extensively in radio transmission. The wave which is altered is called the *carrier wave* and the wave responsible for the change is called the *modulating wave*. Information is transmitted in this way.

There are several ways in which modulation can be performed. In *amplitude modulation* the amplitude of the carrier wave rises and falls as the amplitude of the modulating wave rises and falls.

If the carrier wave is regarded as a sine wave then amplitude modulation alters the amplitude of the sine wave but does not change the angular aspect of the carrier wave, either as regards phase angle or frequency. Modulation which involves change of the angle is known as *angle modulation*. There are two types of angle modulation: *frequency modulation* and *phase modulation*. In frequency modulation the frequency of the carrier wave rises and falls as the amplitude of the modulating wave rises and falls. In phase modulation the relative phase of the carrier wave changes by an amount proportional to the amplitude of the modulating wave. In angle modulation the amplitude of the carrier wave is not altered.

There exist several other types of modulation, including some which make use of pulses (*pulse modulation*).

**modulus** /moj-ŭ-**lŭs**/ (*pl.* moduli) The absolute value of a quantity, not considering its sign or direction. For example, the modulus of minus five, written |–5|, is 5. The modulus of a vector quantity corresponds to the length or magnitude of the vector. The modulus of a COMPLEX NUMBER $x + iy$

is $\sqrt{(x^2 + y^2)}$. If the number is written in the form $r(\cos\theta + i\sin\theta)$, the modulus is $r$. *See also* argument.

**mole** Symbol: mol The SI base unit of amount of substance, defined as the amount of substance that contains as many elementary entities as there are atoms in 0.012 kilogram of carbon-12. The elementary entities may be atoms, molecules, ions, electrons, photons, etc., and they must be specified. One mole contains $6.022\ 52 \times 10^{23}$ entities. One mole of an element with relative atomic mass $A$ has a mass of $A$ grams (this was formerly called one *gram-atom*). One mole of a compound with relative molecular mass $M$ has a mass of $M$ grams (this was formerly called one *gram-molecule*).

**moment** (of a force) The turning effect produced by a force about a point. If the point lies on the line of action of the force the moment of the force is zero. Otherwise it is the product of the force and the perpendicular distance from the point to the line of action of the force. If a number of forces are acting on a body, the resultant moment is the algebraic sum of all the individual moments. For a body in equilibrium, the sum of the clockwise moments is equal to the sum of the anticlockwise moments (this law is sometimes called the *law of moments*). *See also* couple; torque.

**moment generating function** (m.g.f.) A function that is used to calculate the statistical properties of a random variable, $x$. It is defined in terms of a second variable, $t$, such that the m.g.f., $M(t)$, is the expectation value of $e^{tx}$, $E(e^{tx})$. For a discrete random variable
$$M(t) = \Sigma e^{tx} p$$
and for a continuous random variable
$$M(t) = \int e^{tx} f(x) dx$$
Two distributions are the same if their m.g.f.s are the same. The mean and variance of a distribution can be found by differentiating the m.g.f. The mean $E(x) = M'(0)$ and the variance, $Var(x) = M''(0) - (M'(0))^2$.

**moment of area** For a given surface, the moment of area is the moment of mass that the surface would have if it had unit mass per unit area.

**moment of inertia** Symbol: $I$ The rotational analog of mass. The moment of inertia of an object rotating about an axis is given by
$$I = mr^2$$
where $m$ is the mass of an element a distance $r$ from the axis. *See also* radius of gyration; theorem of parallel axes.

**moment of mass** The moment of mass of a point mass about a point, line, or plane is the product of the mass and the distance from the point or of the mass and the perpendicular distance from the line or plane. For a system of point masses, the moment of mass is the sum of the mass-distance products for the individual masses. For an object the integral must be used over the volume of the object.

**momentum, conservation of** /mŏ-**men**-tŭm/ *See* constant linear momentum; law of.

**momentum, linear** Symbol: $p$ The product of an object's mass and its velocity: $p = mv$. The object's momentum cannot change unless a net outside force acts. This relates to Newton's laws and to the definition of force. It also relates to the principle of constant momentum. *See also* angular momentum.

**monotonic** /mon-ŏ-**tonn**-ik/ Always changing in the same direction. A *monotonic increasing function* of a variable $x$ increases or stays constant as $x$ increases, but never decreases. A *monotonic decreasing function* of $x$ decreases or stays constant as $x$ increases, but never increases. Each term in a *monotonic series* is either greater than or equal to the one before it (monotonic increasing) or less than or equal to the one before it (monotonic decreasing). *Compare* alternating series.

**mouse** A computer INPUT DEVICE that is held under the palm of the hand and moved on a flat surface to control the movements of a cursor (pointer) on the computer screen. Instructions can be sent to the computer by pressing ('clicking on') one or more buttons on the mouse.

**multiple** A number or expression that has a given number or expression as a factor. For example, 26 is a multiple of 13.

**multiple integral** *See* iterated integral.

**multiple point** A point on the curve of a function at which several arcs intersect, or which forms an isolated point, and where a simple derivative of the function does not exist. If the equation of the curve is written in the form:

$$(a_1x + b_1y) + (a_2x^2 + b_2xy + c_2y^2)$$
$$+ (a_3x^3 + ...) + ... = 0$$

in which the multiple point is at the origin of a Cartesian-coordinate system, the values of the coefficients of $x$ and $y$ indicate the type of multiple point. If $a_1$ and $b_1$ are zero, that is, if all the first degree terms are zero, then the origin is a singular point. If the terms $a_2$, $b_2$, and $c_2$ are also zero it is a double point. If, in addition, the terms $a_3$, $b_3$, etc., of the third degree terms are zero, it is a *triple point*, and so on. *See also* double point; isolated point.

**multiplicand** /mul-ti-plă-**kand**/ A number or term that is multiplied by another (the *multiplier*) in a multiplication.

**multiplication** Symbol: × The operation of finding the product of two or more quantities. In arithmetic, multiplication of one number, $a$, by another, $b$, consists of adding $a$ to itself $b$ times. This kind of multiplication is commutative, that is, $a × b = b × a$. The identity element for arithmetic multiplication is 1, i.e. multiplication by 1 produces no change. In a series of multiplications, the order in which they are carried out does not change the result. For exam-

ple, $2 × (4 × 5) = (2 × 4) × 5$. This is the associative law for arithmetic multiplication.

Multiplication of vector quantities and matrices do not follow the same rules.

**multiplication of complex numbers** The rules of multiplication which emerge from the definition of a complex number. If $z_1 = a + ib$ and $z_2 = c + id$, then the product $z_1z_2$ is given by: $z_1z_2 = (ac - bd) + i(ad + bc)$.

The product of two complex numbers can be written neatly in terms of their polar forms:

$z_1 = r_1 \exp (i\theta_1) = r_1 (\cos\theta_1 + i \sin\theta_1)$ and
$z_2 = r_2 \exp (i\theta_2) = r_2 (\cos\theta_2 + i \sin\theta_2)$.
This gives
$z_1z_2 = r_1r_2 \exp [i(\theta_1 + \theta_2)] = r_1r_2 [\cos (\theta_1 + \theta_2) + i \sin (\theta_1 + \theta_2)]$. In addition to being very convenient for many purposes, the expression for the product of two complex numbers in terms of polar forms has a simple geometrical interpretation that the modulus of the product is the product of the moduli of the two complex numbers being multiplied together and the argument of the product is the sum of the arguments of the two complex numbers.

**multiplication of fractions** *See* fractions.

**multiplication of matrices** *See* matrix.

**multiplication of vectors** *See* scalar product; vector product.

**multiplier** *See* multiplicand.

**mutually exclusive events** Two events that cannot occur together. If the two events are denoted by $M_1$ and $M_2$ respectively then the two events being mutually exclusive is denoted by $M_1 \cap M_2 = \varnothing$ (the null event).

**myria-** Symbol: my A prefix used in France to denote $10^4$.

**NAND gate** *See* logic gate.

**nano-** Symbol: n A prefix denoting $10^{-9}$. For example, 1 nanometer (nm) = $10^{-9}$ meter (m).

**Napierian logarithm** /nă-**peer**-ee-ăn/ *See* logarithm.

**Napier's formulae** /**nay**-pee-erz/ A set of equations used in spherical trigonometry to calculate the sides and angles in a spherical triangle. In a spherical triangle with sides $a$, $b$, and $c$, and angles opposite these of $\alpha$, $\beta$, and $\gamma$ respectively:

$$\sin\tfrac{1}{2}(a - b)/\sin\tfrac{1}{2}(a + b) = \tan\tfrac{1}{2}(\alpha - \beta)/\tan\tfrac{1}{2}\gamma$$
$$\cos\tfrac{1}{2}(a - b)/\cos\tfrac{1}{2}(a + b) = \tan\tfrac{1}{2}(\alpha + \beta)/\tan\tfrac{1}{2}\gamma$$
$$\sin\tfrac{1}{2}(\alpha - \beta)/\sin\tfrac{1}{2}(\alpha + \beta) = \tan\tfrac{1}{2}(a - b)/\cot\tfrac{1}{2}c$$
$$\cos\tfrac{1}{2}(\alpha - \beta)/\cos\tfrac{1}{2}(\alpha + \beta) = \tan\tfrac{1}{2}(a + b)/\cot\tfrac{1}{2}c$$

The formulae are named for the Scottish mathematician John Napier (1550–1617). *See also* spherical triangle.

**nappe** /nap/ One of the two parts of a conical surface that lie either side of the vertex. *See* cone.

**natural frequency** The frequency at which an object or system will vibrate freely. A free vibration occurs when there is no external periodic force and little resistance. The amplitude of free vibrations must not be too great. For instance, a pendulum swinging with small swings under its own weight moves at its natural frequency. Normally, an object's natural frequency is its fundamental frequency.

**natural logarithm** *See* logarithm.

**natural numbers** Symbol: N The set of numbers {1,2,3, …} used for counting separate objects.

**nautical mile** A unit of length equal to 6080 feet (about 1.852 km), equivalent to 1 minute of arc along a great circle on the Earth's surface. A speed of 1 nautical mile per hour is a knot.

**necessary condition** *See* condition.

**negation** Symbol: ~ or ¬ In logic, the operation of putting *not* or *it is not the case that* in front of a proposition or statement, thus reversing its truth value. The negation of a proposition $p$ is false if $p$ is true and vice versa. The truth-table definition for negation is shown in the illustration. *See also* truth table.

| $P$ | | $\sim P$ |
|-----|---|----------|
| F | | T |
| T | | F |

Negation

**negative** Denoting a number of quantity that is less than zero. Negative numbers are also used to denote quantities that are below some specified reference point. For example, in the Celsius temperature scale a temperature of –24°C is 24° below the freezing point of water. *Compare* positive.

**negative binomial distribution** *See* Pascal's distribution.

**neighborhood** *See* topology.

**nested intervals** A sequence of intervals

such that each interval contains the previous one. The *nested interval theorem* states that for any sequence of bounded and closed nested intervals there is at least one point that belongs to all the intervals. If the lengths of the intervals tend to zero as one goes through the sequence then there is exactly one such point.

**nesting** The embedding of a computer subroutine or a loop of instructions within another subroutine or loop, which in turn may lie within yet another, and so on.

**net** 1. Denoting a weight of goods excluding the weight of the containers or packing. 2. Denoting a profit calculated after deducting all overhead costs, expenses, and taxes. *Compare* gross.
3. A surface that can be folded to form a solid.
4. A network.

**network** A graph consisting of VERTICES that are joined by *arcs* (*directed edges*), with each arc having an arrow on it to indicate its direction, and each arc is associated with a non-negative number called its *weight*. There are many physical applications of networks including electrical circuits and representations of streets in towns, with the physical significance of the weight depending on the physical problem. For example, in some applications there is physical flow (transport) of something between the vertices, such as electrical current in an electrical circuit or cars moving along a street, with the weight being the capacity of the arc. Another type of example is the one in which a network represents some process, with the vertices representing the steps in the process and the weight of an arc joining two vertices represents the time between the two steps.

**neutral equilibrium** Equilibrium such that if the system is disturbed a little, there is no tendency for it to move further nor to return. *See* stability.

**newton** Symbol: N The SI unit of force, equal to the force needed to accelerate one kilogram by one meter per second. 1 N = 1

kg m s$^{-2}$. The unit is named for the English physicist and mathematician Sir Isaac Newton (1642–1727).

**Newtonian mechanics** /new-**toh**-nee-ăn/ Mechanics based on Newton's laws of motion; i.e. relativistic or quantum effects are not taken into account.

**Newton's law of universal gravitation** The force of gravitational attraction between two point masses ($m_1$ and $m_2$) is proportional to each mass and inversely proportional to the square of the distance ($r$) between them. The law is often given in the form

$$F = Gm_1m_2/r^2$$

where $G$ is a constant of proportionality called the *gravitational constant*. The law can also be applied to bodies; for example, spherical objects can be assumed to have their mass acting at their center. *See also* relativity; theory of.

**Newton's laws of motion** Three laws of mechanics formulated by Sir Isaac Newton in 1687. They can be stated as:
1. An object continues in a state of rest or constant velocity unless acted on by an external force.
2. The resultant force acting on an object is proportional to the rate of change of momentum of the object, the change of momentum being in the same direction as the force.
3. If one object exerts a force on another then there is an equal and opposite force (REACTION) on the first object exerted by the second.

The first law was discovered by Galileo, and is both a description of inertia and a definition of zero force. The second law provides a definition of force based on the inertial property of mass. The third law is equivalent to the law of conservation of linear momentum.

**Newton's method** A technique for obtaining successive approximations (iterations) to the solution of an equation, each more accurate than the preceding one. The equation in a variable $x$ is written in the form $f(x) = 0$, and the general formula or algorithm:

$$x_{n+1} = x_n - f(x_n)/f'(x_n)$$

is applied, where $x_n$ is the $n$th approximation. Newton's method can be thought of as repeated estimates of the position on a graph of $f(x)$ against $x$ at which the curve crosses the $x$-axis, by extrapolation of the tangent to the curve. The slope of the tangent at $(x_1, f(x_1))$ is $df/dx$ at $x = x_1$, that is

$$f'(x_1) = f(x_1)/(x_2 - x_1)$$

$x_2 = x_1 - f(x_1)/f'(x_1)$ is therefore the point where the tangent crosses the $x$-axis, and is a closer approximation to $x$ at $f(x) = 0$ than $x_1$ is. Similarly,

$$x_3 = x_2 - f(x_2)/f'(x_2)$$

is a better approximation still. For example, if $f(x) = x^2 - 3 = 0$, then $f'(x) = 2x$ and we obtain the algorithm

$$x_{n+1} = x_n - (x^2_n - 3)/2x_n = \tfrac{1}{2}(x_n + 3/x_n)$$

*See also* iteration.

**node** A point of minimum vibration in a stationary wave pattern, as near the closed end of a resonating pipe. *Compare* antinode. *See also* stationary wave.

**Noether's theorem** /noh-terz/ A fundamental result in physics which relates symmetry to conservation laws. It states that for each continuous symmetry under which a physical system is invariant there is a conserved quantity. For example, invariance of a system under rotational symmetry is associated with the conservation of angular momentum. It does not follow that if a certain symmetry and conservation law associated with Noether's theorem exists in a system described by classical physics then the symmetry and conservation law necessarily have to exist in the corresponding system described by QUANTUM MECHANICS.

**nominal value** (**per value**) The value given to a stock or share by the government or corporation that offers it for sale. Stocks have a nominal value of $100. Shares, however, may have any nominal value. For example, a corporation wishing to raise $100 000 by an issue of shares may issue 100 000 $1 shares or 200 000 50¢ shares, or any other combination. The *issue price*, i.e. the price paid by the first buyers of the shares, may not be the same as the nominal value, although it is likely to be close to it. A share with a nominal value of 50¢ may be offered at an issue price of 55¢; it is then said to be offered at a premium of 5¢. If offered at an issue price of 45¢ it is said to be offered at a discount of 5¢. Once established as a marketable share on a stock exchange, the nominal value has little importance and it is the *market price* at which it is bought and sold. However, the dividend is always expressed as a percentage of the nominal value.

**nomogram** /nom-ŏ-gram/ A graph that consists of three parallel lines, each one a scale for one of three related variables. A straight line drawn between two points, representing known values of two of the variables, crosses the third line at the corresponding value of the third variable. For example, the lines might show the temperature, volume, and pressure of a known mass of gas. If the volume and pressure are known, the temperature can be read off the nomogram.

**nonagon** /non-ă-gon/ A plane figure with nine straight sides. A *regular nonagon* has nine equal sides and nine equal angles.

**non-Cartesian coordinates** Coordinates which are not Cartesian coordinates. POLAR COORDINATES and SPHERICAL POLAR COORDINATES are examples of non-Cartesian coordinates.

In many physical problems which are specified in terms of a partial differential equation it is frequently the case that the equation can only be solved exactly using a particular set of non-Cartesian coordinates, with the coordinate system being used depending on the geometry or symmetry of the problem. For example, spherical polar coordinates are convenient for problems with spherical symmetry.

**non-contradiction, law of** *See* laws of thought.

**non-Euclidean geometry** Any system of geometry in which the parallel postulate of

Euclid does not hold. This postulate can be stated in the form that, if a point lies outside a straight line, only one line parallel to the straight line can be drawn through the point. In the early nineteenth century it was shown that it is possible to have a whole self-consistent formal system of geometry without using the parallel postulate at all. There are two types of non-Euclidean geometry. In one (called *elliptic geometry*) there are no parallel lines through the point. An example of this is a system describing the properties of lines, figures, angles, etc., on the surface of a sphere in which all lines are parts of great circles (i.e. circles that have the same center as the center of the sphere). Since all great circles intersect, no parallel can be drawn through the point. Note also that the angles of a triangle on such a sphere do not add up to 180°. The other type of non-Euclidean geometry is called a *hyperbolic geometry* – here an infinite number of parallels can be drawn through the point.

Note that a type of geometry is not in itself based on 'experiment' – i.e. of measurements of distance, angles, etc. It is a purely abstract system based on certain assumptions (such as Euclid's axioms). Mathematicians study such systems for their own sake – without necessarily looking for practical applications. The practical applications come in when a particular mathematical system gives an accurate description of physical properties – i.e. the properties of the 'real world'. In practical uses (in architecture, surveying, engineering, etc.) it is assumed that Euclidean geometry applies. However, it is found that this is only an approximation, and that the space-time continuum of relativity theory is non-Euclidean in its properties.

**non-isomorphism** *See* isomorphism.

**non-linear oscillations** Oscillations in which the force is not proportional to the displacement from the equilibrium position; i.e. the oscillations are not simple harmonic motion. The general motion of a pendulum is an example of a non-linear oscillation. The motion of a pendulum is a case of a non-linear oscillation that can be

solved exactly but, in general, non-linear oscillations require approximate methods. Associated with this, it is possible for chaotic motion to occur for non-linear oscillators.

**non-linear waves** Waves that are associated with NON-LINEAR OSCILLATIONS. The form of a non-linear wave is more complicated than a sine wave and only approximates to a sine wave in the case of small oscillations. It has been suggested that very large 'freak waves' in oceans are non-linear waves. Other physical examples exist. A SOLITON is a particular type of non-linear wave.

**non-uniform motion** Motion in which the velocity is not constant. This can mean that there is acceleration (including the case of negative acceleration, i.e. deceleration) and/or the direction of the motion is not constant. For example, a body moving in a circle is an example of non-uniform motion since the direction of the body is always changing, which means that there is always an acceleration, even if the speed of the body is constant. If the speed of a body moving in a circle is not uniform then there is said to be *non-uniform circular motion*. The analysis of non-uniform circular motion requires a slight generalization of the analysis of uniform circular motion. In the case of non-uniform circular motion there is a component of the acceleration directed along the tangent to the circle as well as a component directed towards the center of the circle.

**NOR gate** *See* logic gate.

**norm** A generalization of the concept of magnitude to any vector space. The norm of a vector $x$ is usually written $\|x\|$. It is a real number associated with the vector and is positive or zero (for the zero vector). If $a$ is a real number, then
$$\|ax\| = a\|x\|$$
and
$$\|x + y\| \le \|x\| + \|y\|$$

**normal** Denoting a line or plane that is perpendicular to another line or plane. A

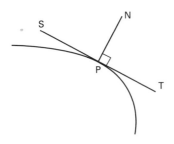

Normal: NP is the normal to the curve

line or plane is said to be normal to a curve if it is perpendicular to the tangent to the curve at the point at which the line and the curve meet. A radius of a circle, for example, is normal to the circumference. A plane passing through the center of a sphere is normal to the surface at all points at which they meet.

**normal chord** A straight line joining two points on a curve that is also a normal to the curve at one or both of the two points. The diameter of a circle can be regarded as a normal chord. It is also possible to construct normal chords of a hyperbola that connect the two branches of the hyperbola.

**normal distribution** (**Gaussian distribution**) The type of statistical distribution followed by, for example, the same measurement taken several times, where the variation of a quantity ($x$) about its mean value ($\mu$) is entirely random. A normal distribution has the probability density function

$$f(x) = \exp[-(x - \mu)^2/2\sigma^2]/\sigma\sqrt{2\pi}$$

where $\sigma$ is known as the *standard deviation*. The distribution is written $N(\mu,\sigma^2)$. The graph of $f(x)$ is bell-shaped and symmetrical about $x = \mu$. The standard normal distribution has $\mu = 0$ and $\sigma^2 = 1$. $x$ can be standardized by letting $z = (x - \mu)/\sigma$. The values $z_\alpha$, for which the area under the curve from $-\infty$ to $z_\alpha$ is $\alpha$, are tabulated; i.e. $z$ is such that $P(z \leq z_\alpha) = \alpha$. Hence

$$P(a<x\leq b) =$$
$$P(a - \mu)/\sigma < z \leq (b - \mu)/\sigma$$

can be found. The alternative term 'Gaussian distribution' is named for the German mathematician Karl Friedrich Gauss (1777–1855).

**normal form** *See* canonical form.

**normalize** To multiply a quantity (e.g. a vector or matrix) by a suitable constant so that its norm is equal to one.

**normal subgroup** A subgroup $H$ of a group $G$ is *normal* if and only if for any element $h$ in $H$, $h^{-1}gh$ is in $H$ for all elements $g$ of $G$.

**NOT gate** *See* logic gate.

**NP-problem** A type of problem where the size of the problem is characterized by some number $n$ and the number of steps which an algorithm would need to solve this problem is $N$, and the dependence of $N$ on $n$ is such that $N$ increases with $n$ more rapidly than any polynomial of $n$. Such problems are called NP-problems because the time it takes to solve them increases more rapidly than any polynomial of $n$. Examples of NP-problems include the TRAVELING SALESMAN PROBLEM and the factorization of large integers. NP-problems are more difficult to solve than P-PROBLEMS. It may be the case that some NP-problems, such as the factorization of large integers, could be solved much more rapidly using a quantum computer.

**n-th root of unity** A complex number $z$ that satisfies the relation $z^n = 1$. There are $n$ such $n$-th roots of unity. They are given by $z = \exp(i2\pi l/n)$, where $l$ takes all the integer values from 0 to $n-1$. The complex numbers that are the $n$-th root of unity can be represented in an ARGAND DIAGRAM for the complex plane as vertices of a regular polygon with $n$ sides. These points lie on a circle with its center at the origin of the complex plane and a radius of one unit. If $n$ is an even integer then the roots must include both 1 and $-1$. If $n$ is an odd integer the roots that are not real numbers have to be pairs of complex numbers that are complex conjugates of each other. The $n$-th roots of unity can also be written as: $z = \cos(2\pi l/n) + i \sin(2\pi l/n)$, with $l$ taking all integer values from 0 to $n-1$.

**NTP** Normal temperature and pressure. *See* STP.

**null hypothesis** *See* hypothesis test.

**null matrix** (**zero matrix**) A MATRIX in which all the elements are equal to zero.

**null set** *See* empty set.

**number line** A straight horizontal line on which each point represents a real number. Integers are points marked at unit distance apart.

**numbers** Symbols used for counting and measuring. The numbers now in general use are based on the Hindu-Arabic system, which was introduced to Europe in the 14th and 15th centuries. The Roman numerals used before this made simple arithmetic very difficult, and most calculations needed an abacus. Hindu-Arabic numerals (0, 1, 2, … 9) enabled calculations to be performed with far greater efficiency because they are grouped systematically in units, tens, hundreds, and so on. *See also* integers; irrational numbers; natural numbers; rational numbers; real numbers; whole numbers.

**numeral** /**new**-mĕ-răl/ A symbol that de-

notes a number. Examples include 0, 1, 2, 3, 4, 5, 6, 7, 8, 9 of Arabic numerals and I, V, X, L, C, D, M of ROMAN NUMERALS.

**numerator** /**new**-mĕ-ray-ter/ The top part of a fraction. For example, in the fraction ¾, 3 is the numerator and 4 is the denominator. The numerator is the dividend.

**numerical analysis** /new-**me**-ră-kăl/ The study of methods of calculation that involve approximations, for example, iterative methods. *See also* iteration.

**numerical integration** A procedure for calculating approximate values of integrals. Sometimes a function is known only as a set of values for corresponding values of a variable and not as a general formula that can be integrated. Also, many functions cannot be integrated in terms of known standard integrals. In these cases, numerical integration methods, such as the TRAPEZIUM RULE and SIMPSON'S RULE, can be used to calculate the area under a graph corresponding to the integral. The area is divided into vertical columns of equal width, the width of each column representing an interval between two values of $x$ for which $f(x)$ is known. Usually a calculation is first carried out with a few columns; these are further subdivided until the desired accuracy is attained, i.e. when further subdivision makes no significant difference to the result.

Number line: this shows an open interval of the real numbers between −1 and +2 and a closed interval from 4 to 6 (including 4 and 6)

**object** The set of points that undergoes a geometrical transformation or mapping. *See also* projection.

**oblate** /**ob**-layt, ŏ-**blayt**/ Denoting a spheroid that has a polar diameter that is smaller than the equatorial diameter. The Earth, for example, is not a perfect sphere but is an oblate spheroid. *Compare* prolate. *See also* ellipsoid.

**oblique** /ŏ-**bleek**/ Forming an angle that is not a right angle.

**oblique coordinates** *See* Cartesian coordinates.

**oblique solid** A solid geometrical figure that is 'slanted'; for example, a cone, cylinder, pyramid, or prism with an axis that is not at right angles to its base. *Compare* right solid.

**oblique spherical triangle** *See* spherical triangle.

**oblique triangle** A triangle that does not contain a right angle.

**oblong** /**ob**-long/ An imprecise term for a RECTANGLE.

**obtuse** /ŏb-**tewss**/ Denoting an angle that is greater than 90° but less than 180°. *Compare* acute; reflex.

**OCR** (optical character recognition) A system used to input information to a computer. The information, usually in the form of letters and numbers, is printed, typed, or sometimes hand-written. The characters used can be read and identified optically by an *OCR reader*. This machine interprets each character and translates it into a series of electrical pulses.

**octagon** /**ok**-tă-gon/ A plane figure with eight straight sides. A *regular octagon* has eight equal sides and eight equal angles.

**octahedron** /ok-tă-**hee**-drŏn/ (*pl.* **octahedrons** or **octahedra**) A POLYHEDRON that has eight faces. A *regular octahedron* has eight faces, each one an equilateral triangle.

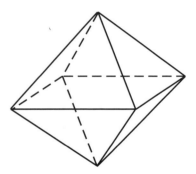

Octahedron: a regular octahedron

**octal** /**ok**-tăl/ Denoting or based on the number eight. An octal number system has eight different digits instead of the ten in the decimal system. Eight is written as 10, nine as 11, and so on. *Compare* binary; decimal; duodecimal; hexadecimal.

**octant** /**ok**-tănt/ 1. One of eight regions into which space is divided by the three axes of a three-dimensional Cartesian coordinate system. The first octant is the one in which $x$, $y$, and $z$ are all positive. The second, third, and fourth octants are num-

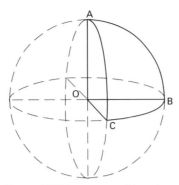

Octant: ABCO is an octant of the sphere.

bered anticlockwise around the positive $z$-axis. The fifth octant is underneath the first, the sixth under the second, etc.

**2.** A unit of plane angle equal to 45 degrees ($\pi/4$ radians).

**odd** Not divisible by two. The set of odd numbers is {1, 3, 5, 7, ...}. *Compare* even.

**odd function** A function f($x$) of a variable $x$ for which f($-x$) = $-$f($x$). For example, sin$x$ and $x^3$ are odd functions of $x$. *Compare* even function.

**odds** When bets are placed on some event the odds are the probability of it happening.

**oersted** Symbol Oe A unit of magnetic field strength in the c.g.s. system. It is equal to $10^3/4\pi$ amperes per meter ($10^3/4\pi$ A m$^{-1}$). The unit is named for the Danish physicist Hans Christian Oersted (1777–1851).

**ohm** /ohm/ Symbol: $\Omega$ The SI unit of electrical resistance, equal to a resistance through which a current of one ampere flows when there is an electric potential difference of one volt across it. 1 $\Omega$ = 1 V A$^{-1}$. The unit is named for the German physicist Georg Simon Ohm (1787–1854).

**one-to-one correspondence** A function or mapping between two sets of things or numbers, such that each element in the first set maps into only one element in the sec-

ond, and vice versa. *See also* function; homomorphism.

**one-to-one mapping** A mapping f from a set $M$ to a set $N$ such that if $m_1$, and $m_2$ are different elements of $M$ then their images f($m_1$) and f($m_2$) are different elements of $N$. This definition means that if f is a one-to-one mapping then the result f($m_1$) = f($m_2$) means that $m_1$ = $m_2$.

**onto** A mapping from one set $S$ to another set $T$ is *onto* if every member of $T$ is the image of some member of $S$ under the mapping. *Compare* into.

**open curve** A curve in which the ends do not meet, for example, a parabola or a hyperbola. *Compare* closed curve.

**open interval** A set consisting of the numbers between two given numbers (end points), not including the end points, for example, all the real numbers greater than 1 and less than 4.5 constitute an open interval. The open interval between two real numbers $a$ and $b$ is written $(a,b)$. Here, the round brackets indicate that the points $a$ and $b$ are not included in the INTERVAL. On a number line, the end points of an open interval are circled. *Compare* closed interval.

**open sentence** In formal logic, a sentence that contains one or more free variables.

**open set** A set defined by limits that are not included in the set itself. The set of all rational numbers greater than 0 and less than ten, written $\{x:0<x<10; x \in R\}$, and the set of all the points inside a circle, but not including the circle itself, are examples of open sets. *Compare* closed set.

**operating system** (OS) The collection of programs used in the control of a computer system. It is generally supplied by the computer manufacturer. An operating system has to decide at any instant which of the many demands on the attention of the central processor to satisfy next. These demands include input from and output to various devices, the execution of a number of programs, and accounting and timing. Large computers, in which many jobs can be run simultaneously, will have a highly complex operating system.

**operation** Any process that combines together members of a set. Combining two members to produce a third is a binary operation. Chief of these are addition, subtraction, multiplication, and division, the main operations in arithmetic. *See* operator.

**operator** 1. A mathematical function, such as addition, subtraction, multiplication, or taking a square root or a logarithm, etc. *See* function.
2. The symbol denoting a mathematical operation or function, for example: $+$, $-$, $\times$, $\sqrt{}$, $\log_{10}$.

**opposite** Denoting the side facing a given angle in a triangle, i.e. the side not forming one arm of the angle. In trigonometry, the ratios of the length of the opposite side to the other side lengths in a right-angled triangle are used to define the sine and tangent functions of the angle.

**optical character recognition** *See* OCR.

**or** *See* disjunction.

**orbit** The curved path or trajectory along which a moving object travels under the influence of a gravitational field. An object with a negligible mass moving under the influence of a planet or other body has an orbit that is a conic section; i.e. a parabola, ellipse, or hyperbola.

**order** 1. (of a matrix) The number of rows and columns in a matrix. *See* matrix.
2. (of a derivative) The number of times a variable is differentiated. For example, $dy/dx$ is a first-order derivative; $d^2y/dx^2$ is second-order; etc.
3. (of a DIFFERENTIAL EQUATION) The order of the highest derivative in an equation. For example,
$$d^3y/dx^3 + 4xd^2y/dx^2 = 0$$
is a third-order differential equation.
$$d^2y/dx^2 - 3x(dy/dx)^3 = 0$$
is a second-order differential equation. *Compare* degree.

**ordered pair** Two numbers indicating values of two variables in a particular order. For example, the $x$- and $y$-coordinates of points in a two-dimensional Cartesian coordinate system form a set of ordered pairs $(x,y)$.

**ordered set** A set of entities in a particular order. *See* sequence.

**ordered triple** Three numbers indicating values of three variables in a particular order. The $x$-, $y$-, and $z$-coordinates of a point in a three-dimensional coordinate system form an ordered triple $(x,y,z)$.

**order of rotational symmetry** the value of $n$ in an $n$-fold axis of symmetry, i.e. the value of $n$ when rotation about this axis by $(360/n)$ degrees gives a result indistinguishable from the original position of the object which is being rotated. For example, if a body has a threefold axis of symmetry then the order of rotational symmetry is three.

**ordinal numbers** Whole numbers that denote order, as distinct from number or quantity. That is, first, second, third, and so on. *Compare* cardinal numbers.

**ordinary differential equation** An equation that contains total derivatives but no partial derivatives. *See* differential equation.

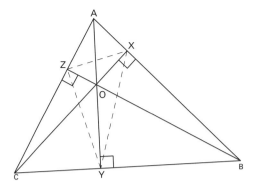

Orthocenter of a triangle ABC: the pedal triangle is XYZ.

**ordinate** The vertical coordinate ($y$-coordinate) in a two-dimensional rectangular Cartesian coordinate system. *See* Cartesian coordinates.

**OR gate** *See* logic gate.

**origin** The fixed reference point in a coordinate system, at which the values of all the COORDINATES are zero and at which the axes meet.

**orthocenter** /or-thoh-sen-ter/ A point in a triangle that is the point of intersection of lines from each vertex perpendicular to the opposite sides. The triangle formed by joining the feet of these vertices is the *pedal triangle*.

**orthogonal circles** /or-**thog**-ŏ-năl/ Two circles that intersect at right angles to each other. Circles which intersect in this way are said to cut *orthogonally*. If A and B are the centers of two circles that cut orthogonally at the points P and Q then the tangent at P to the circle with center A is perpendicular to the radius AP. In a similar way the tangent at P to the circle with center B is perpendicular to the radius BP. Analogous results hold for the tangents at Q. Since the circles are orthogonal the tangents at P are perpendicular, as are the tangents at Q. This means that each of the tangents at both P and Q goes through the center of the other circle. This means that $AP^2 + BP^2 = AQ^2 + BQ^2 = AB^2$, i.e. the

square of the distance between two orthogonal circles is equal to the sum of the squares of their two radii.

**orthogonal group** *See* orthogonal matrix.

**orthogonal matrix** A matrix for which its transpose is also its inverse. Thus, if $A$ is a matrix and its transpose is denoted by $\tilde{A}$ then $A$ is an orthogonal matrix if $A\tilde{A} = I$, where $I$ is the unit matrix. The set of all $n \times n$ orthogonal matrices forms a group called the *orthogonal group*. The set of all $n \times n$ orthogonal matrices with determinants that have the value 1 is a subgroup of the orthogonal group called the *special orthogonal group*. The orthogonal group is denoted $O(n)$ and the special orthogonal group is denoted $SO(n)$. The groups $O(n)$ and $SO(n)$ have several important physical applications, including the description of rotations.

**orthogonal projection** /or-**thog**-ŏ-năl/ A geometrical transformation that produces an image on a line or plane by perpendicular lines crossing the plane. If a line of length $l$ is projected orthogonally from a plane at angle $\theta$ to the image plane, its image length is $l\cos\theta$. The image of a circle is an ellipse. *See also* projection.

**orthogonal vectors** Vectors which are perpendicular. This means that if $a$ and $b$ are non-zero vectors then they are orthogonal vectors if and only if their SCALAR PRODUCT $a.b = 0$. If there is a set of mutually orthogonal unit vectors then there is said to be an *orthonormal* set of vectors. In three-dimensional space the UNIT VECTORS $i$, $j$, and $k$ along the $x$, $y$, and $z$ axes respectively form a set of orthonormal vectors.

**orthonormal** /or-thŏ-**nor**-măl/ *See* orthogonal vectors.

**OS** *See* operating system.

**Osborn's rule** /**oz**-born/ A rule that can be used to convert relations between trigonometric identities and the analogous relations between hyperbolic functions. It states that a $\cos^2$ term becomes a $\cosh^2$ term without change of sign but that a $\sin^2$ term (or more generally a product of two sine terms) becomes a $-\sinh^2$ term, or, more generally, the sign of the term is changed. For example, $\cos^2 x + \sin^2 x = 1$ becomes $\cosh^2 x - \sinh^2 x = 1$ and $\cos^2 x - \sin^2 x = \cos^2 x$ becomes $\cosh^2 x + \sinh^2 x = \cosh^2 x$. Osborn's rule has to be used carefully since there are some trigonometric functions, such as $\tan^2 x$, that can be expressed in terms of sines and cosines in several ways. Although Osborn's rule can be used to find an identity for hyperbolic functions, finding such an identity in this way does not constitute a proof for the identity.

**oscillating series** /**os**-ă-lay-ting/ A special type of nonconvergent series for which the sum does not approach a limit but continually fluctuates. Oscillating series can either fluctuate between bounds, for example the series $1 - 1 + 1 - 1 + ...$, or it can be unbounded, for example $1 - 2 + 3 - 4 + 5 - ...$.

**oscillation** /os-ă-**lay**-shŏn/ A regularly repeated motion or change. *See* vibration.

**ounce 1.** A unit of mass equal to one sixteenth of a pound. It is equivalent to 0.0283 49 kg.
**2.** A unit of capacity, often called a *fluid ounce*, equal to one sixteenth of a pint. It is equivalent to $2.057\ 3 \times 10^{-5}$ m³. In the UK,

it is one twentieth of a UK pint, equivalent to $2.841\ 3 \times 10^{-5}$ m³. 1 UK fluid ounce is equal to 0.960 8 US fluid ounce.

**output 1.** The signal or other form of information obtained from an electrical device, machine, etc. The output of a computer is the information or results derived from the data and programmed instructions fed into it. This information is transferred as a series of electrical pulses from the central processor of the computer to a selected *output device*. Some of these output units convert the pulses into a readable or pictorial form; examples include the printer, plotter, and visual display unit (which can also be used as an input device). Other output devices translate the pulses into a form that can be fed back into the computer at a later stage; the magnetic tape unit is an example.
**2.** The process or means by which output is obtained.
**3.** To deliver as output.
*See also* input; input/output.

**overdamping** *See* damping.

**overflow** The situation arising in computing when a number, such as the result of an arithmetical operation, has a greater magnitude than can be represented in the space allocated to it in a register or a location in store.

**overlay** A technique used in computing when the total storage requirements for a lengthy program exceeds the space available in the main store. The program is split into sections so that only the section or sections required at any one time will be transferred. These program segments (or *overlays*) will all occupy the same area of the main store.

**overtones** The wave patterns which are present in sound and music in which the frequency is greater than the fundamental frequency. It is possible to analyze the overtones of a wave pattern using FOURIER SERIES. The overtones which are heard along with the main note of a specific frequency $f$ vary in different musical instruments. This means that the waveform associated with each note varies in different musical instruments. The *quality* or *timbre* of a note, i.e. how much the waveform for a note deviates from being a sine wave, is therefore dependent on the presence or absence of overtones and hence depends on the instrument.

**palindrome** A number that is the same sequence of integers forward and backward. An example of a palindrome is the number 35753. A conjecture concerning palindromes is that starting from any number, if the integers of that number are reversed and the resulting number is added to the original number, then if this process is repeated long enough a palindrome will result. However, this conjecture has not been proved.

**pandigital number** /pan-**dij**-i-tăl/ A number which contains all of the numbers 0, 1, 2, 3, 4, 5, 6, 7, 8, 9 only once.

**paper tape** A long strip of paper, or sometimes thin flexible plastic, on which information can be recorded as a pattern of round holes punched in rows across the tape, once used extensively in data processing.

**Pappus' theorems** Two theorems concerning the rotation of a curve or a plane shape about a line that lies in the same plane. The first theorem states that the surface area generated by a curve revolving about a line that does not cross it, is equal to the length of the curve times the circumference of the circle traced out by its centroid. The second theorem states that the volume of a solid of revolution generated by a plane area that rotates about a line not crossing it, is equal to the area times the circumference of the circle traced out by the centroid of the area. (Note that the plane area and the line both lie in the same plane.) The theorems are named for the Greek mathematician Pappus of Alexandria (*fl.* AD 320). The second Pappus theorem is sometimes known as the *Guldinus theorem* as it was rediscovered by the Swiss mathematician and astronomer Paul Guldin (1577–1643).

**parabola** /pă-**rab**-ŏ-lă/ A conic with an eccentricity of 1. The curve is symmetrical about an axis through the focus at right angles to the directrix. This axis intercepts the parabola at the *vertex*. A chord through the focus perpendicular to the axis is the *latus rectum* of the parabola.

In Cartesian coordinates a parabola can be represented by an equation:
$$y^2 = 4ax$$
In this form the vertex is at the origin and the $x$-axis is the axis of symmetry. The focus is at the point $(0,a)$ and the directrix is the line $x = -a$ (parallel to the $y$-axis). The *latus rectum* is $4a$.

If a point is taken on a parabola and two lines drawn from it – one parallel to the axis and the other from the point to the focus – then these lines make equal angles with the tangent at that point. This is known as the *reflection property* of the parabola, and is utilized in parabolic reflectors and antennas. *See* paraboloid.

The parabola is the curve traced out by a projectile falling freely under gravity. For example, a tennis ball projected horizontally with a velocity $v$ has, after time $t$, traveled a distance $d = vt$ horizontally, and has also fallen vertically by $h = gt^2/2$ because of the acceleration of free fall $g$. These two equations are *parametric equations* of the parabola. Their standard form, corresponding to $y^2 = 4ax$, is
$$x = at^2$$
$$y = 2at$$
where $x$ represents $h$, the constant $a$ is $g/2$, and $y$ represents $d$. *See illustration overleaf. See also* conic.

**paraboloid** /pă-**rab**-ŏ-loid/ A curved sur-

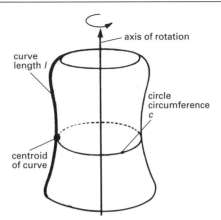

Pappus' theorem: the curved surface area $A = l \times c$.

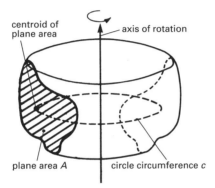

Pappus' theorem: the volume enclosed by the curved surface $V = A \times c$.

Pappus' theorems

face in which the cross-sections in any plane passing through a central axis is a parabola. A *paraboloid of revolution* is formed when a parabola is rotated around its axis of symmetry. Parabolic surfaces are used in telescope mirrors, searchlights, radiant heaters, and radio antennas on account of the focusing property of the parabola.

Another type of paraboloid is the *hyperbolic paraboloid*. This is a surface with the equation:
$$x^2/a^2 - y^2/b^2 = 2cz$$
where $c$ is a positive constant. Sections parallel to the $xy$ plane ($z = 0$) are hyperbolas.

Sections parallel to the other two planes ($x = 0$ or $y = 0$) are parabolas. *See illustration overleaf. See also* conicoid.

**paradox** /pa-ră-doks/ (**antinomy**) A proposition or statement that leads to a contradiction if it is asserted *and* if it is denied.

A famous example of a paradox is *Russell's paradox* in set theory. A set is a collection of things. It is possible to think of sets as belonging to two groups: sets that contain themselves and sets that do not contain themselves. A set that contains itself is itself a member of the set. For in-

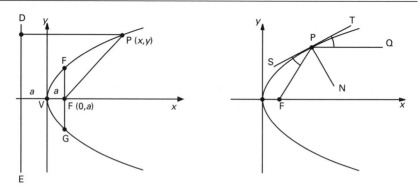

Parabola: F is the focus and DE is the directrix. FG is the latus rectum. If PQ is parallel to the *x*-axis, angles SPF and TPQ are equal (ST is a tangent at P).

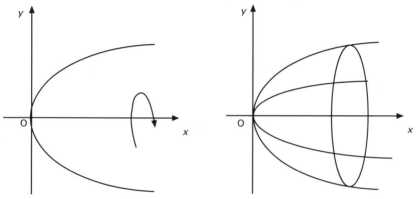

Paraboloid of revolution

stance, the set of all sets is itself a set, so it contains itself. Other examples are the 'set of all things that one can think about' and 'the set of all abstract ideas'. On the other hand the set of all things with four legs does not itself have four legs – it is an example of a set that does not contain itself. Other examples of sets that do not contain themselves (or are not members of themselves) are 'the set of all oranges' and 'the set of all things that are green'.

Consider now the set of all sets that do not contain themselves. The paradox arises from the question, 'Does the set of all sets that do not contain themselves contain itself or not?' If one asserts that it does contain itself then it cannot be one of the things that do not contain themselves. This means that it does not belong to the set, so

it doesn't contain itself. On the other hand, if it does not contain itself it must be one of all the things that do not contain themselves, so it must belong to the set. The short answer to the question is, 'If it does, it doesn't; if it doesn't, it does!' – hence the paradox. Paradoxes like this can be used in investigating fundamental questions in set theory.

**parallax** /pa-ră-laks/ The angle between the direction of an object, for example a star or a planet, from a point on the surface of the Earth and the direction of the same object from the center of the Earth. The measurement of parallax is used to find the distance of an object from the Earth. The object is viewed from two widely separated points on the Earth and the distance be-

Parallelepiped

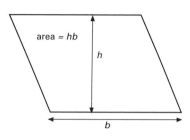

Parallelogram

tween these two points and the direction of the object as seen from each is measured. The parallax at the observation points and the distance of the object from the observer can be found by simple trigonometry.

**parallel** Extending in the same direction and remaining the same distance apart. *Compare* antiparallel.

**parallel axes, theorem of** *See* theorem of parallel axes.

**parallelepiped** /pa-ră-lel-ă-**pÿ**-ped/ A solid figure with six faces that are parallelograms. In a *rectangular parallelepiped* the faces are rectangles. If the faces are squares, the parallelepiped is a cube. If the lateral edges are not perpendicular to the base, it is an *oblique parallelepiped*.

**parallel forces** When the forces on an object pass through one point, their resultant can be found by using the parallelogram of vectors. If the forces are parallel the resultant is found by addition, taking sign into account. There may also be a turning effect in such cases, which can be found by the principle of moments.

**parallelogram** /pa-ră-**lel**-ŏ-gram/ A plane figure with four straight sides, and with opposite sides parallel and of equal length. The opposite angles of a parallelogram are also equal. Its area is the product of the

length of one side and the perpendicular distance from that side to the side opposite. In the special case in which the angles are all right angles the parallelogram is a rectangle; when all the sides are equal it is a rhombus.

**parallelogram (law) of forces** *See* parallelogram of vectors.

**parallelogram (law) of velocities** *See* parallelogram of vectors.

**parallelogram of vectors** A method for finding the resultant of two vectors acting at a point. The two vectors are shown as two adjacent sides of a parallelogram: the resultant is the diagonal of the parallelogram through the starting point. The technique can be used either with careful scale drawing or with trigonometry. The trigonometrical relations give:
$$F = \sqrt{(F_1^2 + F_2^2 + 2F_1F_2\cos\theta)}$$
$$\alpha = \sin^{-1}[(F_2/F)\sin\theta]$$
where $\theta$ is the angle between $F_1$ and $F_2$ and $\alpha$ is the angle between $F$ and $F_1$. *See* vector.

**parallel postulate** *See* Euclidean geometry.

**parameter** /pă-**ram**-ĕ-ter/ A quantity that, when varied, affects the value of another. For example, if a variable $z$ is a function of variables $x$ and $y$, that is $z = f(x,y)$, then $x$ and $y$ are the parameters that determine $z$.

**parametric equations** Equations that, in an implicit function (such as f($x,y$) = 0) express $x$ and $y$ separately in terms of a quantity, which is an independent variable or parameter. For example, the equation of a circle can be written in the form
$$x^2 + y^2 = r^2$$
or as the parametric equations
$$x = r\cos\theta$$
$$y = r\sin\theta$$
where $r$ is the radius. *See also* parabola.

**parametrization** /pa-ră-met-ri-**zay**-shŏn/ A method for associating a parameter $t$ with a point P, which lies on a curve so that each point on the curve is associated with a value of $t$, with $t$ lying in some interval of the real numbers. This is frequently done by expressing the $x$ and $y$ coordinates of P in terms of $t$. The resulting equations for $x$ and $y$ in terms of $t$ are called the PARAMETRIC EQUATIONS for the curve. It is possible to find the expression for d$y$/d$x$ at any point on the curve in terms of the parameter $t$ since d$y$/d$x$ = (d$y$/d$t$)/(d$x$/d$t$).

Perhaps the simplest parametrization of a curve is given by a circle of unit radius. The points on the circle are parametrized by an angle $\theta$, with (in radians) $0 \leq \theta \leq 2\pi$. One has $x = \cos\theta$ and $y = \sin\theta$. In the case of an ellipse $x^2/a^2 + y^2/b^2 = 1$ the parametrization is given by $x = a\cos\theta$ and $y = b\sin\theta$ ($0 \leq \theta \leq 2\pi$). In the case of a parabola $y^2 = 4ax$ the curve is parametrized by $t$ with $x = at^2$ and $y = 2at$.

**paraplanar** /pa-ră-**play**-ner/ Describing a set of vectors that are all parallel to a plane but are not necessarily contained in that plane.

**parsec** /**par**-sek/ Symbol: pc A unit of distance used in astronomy. A star that is one parsec away from the earth has a parallax (apparent shift), due to the Earth's movement around the Sun, of one second of arc. One parsec is approximately 3.085 61 × $10^{16}$ meters.

**partial derivative** The rate of change of a function of several variables as one of the variables changes and the others are held constant. For example, if $z$ = f($x,y$) the partial derivative $\partial z/\partial x$ is the rate of change of $z$ with respect to $x$, with $y$ held constant. Its value will depend on the constant value of $y$ chosen. In three-dimensional Cartesian coordinates, $\partial z/\partial x$ is the gradient of a line at a tangent to the curved surface f($x,y$) and parallel to the $x$-axis. *See also* total derivative.

**partial differential** The infinitesimal change in a function of two or more variables resulting from changing only one of the variables while keeping the others constant. The sum of all the partial differentials is the total differential. *See* differential.

**partial differential equation** An equation that contains partial derivatives of a function with respect to a number of variables. General methods of solution are available only for certain types of linear partial differential equation. Many partial differential equations occur in physical problems. *Laplace's equation,* for example, is
$$\partial^2\phi/\partial x^2 + \partial^2\phi/\partial y^2 + \partial^2\phi/\partial z^2 = 0$$
It occurs in the study of gravitational and electromagnetic fields. The equation is named for the French mathematician, astronomer, and physicist Marquis Pierre Simon de Laplace (1749–1827). *See also* differential equation.

**partial fractions** A sum of fractions that is equal to a particular fraction, for example, ½ + ¼ = ¾. Writing a ratio in terms of partial fractions can be useful in solving equations or calculating integrals. For example
$$1/(x^2 + 5x + 6)$$
can be written as partial fractions by factorizing the denominator into $(x + 2)(x + 3)$, and writing the fraction in the form
$$[A/(x + 2)] + [B/(x + 3)] =$$
$$1/(x^2 + 5x + 6)$$
This means that
$$A(x + 3) + B(x + 2) = 1$$
i.e.
$$Ax + 3A + Bx + 2B = 1$$
The values of $A$ and $B$ can be found by comparing the coefficients of powers of $x$, i.e.

$3A + 2B = 1$ (constant term)
$A + B = 0$ (coefficient of $x$)
Solving these simultaneous equations gives $A = 1$ and $B = -1$. So the fraction $1/(x^2 + 5x + 6)$ can be expressed as partial fractions $1/(x + 2) - 1/(x + 3)$. This is a form that can be integrated as a sum of two simpler integrals.

**partially ordered** A *partially ordered set* is a set with a relation $x < y$ defined for some elements $x$ and $y$ of the set satisfying the conditions:
1. if $x < y$ then $y < x$ is false and $x$ and $y$ are not the same element;
2. if $x < y$ and $y < z$ then $x < z$.
    It need not be the case that $x < y$ or $y < x$ for any two elements $x$ and $y$. An example of a partially ordered set is the set of subsets of a given set where we define $A < B$ for sets $A$ and $B$ to mean that $A$ is a proper subset of $B$.

**partial product** A product of the first $n$ terms $a_1\ a_2\ ...\ a_n$, called the *n-th partial product*, where these $n$ terms are the first $n$ terms of the infinite product.

**partial sum** The sum of a finite number of terms from the start of an infinite series. In a convergent series, the partial sum of the first $r$ terms, $S_r$, is an approximation to the sum to infinity. *See* series.

**particle** An abstract simplification of a real object – the mass is concentrated at the object's center of mass; its volume is zero. Thus rotational aspects can be ignored.

**particular integral** A name given to the PARTICULAR SOLUTION in the case of a second-order linear differential equation of the form $ad^2y/dx^2 + bdy/dx + cy = f(x)$, where $a$, $b$, and $c$ are constants and f is some function of $x$. A way of finding a particular integral which frequently works is to use a function which is similar to f($x$). For example, if f($x$) is a polynomial of some degree in $x$ then the particular integral to be tried should also be a polynomial of the same degree. Similarly, if f($x$) is a trigonometric function such as sin$lx$ or cos$mx$ or an exponential function exp($nx$)

then a similar function can be tried. In all these examples, unknown coefficients can be found by substitution into the original differential equation.

**particular solution** A solution of a differential equation that is given by some particular values of the arbitrary constants that appear in the general solution of the differential equation. A particular solution can be found if it satisfies both the differential equation and the BOUNDARY CONDITIONS that apply.

**partition** A partition of a set $S$ is a finite collection of disjoint sets whose union is $S$.

**pascal** /**pas**-kăl/ Symbol: Pa The SI unit of pressure, equal to a pressure of one newton per square meter (1 Pa = 1 N m$^{-2}$). The pascal is also the unit of stress. It is named for the French mathematician, physicist and religious philosopher Blaise Pascal (1623–62).

**Pascal's distribution** (**negative binomial distribution**) The distribution of the number of independent Bernoulli trials performed up to and including the $r$th success. The probability that the number of trials, $x$, is equal to $k$ is given by
$$P(x=k) = {}^{k-1}C_{r-1}p^r q^{k-r}$$
The mean and variance are $r/p$ and $rq/p^2$ respectively. *See also* geometric distribution.

**Pascal's triangle** A triangle array of numbers in which each row starts and ends with 1 and that is built up by summing two adjacent numbers in a row to obtain the number directly below them in the next

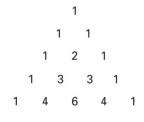

Pascal's triangle

row. Each row in Pascal's triangle is set of binomial coefficients. In the expansion of $(x + y)^n$, the coefficients of $x$ and $y$ are given by the $(n + 1)^{th}$ row of Pascal's triangle.

**pedal curve** The *pedal curve* of a given curve C with respect to a fixed point P is the locus of the foot of the perpendicular from P to a variable tangent to the curve C. The point P is called the *pedal point*. For example, if C is a circle and P is a point on its circumference the pedal curve is a cardioid.

**pedal triangle** *See* orthocenter.

**pencil** A family of geometric objects that share a common property. For example, a pencil of circles consists of all the circles in a given plane that pass through two given points, a pencil of lines consists of all the lines in a given plane passing through a given point, and a pencil of parallel lines consists of all the lines parallel to a given direction.

**pendulum** A body that oscillates freely under the influence of gravity. A *simple pendulum* consists of a small mass oscillating to and fro at the end of a very light string. If the amplitude of oscillation is small (less than about 10°), it moves with simple harmonic motion; the period does not depend on amplitude. There is a continuous interchange of potential and kinetic energy through the motion; at the ends of the swings the potential energy is a maximum and the kinetic energy zero. At the mid-point the kinetic energy is a maximum and the potential energy is zero. The period is given by
$$T = 2\pi\sqrt{(l/g)}$$
Here $l$ is the length of the pendulum (from support to center of the mass) and $g$ is the acceleration of free fall. If the amplitude of oscillation is not small it does not move with simple harmonic motion but the problem of its motion can still be solved exactly.

A *compound pendulum* is a rigid body swinging about a point. The period of a compound pendulum depends on the moment of inertia of the body. For small oscillations it is given by the same relationship as that of the simple pendulum with $l$ replaced by $[\sqrt{(k^2 + h^2)}]/h$. Here, $k$ is the RADIUS OF GYRATION about an axis through the center of mass and $h$ is the distance from the pivot to the center of mass.

**Penrose pattern** /**pen**-rohz/ A two-dimensional pattern of a tiling that has five-fold rotational axes of symmetry and also long-range order, in spite of it being impossible to have a two-dimensional crystal in which five-fold rotational axes of symmetry can occur. It is possible to obtain this type of pattern by combining two sets of rhombuses, with one of the sets being 'thin' rhombuses and the other set being 'fat' rhombuses, in specific ways. A three-dimensional version of Penrose patterns occurs in materials that have QUASICRYSTALLINE SYMMETRY. The pattern is named for the British mathematician Sir Roger Penrose (1931– ).

**pentagon** /**pen**-tă-gon/ A plane figure with five straight sides. In a *regular pentagon*, one with all five sides and angles equal, the angles are all 108°. A regular pentagon can be superimposed on itself after rotation through 72° ($2\pi/5$ radians).

**percentage** A number expressed as a fraction of one hundred. For example, 5 percent (or 5%) is equal to 5/100. Any fraction or decimal can be expressed as a percentage by multiplying it by 100. For example, $0.63 \times 100 = 63\%$ and $\frac{1}{4} \times 100 = 25\%$.

**percentage error** The ERROR or uncertainty in a measurement expressed as a percentage. For example, if, in measuring a length of 20 meters, a tape can measure to the nearest four centimeters, the measurement is written as $20\pm 0.04$ meters and the percentage error is $(0.04/20) \times 100 = 0.2\%$.

**percentile** /per-**sen**-tÿl, -tăl/ One of the set of points that divide a set of data arranged in numerical order into 100 parts. The $r^{th}$ percentile, $P_r$, is the value below and including which $r\%$ of the data lies and above which $(100 - r)\%$ lies. $P_r$ can be found from the cumulative frequency graph. *See also* quartile; range.

**perfect number** A number that is equal to the sum of all its factors except itself. 28 is a perfect number since its factors are 1, 2, 4, 7, and 14, and $1 + 2 + 4 + 7 + 14 = 28$. *See also* Mersenne prime.

**perimeter** /pĕ-**rim**-ĕ-ter/ The distance round the edge of a plane figure. For example, the perimeter of a rectangle is twice the length plus twice the breadth. The perimeter of a circle is its circumference $(2\pi r)$.

**period** Symbol: $T$ The time for one complete cycle of an oscillation, wave motion, or other regularly repeated process. It is the reciprocal of the frequency, and is related to pulsatance, or angular frequency, $(\omega)$ by $T = 2\pi/\omega$.

**periodic function** A function that repeats itself at regular intervals of the variable. For example $\sin x$ is a periodic function of $x$ because $\sin x = \sin (x + 2\pi)$ for all values of $x$.

**periodic motion** Any kind of regularly repeated motion, such as the swinging of a pendulum, the orbiting of a satellite, the vibration of a source of sound, or an electromagnetic wave. If the motion can be represented as a pure sine wave, it is a simple harmonic motion. Harmonic motions in general are given by the sum of two or more pure sine waves.

**period of investment** The length of time for which a fixed amount of capital remains invested. In times of historically low interest rates, an investor prepared to commit his or her money for a long period, such as five or ten years, will gain a higher rate of interest than can be expected for a short-term investment. However, if interest rates are historically high this will not be the case and long-term rates may be lower than short-term rates.

**peripheral unit** (**peripheral**) A device connected to and controlled by the central processor of a computer. Peripherals include input devices, output devices, and backing store. Some examples are visual display units, printers, magnetic tape units, and disk units. *See also* input; output.

**permutation** /per-myŭ-**tay**-shŏn/ An ordered subset of a given set of objects. For three objects, A, B, and C, there are six possible permutations: ABC, ACB, BAC, BCA, CAB, and CBA. The total number of permutations of $n$ objects is $n!$

The total number of permutations of $r$ objects taken from $n$ objects is given by $n!(n - r)!$, assuming that each object can be selected only once. This is written $^nP_r$. For example, the possible permutations of two objects from the set of three objects A, B, and C would be AB, BA, AC, CA, BC, CB. Note that each object is selected only once in this case – if the objects could occur any number of times, the above set of permutations would include AA, BB, and CC. The number of permutations of $r$ objects selected from $n$ objects when each can occur any number of times is $n^r$.

Note also the difference between permutations and combinations: permutations are different if the order of selection is different, so AB and BA are different permutations but the same combination. The number of combinations of $r$ objects from $n$ objects is written $^nC_r$, and $^nP_r = {}^nC_r \times r!$

**permutation group** The set of permutations of a number of indistinguishable objects. The permutation group is important in theoretical physics. It is used to study the energy levels of electrons in atoms and molecules and is also used in nuclear theory.

**perpendicular** /per-pĕn-**dik**-yŭ-ler/ At right angles. The perpendicular bisector of a line crosses it half way along its length and forms a right angle. A vertical surface is perpendicular to a horizontal surface.

**personal computer** (**PC**) *See* microcomputer.

**perturbation** /per-ter-**bay**-shŏn/ A way of obtaining approximate solutions of equations that represent the behavior of a system, by making a slight change in some of the basic parameters. It is an important technique in quantum mechanics and celestial mechanics.

**phase** /fayz/ The stage in a cycle that a WAVE (or other periodic system) has reached at a particular time (taken from some reference point). Two waves are *in phase* if their maxima and minima coincide.

For a simple wave represented by the equation

$$y = a\sin 2\pi(ft - x/\lambda)$$

The phase of the wave is the expression

$$2\pi(ft - x/\lambda)$$

The *phase difference* between two points distances $x_1$ and $x_2$ from the origin is

$$2\pi(x_1 - x_2)/\lambda$$

A more general equation for a progressive wave is

$$y = a\sin 2\pi[ft - (x/\lambda) - \phi]$$

Here, $\phi$ is the *phase constant* – the phase when $t$ and $x$ are zero. Two waves that are out of phase have different stages at the origin. The phase difference is $\phi_1 - \phi_2$. It is equal to $2\pi x/\lambda$, where $x$ is the distance between corresponding points on the two waves. It is the *phase angle* between the two waves; the angle between two rotating vectors (phasors) representing the waves.

**phase angle** *See* phase.

**phase constant** *See* phase.

**phase difference** *See* phase.

**phase portrait** A picture which plots the evolution of possible paths in PHASE SPACE that a point in that space can have, starting from a certain set of initial conditions. In the case of SIMPLE HARMONIC MOTION the phase portrait consists of a set of ellipses which all have the same center. In the case of NON-LINEAR OSCILLATIONS more complicated patterns appear in the phase portrait. The phase portrait is a convenient way of showing the complicated behavior that can occur in dynamical systems, including the presence of ATTRACTORS, chaotic behavior, and LIMIT CYCLES.

**phase space** A multi-dimensional space that can be used to define the state of a system. Phase space has coordinates $(q_1, q_2,\dots p_1, p_2,\dots)$, where $q_1, q_2$, etc., are degrees of freedom of the system and $p_1, p_2$, are the momenta corresponding to these degrees of freedom. For example, a single particle has three degrees of freedom (corresponding to the three coordinates defining its position). It also has three components of momentum corresponding to these degrees of freedom. This means that the state of the particle can be defined by six numbers $(q_1, q_2, q_3, p_1, p_2, p_3)$ and it is thus defined by a point in six-dimensional phase space. If the system changes with time (i.e. the particle changes its position and momentum), then the point in phase space traces out a path (known as the *trajectory*). The system may consist of more than one particle. Thus, if there are $N$ particles in the system then the state of the system is specified by a point in a phase space of $6N$ dimensions. The idea of phase space is useful in chaos theory. *See also* attractor.

**phase speed** The speed with which the phase in a traveling wave is propagated. It is equal to $\lambda/T$, where $T$ is the period. *Compare* group speed.

**phasor** /**fay**-zer/ *See* simple harmonic motion.

**pi** /pÿ/ ($\pi$) The ratio of the circumference of any circle to its diameter. $\pi$ is approximately equal to 3.14159... and is a transcendental number (its exact value cannot be written down, but it can be stated to any degree of accuracy). There are several expressions for $\pi$ in terms of infinite series.

**pico-** /**pÿ**-koh/ Symbol: p A prefix denoting $10^{-12}$. For example, 1 picofarad (pF) = $10^{-12}$ farad (F).

**pictogram** /**pik**-tŏ-gram/ (**pictograph**) A diagram that represents statistical data in a pictorial form. For example, the proportions of pink, red, yellow, and white flowers that grow from a packet of mixed seeds can be shown by rows of the appropriate relative numbers of colored flower shapes.

**piecewise** A function is *piecewise continuous* on *S* if it is defined on *S* and can be separated into a finite number of pieces such that the function is continuous on the interior of each piece. Terms such as *piecewise differentiable* and *piecewise linear* are similarly defined.

**pie chart** A diagram in which proportions are illustrated as sectors of a circle, the relative areas of the sectors representing the different proportions. For example, if, out of 100 workers in a factory, 25 travel to work by car, 50 by bus, 10 by train, and the rest walk, the bus passengers are represented by half of the circle, the car passengers by a quarter, the train users by a 36° sector, and so on.

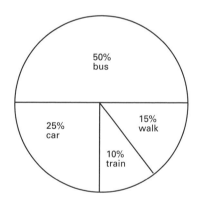

Pie chart showing how people travel to work.

**pint** A unit of capacity. The US liquid pint is equal to one eighth of a US gallon and is equivalent to $4.731\ 8 \times 10^{-4}$ m$^3$. In the UK it is equal to one eighth of a UK gallon and is equivalent to $5.682\ 6 \times 10^{-4}$ m$^3$. The US dry pint is equal to one sixty-fourth of a US bushel and is equivalent to $5.506\ 1 \times 10^{-4}$ m$^3$.

**pixel** /**piks**-ĕl/ *See* computer graphics.

**place value** the position of an integer in a number. For example, in the number 375 the 5 represents 5 units, the 7 represents 7 tens, and the 3 represents 3 hundreds.

**plan** An illustration that shows the appearance of a solid object as viewed from above (vertically downward). *See also* elevation.

**planar** /**play**-ner/ Describing something that occupies a plane or is flat.

**plane** A flat surface, either real or imaginary, in which any two points are joined by a straight line lying entirely on the surface. *Plane geometry* involves the relationships between points, lines, and curves lying in the same plane. In Cartesian coordinates, any point in a plane can be defined by two coordinates, *x* and *y*. In three-dimensional coordinates, each value of *z* corresponds to a plane parallel to the plane in which the *x* and *y* axs lie. For any three points, there exists only one plane containing all three. A particular plane can also be specified by a straight line and a point.

**plane shape** A shape that exists in a plane, i.e. a two-dimensional shape that does not have a depth but has a width and height. Examples of plane shapes include triangles, squares, rectangles, kites, rhombuses, and parallelograms. Plane shapes also include curves such as circles and ellipses.

**Platonic solids** /plă-**tonn**-ik/ A name given to the set of five regular polyhedra. *See* polyhedron.

**plot** To draw on a graph. A series of individual points plotted on a graph may show a general relationship between the variables represented by the horizontal and vertical axs. For example, in a scientific experiment one quantity can be represented by *x* and another by *y*. The values of *y* at different values of *x* are then plotted as a series of points on a graph. If these fall on a line or curve, then the line or curve drawn through the points is said to be a plot of $y$ against $x$.

**plotter** An output device of a computer system that produces a permanent record of the results of some program by drawing lines on paper. One pen, or maybe two or more pens with different colored ink, are moved over the paper according to instructions sent from the computer or from a backing store. Plotters are used for drawing graphs, contour maps, etc.

**plotting** The process of marking points on a system of coordinates, or of drawing a graph by marking points.

**Poincaré conjecture** /pwank-ka-**ray**/ A conjecture in TOPOLOGY that can be stated in the form that there is a HOMEOMORPHISM between any $n$-dimensional MANIFOLD which is *closed*, i.e. it is like a loop and has no end points, and is SIMPLY CONNECTED and the sphere of that dimension. The French mathematician Henri Poincaré (1854–1912) originally made in conjecture in three dimensions ($n = 3$). The more general version of the conjecture is sometimes known as the *generalized Poincaré conjecture*.

The generalized Poincaré conjecture is readily proved for $n = 1$ and $n = 2$. It has been proved with more difficulty in the case of $n > 4$ in the 1960s. The case of $n = 4$ was proved, with considerable difficulty in the 1980s. Proving the case of $n = 3$, i.e. the original Poincaré conjecture, has proved to be the hardest of all dimensions and a proof of this conjecture is one of the MAGNIFICENT SEVEN PROBLEMS of mathematics. It has been claimed that the conjecture has been proved in three dimensions but, at the time of writing, this has not been confirmed.

**point** A location in space, on a surface, or in a coordinate system. A point has no dimensions and is defined only by its position.

**point group** A set of symmetry operations on some body in which one of the points of the body is left fixed in space. An important application of point groups is to the symmetry of isolated molecules in chemistry. Examples of the symmetry operations in point groups include rotation about a fixed axis, reflection about a plane of symmetry, inversion through a center of symmetry, i.e. invariance in the appearance of the body when a point in the body with coordinates $(x,y,z)$ is carried to a point with coordinates $(-x,-y,-z)$, and rotation about some axis followed by reflection in a plane perpendicular to that axis.

An important set of point groups is the set of CRYSTALLOGRAPHIC POINT GROUPS, i.e. the 32 point groups in three dimensions that are compatible with the translational symmetry of a crystal lattice (*see* space group). In a crystallographic point group only two-fold, three-fold, four-fold, and six-fold rotational symmetries are possible.

**point of contact** A single point at which two curves, or two curved surfaces touch. There is only one point of contact between the circumference of a circle and tangent to the circle. Two spheres also can have only one point of contact.

**point of inflection** A point on a curved line at which the tangent changes its direction of rotation. Approaching from one side of the point of inflection, the slope of the tangent to the curve increases; and moving away from it on the other side, it decreases. For example, the graph of $y = x^3 - 3x^2$ in rectangular Cartesian coordinates, has a point of inflection at the point $x = 1$, $y = -2$. The second derivative $d^2y/dx^2$ on the graph of a function $y = f(x)$ is zero and changes its sign at a point of inflection. Thus, in the example above, $d^2y/dx^2 = 6x - 6$, which is equal to zero at the point $x = 1$.

**Poisson distribution** A probability distribution for a discrete random variable. It is defined, for a variable ($r$) that can take values in the range 0, 1, 2, ..., and has a mean value $\mu$, as
$$P(r) = e^{-\mu r}/r!$$
It is named for the French mathematician and mathematical physicist Siméon-Denis Poisson (1781–1840).

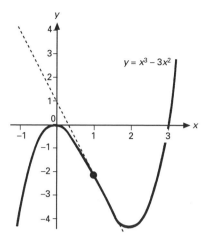

Point of inflection: in this case the point of inflection is $x = 1$, $y = -2$. The derivative $dy/dx = -3$ at this point.

A binomial distribution with a small frequency of success $p$ in a large number $n$ of trials can be approximated by a Poisson distribution with mean $np$.

**Poisson equation** A generalization of the LAPLACE EQUATION in which there is some function $f(x,y,z)$ on the right-hand side of the equation. This means that the Poisson equation has the form $\nabla^2\phi = (x,y,z)$, where $\nabla^2$ is the LAPLACIAN. Here, for example, $\phi$ can be the gravitational potential in a region in which there is matter, the electrostatic potential in a region in which there is electric charge, or the temperature in a region in which there is a source of heat. In general, the function $f(x,y,z)$ is called the *source density*. In the examples given the source density is respectively proportional to the density of mass, the density of electric charge, and the amount of heat generated per unit time, per unit volume. The Poisson equation is the main equation of POTENTIAL THEORY.

**Poisson's ratio** The ratio of the contraction per unit diameter of a rod that is stretched to the elongation per unit length of the rod. If the starting diameter of the rod is $d$ and the contraction is $\Delta d$ the con-

traction per unit diameter, denoted $c$, is $\Delta d/d$. If the starting length of the rod is $l$ and it is extended by $\Delta l$ then the elongation per unit length of the rod, denoted $s$, is given by $s = \Delta l/l$. Poisson's ratio is defined to be $c/s$. If the elasticity of a solid is independent of direction then the ratio of $c/s$ is 0.25. Values of the ratio differ for actual materials. For example, it is found empirically that for steels the ratio is about 0.28. In the case of aluminum alloys it is about 0.33. The value of Poisson's ratio for a material is constant for that material up to its elastic limit. If the value of Poisson's ratio is less than 0.50 for a material then stretching that material increases the net volume of the material. If Poisson's ratio is exactly 0.50 then the volume is constant.

**polar coordinates** A method of defining the position of a point by its distance and direction from a fixed reference point (pole). The direction is given as the angle between the line from the origin to the point, and a fixed line (axis). On a flat surface only one angle, $\theta$, and the radius, $r$, are needed to specify each point. For example, if the axis is horizontal, the point $(r,\theta) = (1,\pi/2)$ is the point one unit length away from the origin in the perpendicular direc-

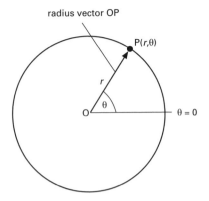

radius vector OP

P(r,θ)

θ = 0

Polar coordinates: coordinates of a point P in two dimensions.

tion. Conventionally, angles are taken as positive in the anticlockwise sense.

In a rectangular Cartesian coordinate system with the same origin and the $x$-axis at $\theta = 0$, the $x$- and $y$-coordinates of the point $(r,\theta)$ are

$$x = r\cos\theta$$
$$y = r\sin\theta$$

Conversely

$$r = \sqrt{(x^2 + y^2)}$$

and

$$\theta = \tan^{-1}(y/x)$$

In three dimensions, two forms of polar coordinate systems can be used. See cylindrical polar coordinates; spherical polar coordinates. See also Cartesian coordinates.

**polar vector** A true vector, i.e. a vector which does not change sign in going from a right-handed coordinate system to a left-handed coordinate system. Compare axial vector.

**pole** 1. One of the points on the Earth's surface through which its axis of rotation passes, or the corresponding point on any other sphere.
2. See stereographic projection.
3. See polar coordinates.

**Polish notation** A notation in computer science in which parentheses are unnecessary since all formulae can be written unambiguously without them. In Polish notation operators precede their operands. Thus, $a + b$ is written $+ab$. The notation was invented by a Polish mathematician, Jan Lukasiewicz (1878–1956).

**polygon** /**pol**-ee-gon/ A plane figure bounded by a number of straight sides. In a regular polygon, all the sides are equal and all the internal angles are equal. In a regular polygon of $n$ sides the exterior angle is $360°/n$.

**polyhedron** /pol-ee-**hee**-drŏn/ A solid figure bounded by plane polygonal faces. The point at which three or more faces intersect on a polyhedron is called a vertex, and a line along which two faces intersect is called an edge. In a regular polyhedron (also known as a Platonic solid), all the faces are congruent regular polygons. There are only five regular polyhedrons: the regular tetrahedron, which has four equilateral triangular faces; the regular hexahedron, or cube, which has six equilateral square faces; the regular octahedron, which has eight equilateral triangular faces; the regular dodecahedron, which has twelve regular pentagonal faces; and the regular icosahedron, which has twenty equilateral triangular faces. These are all convex polyhedrons. That is, all the angles between faces and edges are convex and the polyhedron can be laid down flat on any one of the faces. In a concave polyhedron, there is at least one face in a plane that cuts through the polyhedron. The polyhedron cannot be laid down on this face.

**polynomial** /pol-ee-**noh**-mee-ăl/ A sum of multiples of integer powers of a variable. The general equation for a polynomial in the variable $x$ is

$$a_0x^n + a_1x^{n-1} + a_2x^{n-2} + ...$$

where $a_0, a_1$, etc., are constants and $n$ is the highest power of $x$, called the *degree* of the polynomial. If $n = 1$, it is *linear* expression, for example, $f(x) = 2x + 3$. If $n = 2$, it is *quadratic*, for example, $x^2 + 2x + 4$. If $n = 3$, it is *cubic*, for example, $x^3 + 8x^2 + 2x + 2$. If $n = 4$, it is *quartic*. If $n = 5$, it is *quintic*.

On a Cartesian coordinate graph on which $(n + 1)$ individual points are plotted, there is at least one polynomial curve that passes through all the points. By choosing suitable values of $a_0$ and $a_1$, the straight line

$$y = a_0x + a_1$$

can be made to pass through any two points. Similarly a quadratic

$$y = a_0x^2 + a_1x + a_2$$

can be made to pass through any three points.

A polynomial may have more than one variable:

$$4x^2 + 2xy + y^2$$

is a polynomial of degree 2 (second-degree polynomial) of two variables.

**polytope** /**pol**-ee-tohp/ The analog in $n$ dimensions of point, line, polygon, and polyhedron in 0, 1, 2, and 3 dimensions respectively.

**position vector** The vector that represents the displacement of a point from a given reference origin. If a point P in polar coordinates has coordinates $(r,\theta)$, then $r$ is the position vector of P – a vector of magnitude $r$ making an angle $\theta$ with the axis. *See* vector.

**positive** Denoting a number or quantity that is greater than zero. Numbers that are used in counting things and measuring sizes are all positive numbers. If a change in a quantity is positive, it increases, that is it moves away from zero if it is already posi-tive, and toward it if it is negative. *Compare* negative.

**possibility space** A two-dimensional array that sets out all the possible outcomes of two events. By arranging all the possible outcomes systematically in this way there is no chance of missing any of the possible outcomes. For example if the possible outcomes of rolling two dice are listed as a possibility space then it can be seen that there are 36 possible outcomes, i.e. the possibility space is a 6 by 6 array. This enables probabilities of various outcomes to be calculated. For example, to find the probability of finding a specific total for two dice it is possible to circle the entries corresponding to that total and calculate the probability of that total as the ratio of circled entries to the total number of entries in the array. To find the probability of the total for two dice being 4 the three entries corresponding to a total of 4 are circled, meaning that the probability of a total score of 4 is 3/36 = 1/12.

**postulate** *See* axiom.

**potential energy** Symbol: $V$ The work an object can do because of its position or state. There are many examples. The work an object at height can do in falling is its gravitational potential energy. The ENERGY 'stored' in elastic or a spring under tension or compression is elastic potential energy. Potential difference in electricity is a similar concept, and so on. In practice the potential energy of a system is the energy involved in bringing it to its current state from some reference state; i.e. it is the same as the work that the system could do in moving from its current state back to a reference state.

**potential theory** The branch of mathematical physics that analyzes fields, such as gravitational, magnetic and electric fields, in terms of the potentials of these fields. The main equations used in potential theory are the POISSON EQUATION and the related LAPLACE EQUATION. The development of potential theory had a considerable influence on the development of the theory of differential equations and the CALCULUS OF VARIATIONS. Potential theory is also related to VECTOR CALCULUS.

**pound** A unit of mass now defined as 0.453 592 37 kg.

**poundal** Symbol: pdl The unit of force in the f.p.s. system. It is equal to 0.138 255 newton (0.138 255 N).

**power** 1. The number of times a quantity is to be multiplied by itself. For example, $2^4$ = 2 × 2 × 2 × 2 = 16 is known as the fourth power of two, or two to the power four. *See also* exponent; power series.
2. Symbol: *P* The rate of energy transfer (or work done) by or to a system. The SI unit of power is the watt – the energy transfer in joules per second.

**power series** A series in which the terms contain regularly increasing powers of a variable. For example,

$$S_n = 1 + 2x + 3x^2 + 4x^3 + \ldots + nx^{n-1}$$

is a power series in the variable $x$. In general, a power series has the form

$$a_0 + a_1x + a_2x^2 + \ldots + a_nx^n$$

where $a_0$, $a_1$, etc. are constants.

**P-problem** A type of problem in which the size of the problem is characterized by some number $n$ and the number of steps that an algorithm would need to solve this problem is $N$, where $n$ and $N$ are related by $N \leq cn^p$, where $c$ and $p$ are constants. Such problems are called P-problems because the time it takes to solve them increases as a polynomial of $n$. An example of a P-problem is the problem of multiplying two numbers together. P-problems are easier to solve than NP-PROBLEMS.

**precession** /pri-**sesh**-ŏn/ If an object is spinning on an axis and a force is applied at right angles to this axis, then the axis of rotation can itself move around another axis at an angle to it. The effect is seen in tops and gyroscopes, which 'wobble' slowly while they spin as a result of the force of gravity. The Earth also precesses – the axis of rotation slowly describes a cone.

The precession of Mercury is a movement of the whole orbit of the planet around an axis perpendicular to the orbital plane. Relativistic mechanics needs to be used to describe this precession quantitatively.

**precision** The number of figures in a number. For example 2.342 is stated to a precision of four significant figures, or three decimal places. The precision of a number normally reflects the ACCURACY of the value it represents.

**premiss** In LOGIC, an initial proposition or statement that is known or assumed to be true and on which a logical argument is based.

**premium** 1. The difference between the issue price of a stock or share and its nominal value when the issue price is in excess of the nominal value. *Compare* discount.
2. The amount of money paid each year to an insurance company to purchase insurance cover for a specified risk.

**pressure** Symbol: *p* The pressure on a surface due to forces from another surface or from a fluid is the force acting at right angles to unit area of the surface:

pressure = force/area.

The unit is the pascal (Pa), equal to the newton per square meter.

Objects are often designed to maximize or minimize pressure applied. To give maximum pressure, a small contact area is needed – as with drawing pins and knives. To give minimum pressure, a large contact area is needed – as with snowshoes and the large tires of certain vehicles.

Where the pressure on a surface is caused by the particles of a fluid (liquid or gas), it is not always easy to find the force on unit area. The pressure at a given depth in a fluid is the product of the depth, the average fluid density, and $g$ (the acceleration of free fall):

pressure in a fluid = depth × mean density × $g$

As it is normally possible to measure the mean density of a liquid only, this relation is usually restricted to liquids.

The pressure at a point at a certain depth in a fluid:
1. is the same in all directions;
2. applies force at 90° to any contact surface;
3. does not depend on the shape of the container.

**pressure of the atmosphere** The pressure at a point near the Earth's surface due to the weight of air above that point. Its value varies around about 100 kPa (100 000 newtons per square meter).

**prime** A *prime number* is a positive integer which is not 1 and has no factors except 1 and itself. The set of prime numbers is {2, 3, 5, 7, 11, 13, 17, 19, 23, 29, ...}. There is an infinite number of prime numbers but no general formula for them. The *prime factors* of a number are the prime numbers that divide into it exactly. For example, the prime factors of 45 are 3 and 5 since $45 = 3 \times 3 \times 5$. Each whole number has a unique set of prime factors. *See* Eratosthenes, sieve of.

**prime number theorem** A result concerning the proportion of numbers that are prime numbers that is valid for large numbers. It states that if $\pi(x)$ is the number of prime numbers less than or equal to $x$ then as $x \rightarrow \infty$, $[\pi(x)\ln x]/x \rightarrow 1$. This means that as $x$ becomes a large number it becomes a good approximation that $\pi(x)$ is given by the ratio $x/\ln x$, i.e. the ratio of the number to its natural logarithm. This result, which was first proposed in the late 19th century, has not been given a simple proof in terms of elementary mathematics.

**primitive** A particular type of $n$-th root of unity in which all the $n$ roots of unity are powers of that $n$-th root. For example, in the case of the fourth roots of unity $i$ and $-i$ are primitive fourth roots of unity but the other two fourth roots of unity, 1 and $-1$, are not primitive.

**principal** A sum of money that is borrowed, on which interest is charged. *See* compound interest; simple interest.

**principal diagonal** *See* square matrix.

**principle of equivalence** *See* relativity; theory of.

**principle of moments** The principle that when an object or system is in equilibrium the sum of the moments in any direction equals the sum of the moments in the opposite direction. Because there is no resultant turning force, the moments of the forces can be measured relative to any point in the system or outside it.

**principle of superposition** The result for a system described by a linear equation that if $f$ and $g$ are both solutions to the equation then $f + g$ is also a solution. This principle applies to more than two solutions. An example of the principle of superposition is given in the theory of small (harmonic) oscillations about an equilibrium point. In the case of a vibrating string the behavior of the string can be described by adding together all the harmonic modes of oscillation, with there being no interference between these modes. The principle works in this example because the equation for small harmonic oscillations is linear. However, the principle of superposition does not apply to NON-LINEAR OSCILLATIONS or, more generally, systems governed by non-linear equations. The inapplicability of the principle of superposition in such cases is one of the main reasons why analyzing such systems is much more difficult than for systems described by linear equations.

The principle of superposition also applies to states in quantum-mechanical systems.

**printer** A computer output device that produces hard copy as a printout. There are various types, including daisy-wheel, dot-matrix, ink-jet, and laser.

**printout** The computer output, in the form of characters printed on a continuous sheet of paper, produced by a printer.

**prism** /**priz**-ăm/ A polyhedron with two parallel opposite faces, called *bases*, that

right triangular prism

oblique triangular prism

right hexagonal prism

Prism: examples of prisms

are congruent polygons. All the other faces, called *lateral faces*, are parallelograms formed by the straight parallel lines between corresponding vertices of the bases. If the bases have a center, the line joining the centers is the *axis* of the prism. If the axis is at right angles to the base, the prism is a *right prism* (in which case the lateral faces are rectangles); otherwise it is an *oblique prism*. A *triangular prism* has triangular bases and three lateral faces. This is the shape of many of the glass prisms used in optical instruments. A *quadrangular prism* has a quadrilateral base and four lateral faces. The cube is a special case of this with square bases and square lateral faces.

**probability** The likelihood of a given event occurring. If an experiment has $n$ possible and equally likely outcomes, $m$ of which are event $A$, then the probability of $A$ is $P(A) = m/n$. For example, if $A$ is an even number coming up when a die is thrown, then $P(A) = 3/6$. When the probabilities of the different possible results are not already known, and event $A$ has occurred $m$ times in $n$ trials, $P(A)$ is defined as the limit of $m/n$ as $n$ becomes infinitely large.

In set theory, if $S$ is a set of events (called the *sample space*) and $A$ and $B$ are events in $S$ (i.e. subsets of $S$), the *probability function* P can be represented in set notation. $P(A) = 1$ and $P(0) = 0$ mean that $A$ is 100% certain and the probability of none of the events in $S$ occurring is zero.

Here, $0 \leq P(A) \leq 1$ for all $A$ in $S$. If $A$ and $B$ are separate *independent* events, i.e., if $A \cap B = 0$, then $P(A \cup B) = P(A) + P(B)$. If $A \cap B \neq 0$ then $P(A \cup B) = P(A) + P(B) - P(A \cap B)$.

The *conditional probability* is the probability that $A$ occurs when it is known that $B$ has occurred. It is written as

$$P(A|B) = P(A \cap B)/P(B)$$

If $A$ and $B$ are independent events, $P(A|B) = P(A)$ and $P(A \cap B) = P(A)\,P(B)$. If $A$ and $B$ cannot occur simultaneously, i.e. are mutually exclusive events, $P(A \cap B) = 0$.

**probability density function** *See* random variable.

**probability function** *See* probability.

**procedure** *See* subroutine.

**processor** /**pros**-ess-er/ *See* central processor.

**product** The result obtained by multiplication of numbers, vectors, matrices, etc. *See also* Cartesian product.

**product formulae** *See* addition formulae.

**product notation** Symbol Π. The symbol (a capital Greek letter pi) used to denote the product of either a finite number of terms in the sequence $a_1, a_2, ..., a_n$ or an infinite number of terms in the sequence $a_1, a_2, ...$.

**program** A complete set of instructions to a computer, written in a *programming language*. (The word is also used as a verb, meaning to write such instructions.) These instructions, together with the facts (usually called *data*) on which the instructions operate, enable the computer to perform a wide variety of tasks. For example, there are instructions to do arithmetic, to move data between the main store and the central processor of the computer, to perform logical operations, and to alter the flow of control in the program.

The instructions and data must be expressed in such a way that the central processor can recognize and interpret the instructions and cause them to be carried out on the right data. They must in fact be in binary form, i.e. in a code consisting of the binary digits 0 and 1 (bits). This binary code is known as *machine code* (or *machine language*). Each type of computer has its own machine code.

It is difficult and time-consuming for people to write programs in machine code. Instead programs are usually written in a *source language*, and these *source programs* are then translated into machine code. Most source programs are written in a *high-level language* and are converted into machine code by a complicated program called a *compiler*. High-level languages are closer to natural language and mathematical notation than to machine code, with the instructions taking the form of *statements*. They are fairly easy to use. They are designed to solve particular sorts of problems and are therefore described as 'problem-orientated'.

It is also possible to write source programs in a *low-level language*. These languages resemble machine code more closely than natural language. They are designed for particular computers and are thus described as 'machine-orientated'. *Assembly languages* are low-level languages. A program written in an assembly language is converted into machine code by means of a special program known as an *assembler*. *See also* routine; software; subroutine.

**programming language** *See* program.

**progression** *See* sequence.

**progressive wave** *See* wave.

**projectile** An object falling freely in a gravitational field, having been projected at a speed $v$ and at an angle of elevation $\theta$ to the horizontal. In the special case that $\theta = 90°$, the motion is linear in the vertical direction. It may then be treated using the equations of motion. In all other cases the vertical and horizontal components of velocity must be treated separately. In the absence of friction, the horizontal component is constant and the vertical motion may be treated using the equations of motion. The path of the projectile is then an arc of a parabola. Some useful relations are given below.
Time to reach maximum height:
$$t = v\sin\theta/g$$
Maximum height:
$$h = v^2\sin^2\theta/2g$$
Horizontal range:
$$R = v^2\sin2\theta/g$$
The last result means that the angle for maximum horizontal range is $\theta = 45°$. *See also* orbit.

**projection** A geometrical transformation in which one line, shape, etc., is converted into another according to certain geometrical rules. A set of points (the *object*) is converted into another set (the *image*) by the projection. *See* central projection; Mercator's projection; orthogonal projection; stereographic projection.

**projective geometry** /prŏ-**jek**-tiv/ The study of how the geometric properties of a figure are altered by projection. There is a one-to-one correspondence between points in a figure and points in its projected image, but often the ratios of lengths will be changed. In central projection for example, a triangle maps into a triangle and a quadrilateral into a quadrilateral, but the sides and angles may change. *See also* projection.

**prolate** /**proh**-layt/ Denoting a spheroid that has a polar diameter that is greater than the equatorial diameter. *Compare* oblate. *See* ellipsoid.

**proof** A logical argument showing that a statement, proposition, or mathematical formula is true. A proof consists of a set of basic assumptions, called axioms or premisses, that are combined according to logical rules, to derive as a conclusion the formula that is being proved. A proof of a proposition or formula *P* is just a valid argument from true premises to give *P* as a conclusion. *See also* direct proof; indirect proof.

**proof by contradiction** *See* reductio ad absurdum.

**proper fraction** *See* fraction.

**proper subset** A subset *S* of a set *T* is a *proper subset* if there are elements of *T* that are not in *S*, i.e. *S* has fewer elements than *T* (if *T* is a finite set).

**proportion** The relation between two sets of numbers when the ratio between their corresponding members is constant. For example, the sets {2, 6, 7, 11} and {6, 18, 21, 33} are in proportion (because 2:6 = 6:18 = 7:21 = 11:33). *See also* proportional.

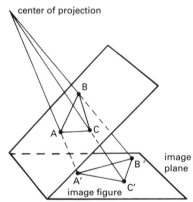

central projection of a triangle ABC onto a triangle A'B'C'

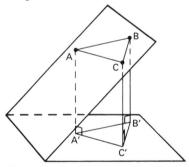

orthogonal projection of a triangle ABC onto a triangle A'B'C'

Projection from one plane onto another

**proportional** Symbol: ∝ Varying in a constant ratio to another quantity. For example, if the length $l$ of a metal bar increases by 1 millimeter for every 10°C rise in its temperature $T$, then the length is proportional to temperature and the constant of proportionality $k$ is 1/10 millimeter per degree Celsius; $l = l_0 + kT$, where $l_0$ is the initial length. If two quantities $a$ and $b$ are *directly proportional* then $a/b = k$, where $k$ is a constant. If they are *inversely proportional* then their product is a constant; i.e. $ab = k$, or $a = k/b$.

**proposition** A sentence or formula in a logical argument. A proposition can have a truth value; that is, it can be either true or false but not both. Any logical argument consists of a succession of propositions linked by logical operations with a proposition as conclusion.

Propositions may be simple or compound. A *compound proposition* is one that is made up of more than one proposition. For example, a proposition $P$ might consist of the constituent parts 'if $R$, then $S$ or $Q$'; i.e. in this case $P = R \rightarrow (S \vee Q)$. A *simple proposition* is one that is not compound. *See also* logic; symbolic logic.

**propositional calculus** *See* symbolic logic.

**propositional logic** *See* symbolic logic.

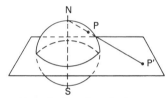

Stereographic projection of a point P on the surface of a sphere onto a plane perpendicular to the line joining the poles N and S. The image of P is P′.

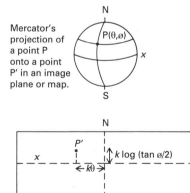

Mercator's projection of a point P onto a point P′ in an image plane or map.

The image P′ of a point P in Mercator's projection.

Projection from a sphere onto a plane

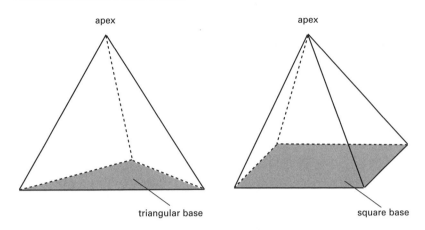

apex        apex

triangular base        square base

Pyramids

**protractor** A drawing instrument used for marking out or measuring angles. It usually consists of a semicircular piece of transparent plastic sheet, marked with radial lines at one-degree intervals.

**pseudoscalar** /soo-doh-**skay**-ler/ A quantity that is similar to a SCALAR quantity but changes sign on going from a right-handed coordinate system to a left-handed coordinate system or if the order of the vectors is interchanged. An example of a pseudoscalar is the quantity formed in the TRIPLE SCALAR PRODUCT. The dot product of any AXIAL VECTOR and a POLAR VECTOR (true vector) results in a pseudoscalar. By contrast, the dot product of two polar vectors or two axial vectors results in a *true scalar*, i.e. a scalar that does not change its sign on going from a right-handed coordinate system to a left-handed coordinate system.

**pseudovector** / soo-doh-**vek**-ter/ *See* axial vector.

**Ptolemy's theorem** /**tol**-ĕ-meez/ *See* cyclic polygon.

**pulley** A class of machine. In any pulley system power is transferred through the tension in a string wound over one or more wheels. The force ratio and distance ratio depend on the relative arrangement of strings and wheels. The efficiency is not usually very high as work must be done to overcome friction in the strings and the wheel bearings and to lift any moving wheels. *See* machine.

**pulsatance** /**pul**-să-tăns/ *See* angular frequency.

**pulse modulation** *See* modulation.

**punched card** *See* card.

**pure mathematics** The study of mathematical theory and structures, without necessarily having an immediate application in mind. For example, the study of the general properties of vectors, considered purely as entities with certain properties, could be considered as a branch of pure mathematics. The use of vector algebra in mechanics to solve a problem on forces or relative velocity is a branch of applied mathematics. Pure mathematics, then, deals with abstract entities, without any necessary reference to physical applications in the 'real world'.

**pyramid** A solid figure in which one of the faces, the base, is a polygon and the others are triangles with the same vertex. If the base has a center of symmetry, a line

Pythagoras' theorem: $c^2 = a^2 + b^2$.

from the vertex to the center is the *axis* of the pyramid. If this axis is at right angles to the base the pyramid is a *right pyramid*; otherwise it is an *oblique pyramid*. A *regular pyramid* is one in which the base is a regular polygon and the axis is at right angles to the base. In a regular pyramid all the lateral faces are congruent isosceles triangles making the same angle with the base. A *square pyramid* has a square base and four congruent triangular faces. The volume of a pyramid is one third of the area of the base multiplied by the perpendicular distance from the vertex to the base.

**Pythagoras' theorem** /pӯ-**thag**-ŏ-răs-iz/ A relationship between the lengths of the sides in a right-angled triangle. The square of the hypotenuse (the side opposite the right angle) is equal to the sum of the squares of the other two sides. There are many proofs of Pythagoras' theorem. It is named for the Greek mathematician and philosopher Pythagoras (*c.* 580 BC–*c.* 500 BC).

**Pythagorean triple** /pӯ-thag-ŏ-**ree**-ăn/ A set of three positive integers $a$, $b$, $c$ that can correspond to the lengths of sides in right-angled triangles and hence, by PYTHAGORAS' THEOREM satisfy the relation $a^2 + b^2 = c^2$. The most familiar example of a Pythagorean triple is (3, 4, 5). Other examples of Pythagorean triples include (5, 12, 13), (7, 24, 25) and (8, 15, 17). It was established by the ancient Greeks that an infinite number of Pythagorean triples exist. If $(a, b, c)$ is a Pythagorean triple then $(ma, mb, mc)$, where $m$ is a positive integer is also a Pythagorean triple.

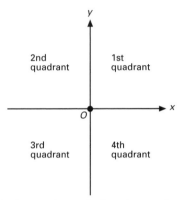

Quadrant: the four quadrants of a Cartesian coordinate system.

**quadrangular prism** /kwod-**rang**-gyŭ-ler/ *See* prism.

**quadrant** /**kwod**-rănt/ **1.** One of four divisions of a plane. In rectangular CARTESIAN COORDINATES, the first quadrant is the area to the right of the $y$-axis and above the $x$-axis, that is, where both $x$ and $y$ are positive. The second quadrant is the area to the left of the $y$-axis and above the $x$-axis, where $x$ is negative and $y$ is positive. The third quadrant is below the $x$-axis and to the left of the $y$-axis, where both $x$ and $y$ are negative. The fourth quadrant is below the $x$-axis and to the right of the $y$-axis, where $x$ is positive and $y$ is negative. In polar coordinates, the first, second, third, and fourth quadrants occur when the direction angle, $\theta$, is 0 to 90° (0 to $\pi/2$); 90° to 180° ($\pi/2$ to $\pi$); 180° to 270° ($\pi$ to $3\pi/2$); and 270° to 360° ($3\pi/2$ to $2\pi$), respectively. *See also* polar coordinates.
**2.** A quarter of a circle, bounded by two

perpendicular radii and a quarter of the circumference.
**3.** A unit of plane angle equal to 90 degrees ($\pi/2$ radians). A quadrant is a right angle.

**quadrantal spherical triangle** /kwod-ran-**tăl**/ *See* spherical triangle.

**quadratic equation** /kwod-**rat**-ik/ A polynomial equation in which the highest power of the unknown variable is two. The general form of a quadratic equation in the variable $x$ is

$$ax^2 + bx + c = 0$$

where $a$, $b$, and $c$ are constants. It is also sometimes written in the reduced form

$$x^2 + bx/a + c/a = 0$$

In general, there are two values of $x$ that satisfy the equation. These solutions (or roots), are given by the formula

$$x = [-b \pm \sqrt{(b^2 - 4ac)}]/2a$$

The quantity $b^2 - 4ac$ is called the DISCRIMINANT. If it is a positive number, there are two real roots. If it is zero, there are two

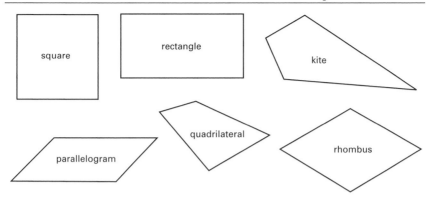

Quadrilateral: six types of quadrilateral.

equal roots. If it is negative there are no real roots. The Cartesian coordinate graph of a quadratic function
$$y = ax^2 + bx + c$$
is a parabola and the points where it crosses the $x$-axis are solutions to
$$ax^2 + bx + c = 0$$
If it crosses the axis twice there are two real roots, if it touches the axis at a turning point the roots are equal, and if it does not cross it at all there are no real roots. In this last case, where the discriminant is negative, the roots are two conjugate complex numbers.

**quadratic graph** A graph of the curve $y = ax^2 + bx + c$, where $a$, $b$, and $c$ are constants. If $a$ is a positive number then the graph has a minimum when $x = -b/2a$. It has a maximum when $x = -b/2a$ if $a$ is a negative number. There is only one stationary point in a quadratic graph. This can either be a minimum or a maximum, as specified. *See also* quadratic equation.

**quadric cone** /**kwod**-rik/ A type of QUADRIC SURFACE that is described by the equation:
$$x^2/a^2 + y^2/b^2 = z^2/c^2,$$
using a particular coordinate system. If a cross-section of this surface is taken parallel to the $xy$-plane an ellipse results in general. In the specific case of $a = b$, a circle results. Cross-sections parallel to the planes of the other axes result in hyperbolas.

**quadric surface** *See* conicoid.

**quadrilateral** /kwod-ră-**lat**-ĕ-răl/ A plane figure with four straight sides. For example, squares, kites, rhombuses, and trapeziums are all quadrilaterals. A square is a regular quadrilateral.

**quantifier** /**kwon**-tă-fÿ-er/ *See* existential quantifier; universal quantifier.

**quantum computer** A computer that operates by using the principles of QUANTUM MECHANICS. In particular, a quantum computer could process alternative pathways simultaneously. It is possible that for many problems, such as factorizing large integers, a quantum computer could perform calculations much more quickly than a conventional computer. At the time of writing, quantum computers have not been developed to the point of being technologically useful. Several different quantum mechanical systems have been proposed and used as quantum computers but they all have practical difficulties such as requiring very low temperatures for their operation.

**quantum mechanics** The theory that governs the behavior of particles at the atomic and sub-atomic level. It bears the same relationship to Newtonian mechanics as wave optics does to geometrical optics. The development of quantum mechanics has had a substantial influence on several branches of mathematics, notably the theory of operators and group theory. Quantum mechanics can be formulated mathematically in several ways including the use of differential equations, matrices, variational principles, and information theory. Characteristic features of quantum mechanics are that many quantities have a set of discrete values rather than a range of continuous values and that particles such as electrons cannot be ascribed precisely defined trajectories like the orbits of planets. A further feature is the existence of non-commutative operators, i.e. pairs of operators in which the final result of the operations depends on the order in which the operators operate. At the time of writing, there is a great deal of activity in developing quantum, i.e. discrete and non-commutative, versions of mathematics which have been previously formulated for continuous, commutative quantities. Some of these branches of 'quantum mathematics' may well turn out to be important in physics.

**quart** A unit of capacity equal to two pints. In the USA a dry quart is equal to two US dry pints.

**quartic equation** /kwor-tik/ A polynomial equation in which the highest power of the unknown variable is four. The general form of a quartic equation in a variable $x$ is
$$ax^4 + bx^3 + cx^2 + dx + e = 0$$
where $a$, $b$, $c$, $d$, and $e$ are constants. It is also sometimes written in the reduced form
$$x^4 + bx^3/a + cx^2/a + dx/a + e/a = 0$$
In general, there are four values of $x$ that satisfy a quartic equation. For example,
$$2x^4 - 9x^3 + 4x^2 + 21x - 18 = 0$$
can be factorized to
$$(2x + 3)(x - 1)(x - 2)(x - 3) = 0$$
and its solutions (or roots) are $-3/2$, 1, 2, and 3. On a Cartesian coordinate graph, the curve
$$y = 2x^4 - 9x^3 + 4x^2 + 21x - 18 = 0$$
crosses the $x$-axis at $x = -3/2$; $x = 1$; $x = 2$; and $x = 3$. *Compare* cubic equation; quadratic equation.

**quartile** One of the three points that divide a set of data arranged in numerical order into four equal parts. The lower (or first) quartile, $Q_1$, is the 25th PERCENTILE ($P_{25}$). The middle (or second) quartile, $Q_2$, is the MEDIAN ($P_{50}$). The upper (or third) quartile, $Q_3$, is the 75th percentile ($P_{75}$).

**quasicrystalline symmetry** /kway-sÿ-kriss-tă-lin, -lÿn, kway-zÿ-, kwah-see-, kwah-zee-/ The symmetry that is the three-dimensional analog of a PENROSE PATTERN, i.e. there is order but not the periodicity of crystals. Quasicrystalline symmetry is realized in certain solids, called *quasicrystals*, such as AlMn. This symmetry is made apparent by the fact that quasicrystals give clear x-ray diffraction images, characteristic of there being order. Quasicrystals such as AlMn have the point group symmetry of an ICOSAHEDRON, a type of point group symmetry that is incompatible with crystalline symmetry.

**quaternions** /kwă-ter-nee-ŏnz/ Generalized complex numbers invented by Hamilton. A quaternion is of the form $a + bi + cj + dk$, where $i^2 = j^2 = k^2 = -1$ and $ij = -ji = k$ and $a$, $b$, $c$, $d$ are real numbers. The most striking feature of quaternions is that multiplication is not commutative. They have applications in the study of the rotations of rigid bodies in space.

**quintic equation** /kwin-tik/ A polynomial equation in which the highest power of the unknown variable is five. The general form of a quintic equation in a variable $x$ is:
$$ax^5 + bx^4 + cx^3 + dx^2 + ex + f = 0$$
where $a$, $b$, $c$, $d$, $e$, and $f$ are constants. It is also sometimes written in the reduced form
$$x^5 + bx^4/a + cx^3/a + dx^2/a + ex/a + f/a = 0$$
In general, there are five values of $x$ that satisfy a quintic equation. For example,
$$2x^5 - 17x^4 + 40x^3 + 5x^2 - 102x + 72 = 0$$
can be factorized to
$$(2x + 3)(x - 1)(x - 2)(x - 3)(x - 4) = 0$$
and its solutions (or roots) are $-3/2$, 1, 2, 3, and 4. On a Cartesian coordinate graph, the curve
$$y = 2x^5 - 17x^4 + 40x^3 +$$

$$5x^2 - 102x + 72$$

crosses the $x$-axis at $x = -3/2$; $x = 1$; $x = 2$; $x = 3$; and $x = 4$. *Compare* cubic equation; quadratic equation; quartic equation.

**quotient** /**kwoh**-shĕnt/ The result of dividing one number by another. There may or may not be a remainder. For example, 16/3 gives a quotient of 5 and a remainder of 1.

**qwerty** /**kwer**-tee/ Describing the standard layout of ALPHANUMERIC characters on a typewriter or computer keyboard (named for the first six letters of the top letter row).

**radial** /**ray**-dee-ăl/ Along the direction of the radius.

**radial symmetry** Symmetry about a line or plane through the center of a shape. *See also* bilateral symmetry; symmetrical.

**radian** Symbol: rad The SI unit for measuring plane angle. It is the angle subtended at the center of a circle by an arc equal in length to the radius of the circle. $\pi$ radians $= 180°$.

**radical** /**rad**-ă-kăl/ An expression for a root. For example, $\sqrt{2}$, where $\sqrt{}$ is the radical sign.

**radical axis** The radical axis of two circles
$$x^2 + y^2 + 2a_1x + 2b_1y + c_1 = 0$$
and
$$x^2 + y^2 + 2a_2x + 2b_2y + c_2 = 0$$
is the straight line obtained by eliminating the square terms between the equations of the circles, i.e.
$$2(a_1 - a_2)x + 2(b_1 - b_2)y + (c_1 - c_2) = 0$$
When the circles intersect, the radical axis passes through their two points of intersection.

**radius** /**ray**-dee-ŭs/ (*pl.* **radii** or **radiuses**) The distance from the center of a circle to any point on its circumference or from the center of a sphere to its surface. In polar coordinates, a radius $r$ (distance from a fixed origin) is used with angular position $\theta$ to specify the positions of points.

**radius of convergence** For a power series $a_0 + a_1(x - a) + a_2(x - a)^2 + \ldots + a_n(x - a)^n + \ldots$ there is a value $R$ such that the series converges for values of $|x - a| < R$. Here, $R$ is the *radius of convergence* of the power series.

**radius of curvature** *See* curvature.

**radius of gyration** Symbol: $k$ For a body of mass $m$ and moment of inertia $I$ about an axis, the radius of gyration about that axis is given by
$$k^2 = I/m$$
In other words, a point mass $m$ rotating at a distance $k$ from the axis would have the same moment of inertia as the body.

**radius vector** The vector that represents the distance and direction of a point from the origin in a polar coordinate system.

**radix** /**ray**-diks/ (*pl.* **radices** or **radixes**) The base number of any counting system, also known as the base. For example, the decimal system has the radix 10, whereas the radix of the binomial system is 2. *See also* base.

**RAM** Random-access memory. *See* random access.

**random access** A method of organizing information in a computer storage device so that one piece of information may be reached directly in about the same time as any other. Main store and disk units operate by random access and are thus known as *random-access memory* (RAM). In contrast a magnetic tape unit operates more slowly by *serial access*: a particular piece of information can only be retrieved by working through the preceding blocks of data on the tape.

**random error** *See* error.

**randomness** The lack of a pattern or order in either a physical system or a set of numbers. Precise mathematical analysis of the concept of randomness is closely associated with CHAOS THEORY, ENTROPY, and INFORMATION THEORY.

**random number table** A table consisting of a sequence of randomly chosen digits from 0 to 9, where each digit has a probability of 0.1 of appearing in a particular position, and choices for different positions are independent. Random numbers are used in statistical random sampling.

**random sampling** *See* sampling.

**random variable** (**chance variable; stochastic variable**) A quantity that can take any one of a number of unpredicted values. A *discrete random variable*, X, has a definite set of possible values $x_1, x_2, x_3, \ldots x_n$, with corresponding probabilities $p_1, p_2, p_3, \ldots p_n$. Since X must take one of the values in this set,

$$p_1 + p_2 + \ldots + p_n = 1$$

If X is a *continuous random variable*, it can take any value over a continuous range. The probabilities of a particular value x occurring is given by a *probability density function* f(x). On a graph of f(x) against x, the area under the curve between two values a and b is the probability that X lies between a and b. The total area under the curve is 1.

**random walk** A succession of movements along line segments where the direction and the length of each move is randomly determined. The problem is to determine the probable location of a point subject to such random motions given the probability of moving some distance in some direction, where the probabilities are the same at each step. Random walks can be used to obtain probability distributions to practical problems. Consider, for example, a drunk man moving a distance of one unit in unit time, the direction of motion being random at each step. The problem is to find the probability distribution of the

distance of the point from the starting point after some fixed time. Technically a random walk is a sequence

$$S_n = X_1 + X_2 + \ldots + X_n$$

where $\{X_i\}$ is a sequence of independent random variables.

**range** **1.** The difference between the largest and smallest values in a set of data. It is a measure of dispersion. In terms of percentiles, the range is $(P_{100} - P_0)$. *Compare* interquartile range.
**2.** A set of numbers or quantities that form possible results of a mapping. In algebra, the range of a function f(x) is the set of values that f(x) can take for all possible values of x. For example, if f(x) is taking the square root of positive rational numbers, then the range would be the set of real numbers. *See also* domain.

**rank** A method of ordering a set of objects according to the magnitude or importance of a variable measured on them, e.g. arranging ten people in order of height. If the objects are ranked using two different variables, the degree of association between the two rankings is given by the coefficient of rank correlation. *See also* Kendall's method.

**raster graphics** *See* computer graphics.

**rate of change** The rate at which one quantity y changes with respect to another quantity x. If y is written as y = f(x) and if f(x) is a differentiable function, the rate of change of y with respect to x is described by differential calculus. This rate of change is denoted by dy/dx *or* f'(x). The value of the rate of change varies with x since dy/dx is a function of x. The value of dy/dx at a specific value of x is found by expressing dy/dx as a function of x and substituting the value of x into the function for the rate of change.

Frequently in physical systems the rate of change of a function with respect to time is of interest. For example, the velocity of a body is the rate of change of the displacement of the body with respect to time and the acceleration of a body is the rate of change of the velocity of the body with respect to time.

**ratio** /ray-shee-oh/ One number or quantity divided by another. The ratio of two variable quantities $x$ and $y$, written as $x/y$ or $x:y$, is constant if $y$ is proportional to $x$. *See also* fraction.

**rational function** A REAL FUNCTION, $f(x)$, of a real variable $x$ in some domain, which is frequently taken to be the set of all real numbers, that can be expressed as the ratio of two polynomial functions $g(x)$ and $h(x)$. In this definition it is assumed that the two polynomials do not have a common factor with a degree that is greater than or equal to 1. A rational function defined in this way is continuous except for values of $x$ for which the value of the denominator $h(x)$ is zero. The concept of a rational function can be extended to a function, $f(z)$, of a complex variable $z$.

**rationalize** /rash-ŏ-nă-lÿz/ To remove radicals (such as square root, or $\sqrt{}$) from an algebraic expression without changing its value, thus making the equation easier to deal with. For example, by squaring both sides, the equation $\sqrt{(2x + 3)} = x$ rationalizes to $2x + 3 = x^2$, equivalent to $x^2 - 2x - 3 = 0$.

**rationalized units** A system of units in which the equations have a logical form related to the shape of the system. SI units form a rationalized system of units. For example, in it formulae concerned with circular symmetry contain a factor of $2\pi$; those concerned with radial symmetry contain a factor of $4\pi$.

**rational numbers** /rash-ŏ-năl/ Symbol: $Q$ The set of numbers that includes integers and fractions. Rational numbers can be written down exactly as ratios or as finite or repeating decimals. For example, $\frac{1}{3}$ ($= 0.333...$) and $\frac{1}{4}$ ($= 0.25$) are rational. The square root of 2 ($= 1.414\ 213\ 6...$) is not. *Compare* irrational numbers.

**ratio theorem** A result concerning vectors stating that if two points A and B are specified by position vectors $a$ and $b$, relative to some origin O, and the point C divides the line AB in the ratio $l:m$ then the position vector $c$ for C is given by: $c = (ma + lb)/(l + m)$. By defining $l$ and $m$ so that $l + m = 1$, the expression $c = ma + (1 - m)b$ is obtained.

**ray** A set consisting of all the points on a given line to the left or to the right of a given point, and including that point itself.

**reaction** Newton's third law of force states that whenever object A applies a force on object B, B applies the same force on A. An old word for force is 'action'; 'reaction' is thus the other member of the pair. Thus in the interaction between two electric charges, each exerts a force on the other. Thus, in general, action and reaction have little meaning. The word 'reaction' is still sometimes used in restricted cases, such as the reaction of a support on the object it supports. In this case the 'action' is the effect of the weight of the object on the support.

**reader** A device used in a computer system to sense the information recorded on some source and convert it into another form.

**read-only memory** (ROM) *See* store.

**read–write head** *See* disk; drum; magnetic tape.

**real analysis** The branch of analysis concerned with REAL FUNCTIONS.

**real axis** An axis of the complex plane on which the points represent the real numbers. It is customary to draw the real axis as the $x$-axis of the complex plane.

**real function** A function f that maps members of the set R of real numbers (or a subset of R) to members of R. This means that if $x$ is a real number in R or its subset then $f(x)$ is also a real number.

**real line** The line that represents the real numbers on the complex plane. The real line is drawn as a horizontal line. A specific point O is chosen to be the origin and another specific point A, which is to the right of O on the horizontal line, is chosen so that the length of OA represents one unit. All positive real numbers are represented by points that lie to the right of O on the real line and all negative real numbers are represented by points that lie to the left of O, with the origin being the real number with value zero.

**real numbers** Symbol: $R$ The set of numbers that includes all rational and irrational numbers.

**real part** Symbol Re$z$. The part $x$ of a complex number $z$ defined by $z = x + iy$, where $x$ and $y$ are both real numbers. *See also* imaginary part.

**real time** The actual time in which a physical process takes place or in which a physical process, machine, etc., is under the direct control of a computer. A *real-time system* is able to react sufficiently rapidly so that it may control a continuing process, making changes or modifications when necessary. Air-traffic control and airline reservations require real-time systems. *Compare* batch processing. *See also* time sharing.

**reciprocal** /ri-**sip**-rŏ-kăl/ The number 1 divided by a quantity. For example, the reciprocal of 2 is ½. The reciprocal of $(x^2 + 1)$ is $1/(x^2 + 1)$. The product of any expression and its reciprocal is 1. For any function, the reciprocal is the multiplicative inverse.

**rectangle** /**rek**-tang-găl/ A plane figure with four straight sides, two parallel pairs of equal length forming four right angles. The area of a rectangle is the product of the two different side lengths, the length times the breadth. A rectangle has two axes of symmetry, the two lines joining the midpoints of opposite sides. It can also be superimposed on itself after rotation through 180° ($\pi$ radians). The two diagonals of a rectangle have equal lengths.

**rectangular hyperbola** /rek-**tang**-gyŭ-ler/ *See* hyperbola.

**rectangular parallelepiped** *See* parallelepiped.

**rectilinear** /rek-tă-**lin**-ee-er/ Describing motion in a straight line.

**recurring decimal** A repeating DECIMAL.

**recursion formula** A formula that relates some quantity $Q_n$, where $n$ is a non-negative integer, to quantities such as $Q_{n-1}$, $Q_{n-2}$, .... Formulas of this type are used extensively to calculate integrals involving powers of trigonometric functions. For example if $I_n = \int \sin^n x \, dx$, there is a recursion formula for $I_n$ which can be found by integrating $I_n$ by parts. This gives: $I_n = [(n-1)/n]I_{n-2}$, where $n \geq 2$. The cases of $I_0$ and $I_1$ are readily found by direct integration. The recursion formula means that $I_n$ can be determined for all even and odd values of $n$. Another example of a recursion formula is the expression $\Gamma(n + 1) = n\Gamma(n)$ for the GAMMA FUNCTION.

A recursion formula is sometimes called a *reduction formula*.

**reduced form** (of a polynomial) The EQUATION of the form
$$x^n + (b/a)x^{n-1} + (c/a)x^{n-2} + \dots = 0$$
that is derived from a POLYNOMIAL of the form
$$ax^n + bx^{n-1} + cx^{n-2} + \dots = 0$$
For example,
$$2x^2 - 10x + 12 = 0$$
is equivalent to the reduced form
$$x^2 - 5x + 6 = 0$$
*See also* quadratic equation.

***reductio ad absurdum*** /ri-**duk**-tee-oh ad ăb-**ser**-dŭm/ A method of proof which proceeds by assuming the falsity of what we wish to prove and showing that it leads to a contradiction. Hence the statement whose falsity we assumed must be true. The following proof that $\sqrt{2}$ is irrational is a simple example of proof by this method.

Assume $\sqrt{2}$ is rational. In that case it can be expressed in the form $a/b$ where $a$ and $b$ are integers. Assume that this fraction is in its lowest terms and so $a$ and $b$ have no common factor. Since $a/b = \sqrt{2}$ then $a^2/b^2 = 2$. Hence $a^2 = 2b^2$. This means that $a^2$ is even and hence $a$ itself is even. In that case we can write $a$ as $2m$ where $m$ is some integer. But then since $a^2 = 2b^2$ we have $(2m)^2 = 2b^2$, or $4m^2 = 2b^2$. Dividing by 2 we get $2m^2 = b^2$. But this means that $b^2$ and hence $b$ is also even. Hence, $a$ and $b$ do have a common factor, namely 2. But we assumed they had no common factor. Since we have reached a contradiction, our starting-point – the assumption that $\sqrt{2}$ is rational – must be false.

**reduction formulae** In trigonometry, the equations that express sine, cosine, and tangent functions of an angle in terms of an angle between 0 and 90° ($\pi/2$). For example:

$$\sin(90° + \alpha) = \cos\alpha$$
$$\sin(180° + \alpha) = -\sin\alpha$$
$$\sin(270° + \alpha) = -\cos\alpha$$
$$\cos(90° + \alpha) = -\sin\alpha$$
$$\tan(90° + \alpha) = -\cot\alpha$$

**re-entrant polygon** A many-sided figure (polygon) with one of its internal angles a reflex angle (i.e. greater than 180°).

**reflection** /ri-**flek**-shŏn/ The geometrical transformation of a point or a set of points from one side of a point, line, or plane to a symmetrical position on the other side. On *reflection in a line*, the image of a point P

would be point P' at the same distance from the line but on the other side. The line, the *axis of reflection*, is the perpendicular bisector of the line PP'. In a symmetrical plane figure there is an axis of reflection, also called the *axis of symmetry*, in which the figure is reflected onto itself. An equilateral triangle, for example, has three axes of symmetry. In a circle, any diameter is an axis of symmetry. Similarly, a solid may undergo *reflection in a plane*. In a sphere, any plane passing through the sphere's center would be a plane of symmetry.

In a Cartesian coordinate system, reflection in the $x$-axis changes the sign of the $y$-coordinate. A point $(a,b)$ would become $(a,-b)$. Reflection in the $y$-axis changes the sign of the $x$-coordinate, making $(a,b)$ become $(-a,b)$. In three dimensions, changing the sign of the $z$-coordinate is equivalent to reflection in the plane of the $x$ and $y$ axes. *Reflection in a point* is equivalent to rotation through 180°. Each point P is moved to a position P' so that the point of reflection bisects the line PP'. *Reflection in the origin* of plane Cartesian coordinates changes the signs of all the coordinates. It is equivalent to reflection in the $x$-axis followed by reflection in the $y$-axis, or vice versa. *See also* rotation.

**reflex** /**ree**-fleks/ Describing an angle that is greater than 180° (but less than 360°). *Compare* acute; obtuse.

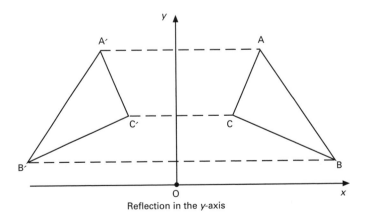

Reflection in the *y*-axis

**reflexive** /ri-**fleks**-iv/ A relation R defined on a given set is said to be reflexive if every member of the set has this relation to itself. Equality is an example of a reflexive relation.

**register** *See* central processor.

**regression line** A line $y = ax + b$, called the regression line of $y$ on $x$, which gives the expected value of a random variable $y$ conditional on the given value of a random variable $x$. The regression line of $x$ on $y$ is not in general the same as that of $y$ on $x$. If a SCATTER DIAGRAM of data points $(x_1,y_1), ..., (x_n,y_n)$ is drawn and a linear relationship is shown up, the line can be drawn by hand. The best line is drawn using the LEAST-SQUARES METHOD. *See also* correlation.

**regular** Describing a figure that has all faces or sides of equal size and shape. *See* polygon; polyhedron.

**relation** A property that holds for ordered pairs of elements of some set, for example being greater than. We can think of a relation abstractly as the set of all ordered pairs in which the two members have the given relation to one another.

**relative** Expressed as a difference from or as a ratio to, some reference level. Relative density, for example, is the mass of a substance per unit volume expressed as a fraction of a standard density, such as that of water at the same temperature. *Compare* absolute.

**relative error** The ERROR or uncertainty in a measurement expressed as a fraction of the measurement. For example, if, in measuring a length of 10 meters, the tape measures only to the nearest centimeter, then the measurement might be written as 10 ± 0.01 meters. The relative error is 0.01/10 = 0.001. *Compare* absolute error.

**relative maximum** *See* local maximum.

**relative minimum** *See* local minimum.

**relative velocity** If two objects are moving at velocities $v_A$ and $v_B$ in a given direction the velocity of A relative to B is $v_A - v_B$ in that direction. In general, if two objects are moving in the same frame at nonrelativistic speeds their relative velocity is the vector difference of the two velocities.

**relativistic mass** /rel-ă-ti-**vis**-tik/ The mass of an object as measured by an observer at rest in a frame of reference in which the object is moving with a velocity $v$. It is given by
$$m = m_0/\sqrt{(1 - v^2/c^2)}$$
where $m_0$ is the REST MASS, $c$ is the velocity of light in free space, and $m$ is the relativistic mass. The equation is a consequence of the special theory of relativity, and is in excellent agreement with experiment. No object can travel at the speed of light in free space because its mass would then be infinite. *See also* relativity, theory of.

**relativistic mechanics** A system of mechanics based on relativity theory. *See also* classical mechanics.

**relativistic speed** (relativistic velocity) Any speed (velocity) that is sufficiently high to make the mass of an object significantly greater than its REST MASS. It is usually expressed as a fraction of $c$, the speed of light in free space. At a speed of $c/2$ the RELATIVISTIC MASS of an object is about 15% greater than the rest mass.

**relativity, theory of** /rel-ă-**tiv**-ă-tee/ A theory put forward in two parts by Albert Einstein. The spcecial theory (1905) referred only to nonaccelerated (inertial) frames of reference. The general theory (1915) is also applicable to accelerated systems.

The *special theory* was based on two postulates:
1. That physical laws are the same in all inertial frames of reference.
2. That the speed of light in a vacuum is constant for all observers, regardless of the motion of the source or observer.
The second postulate seems contrary to 'common sense' ideas of motion. Einstein was led to the theory by considering the

problem of the 'ether' and the relation between electric and magnetic fields in relative motion. The theory accounts for the negative result observed in the MICHELSON–MORLEY EXPERIMENT and shows that the Lorentz–Fitzgerald contraction is only an apparent effect of motion on an object relative to an observer, not a 'real' contraction. It leads to the result that the mass of an object moving at a speed $v$ relative to an observer is given by:

$$m = m_0/\sqrt{(1 - v^2/c^2)}$$

where $c$ is the speed of light and $m_0$ the mass of the object when at rest relative to the observer. The increase in mass is significant at high speeds. Another consequence of the theory is that an object has an energy content by virtue of its mass, and similarly that energy has inertia. Mass and energy are related by the famous equation $E = mc^2$.

The *general theory* of relativity seeks to explain the difference between accelerated and nonaccelerated systems and the nature of the forces acting in both of them. For example, a person in a spacecraft far out in space would not be subject to gravitational forces. If the craft were rotating, he would be pressed against the walls of the craft and would consider that he had weight. There would not be any difference between this force and the force of gravity. To an outside observer the force is simply a result of the tendency to continue in a straight line; i.e. his inertia. This type of analysis of forces led Einstein to a *principle of equivalence* that inertial forces and gravitational forces are equivalent, and that gravitation can be a consequence of the geometrical properties of space. He visualized a four-dimensional space–time continuum in which the presence of a mass affects the geometry – the space-time is 'curved' by the mass with the geometry being non-Euclidean.

**remainder** The number left when one number is divided into another. Dividing 12 into 57 gives 4 remainder 9 ($4 \times 12 = 48$; $57 - 48 = 9$).

**remainder theorem** The theorem expressed by the equation

$$f(x) = (x - a)g(x) + f(a)$$

This means that if a polynomial in $x$, $f(x)$, is divided by $(x - a)$, where $a$ is a constant, the remainder term is equal to the value of the polynomial when $x = a$. For example, if

$$2x^3 + 3x^2 - x - 4$$

is divided by $(x - 4)$, then the remainder term is

$$f(4) = 128 + 48 - 4 - 4 = 168$$

The remainder theorem is useful for finding the factors of a polynomial. In this example,

$$f(1) = 2 + 3 - 1 - 4 = 0$$

Thus, there is no remainder so $(x - 1)$ is a factor.

**repeated root** A root of an equation that occurs more than once.

**repeating decimal** *See* decimal.

**representation of a group** A set of operators that correspond to the elements of the GROUP. The operators are defined in a VECTOR SPACE $V$. The *dimension* of the representation is the dimensionality of $V$. Frequently, the set of operators is expressed in terms of matrices, with this type of representation being called a *matrix representation* of the group. Group representations are of great importance in QUANTUM MECHANICS. In particular, *irreducible representations*, i.e. representations that cannot be reduced to representations of lower dimensions, characterize quantum mechanical systems since the quantum numbers that characterize the energy levels of a system correspond to the irreducible representations of the group that describes the symmetry of the system. For instance, the irreducible representations of the ROTATION GROUP characterize the energy levels of atoms and the irreducible representations of POINT GROUPS characterize the energy levels of molecules.

**representative fraction** A fraction used to express the SCALE of a map in which the numerator represents a distance on the map and the denominator represents the corresponding distance on the ground. As a fraction is a ratio, the units of the numerator and denominator must be the same.

For example, a scale of 1 cm = 1 km would be given as a representative fraction of 1/100 000, because there are 100 000 cm in 1 km.

**residue** /rez-ă-dew/ If there exists an $x$ such that $x^n \equiv a \pmod{p}$, i.e. $x^n$ is congruent to $a$ modulo $p$, then $a$ is called a residue of $p$ of order $n$.

**resolution of vectors** The determination of the components of a vector in two given directions at 90°. The term is sometimes used in relation to finding any pair of components (not necessarily at 90° to each other).

**resonance** The large-amplitude vibration of an object or system when given impulses at its natural frequency. For instance, a pendulum swings with a natural frequency that depends on its length. If it is given a periodic 'push' at this frequency – for example, at each maximum of a complete oscillation – the amplitude is increased with little effort. Much more effort would be required to produce a swing of the same amplitude at a different frequency.

**restitution, coefficient of** /res-tă-tew-shŏn/ Symbol: $e$ For the impact of two bodies, the elasticity of the collision is measured by the coefficient of restitution. It is the relative velocity after collision divided by the relative velocity before collision (with the velocities measured along the line of centers). For spheres A and B:

$$v_A' - v_B' = e(v_A - v_B)$$

$v$ indicates velocity before collision; $v'$ velocity after collision. Kinetic energy is conserved only in a perfectly elastic collision.

**rest mass** Symbol: $m_0$ The mass of an object at rest as measured by an observer at rest in the same frame of reference. *See also* relativistic mass.

**resultant** /ri-zul-tănt/ **1.** A vector with the same effect as a number of vectors. Thus, the resultant of a set of forces is a force that has the same effect; it is equal in magnitude and opposite in direction to the equilibrium. Depending on the circumstances, the resultant of a set of vectors can be found by different methods. *See* parallel forces; parallelogram of vectors; principle of moments.
**2.** *See* eliminant.

**revolution, solid of** A solid generated by revolving a plane area about a line called the *axis of revolution*. For example, rotating a rectangle about an axis joining the midpoints of two opposite sides produces a cylinder as the solid of revolution.

**rhombic dodecahedron** /rom-bik/ A type of polyhedron in which there are 12 faces, each of which is a RHOMBUS.

**rhombohedron** /rom-bŏ-hee-drŏn/ A solid figure bounded by six faces, each one a parallelogram, with opposite faces congruent.

**rhomboid** /rom-boid/ A parallelogram that is neither a rhombus nor a rectangle. *See* parallelogram.

**rhombus** /rom-bŭs/ (*pl.* **rhombuses** or **rhombi**) A plane figure with four straight sides of equal length; i.e. a parallelogram with equal sides. Its area is equal to half the product of the lengths of its two diagonals, which bisect each other perpendicularly. The rhombus is symmetrical about both of its diagonals and also has rotational symmetry, in that it can be superimposed on itself after rotation through 180° ($\pi$ radians).

**Riemannian geometry** /ree-mah-nee-ăn/ A type of geometry that describes the higher-dimensional analogs of curved surfaces. When extended from space to space–time, Riemannian geometry can be used to describe the effect of gravitation in the general theory of relativity. This type of geometry is named for the German mathematician (Georg Friedrich) Bernhard Riemann (1826–66).

**Riemann integral** /ree-mahn/ *See* definite integral; Riemann sum.

**Riemann sum** The series that approximates the area between the curve of a function $f(x)$ and the $x$-axis:

$$\sum_{i=1}^{n} f(\xi_i)\Delta x_i$$

where $\Delta x$ is an increment of $x$, $\xi_i$ is any value of $f(x)$ within that interval, and $n$ is the number of intervals. The definite (or Riemann) integral is the limit of the sum as $n$ becomes infinitely large and $\Delta x$ infinitesimally small.

**Riemann zeta function** /zay-tă/ A function of a complex variable $z = x + iy$, where $x$ and $y$ are real numbers, defined by the infinite series:

$$\zeta(z) = \sum_{n=1}^{\infty} n^{-z}$$

The Riemann zeta function can also be defined for real numbers. There is a conjecture concerning the Riemann zeta function for a complex variable $z$ that has remained unproved since Riemann postulated it in the nineteenth century. If Riemann's conjecture is correct it would establish a firm result about the distribution of prime numbers.

**right angle** An angle that is 90° or $\pi/2$ radians. It is the angle between two lines or planes that are perpendicular to each other. The corner of a square, for example, is a right angle.

**right-handed system** A way of specifying a coordinate system such that rotating the $x$-axis by 90° into the $y$-axis corresponds to rotating a right-handed screw in the positive $z$ direction.

When discussing VECTOR PRODUCTS it is necessary to specify whether a right-handed or left-handed coordinate system is being used.

**right solid** A solid geometrical figure that is upright; for example, a cone, cylinder, pyimid, or prism that has an axis at right angles to the base. *Compare* oblique solid.

**rigid body** In mechanics, a body for which any change of shape produced by forces on the body can be neglected in the calculations.

**ring** A set of entities with two binary operations called addition and multiplication and denoted by + and • respectively, such that:
1. the set is a commutative group under addition;
2. for every pair of elements $a,b$, in the ring the product $a•b$ is unique, multiplication is associative, i.e. $(a•b)•c = a•(b•c)$, and multiplication is distributive with respect to addition, i.e. $a•(b + c) = a•b + a•c$ and $(b + c)•a = b•a + c•a$ for each $a$, $b$, and $c$ in the set.

If multiplication is also commutative, the ring is called a *commutative ring*. For example, the set of real numbers, the set of integers, and the set of rational numbers

Right-handed system

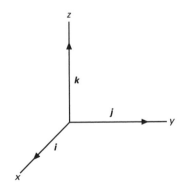

Left-handed system

are rings with respect to ordinary addition and multiplication.

**robot** A feedback-controlled mechanical device. *Robotics* is the study of the design, applications, and control and sensory systems of robots; for example, the design of robot arms that can approach an object from any orientation and grip it. A robot's control system may be simple and consist of only a sequencing device so that the device moves in a repetitive pattern, or more sophisticated so that the robot's movements are generated by computer from data about the environment. The robot's sensory system gathers information needed by the control system, usually visually by using a television camera.

**rod** An object that is considered to have a length but no breadth or depth, with all the mass of the object concentrated along the length. Although this definition is an idealization it is a reasonably good approximation for objects in which the length is much bigger than both the breadth and depth. The pole of a broom can be modelled fairly accurately by regarding it as a rod. In a *uniform rod* equal lengths of the rod have equal masses. A broom pole is an example of a uniform rod if it is made of the same wood and has the same cross-sectional area all along the length of the rod. A *nonuniform rod* is a rod in which equal lengths of the rod do not have equal masses. An example would be an object in two connected

sections made of different woods. A *light rod* is a rod in which the mass of the rod is neglected. The concept of a light rod is useful when considering two bodies connected by a rod, with the masses of both the bodies being much greater than the mass of the rod.

**Rolle's theorem** /rol-ĕz/ A curve that intersects the $x$-axis at two points $a$ and $b$, is continuous, and has a tangent at every point between $a$ and $b$, must have at least one point in this interval at which the tangent to the curve is horizontal. For a curve $y = f(x)$, it follows from Rolle's theorem that the function $f(x)$ has a turning point (a maximum or minimum value) between $f(a)$ and $f(b)$, where the derivative $f'(x) = 0$. The theorem is named for the French mathematician Michel Rolle (1652–1719). *See also* turning point.

**rolling friction** *See* friction.

**Roman numerals** The system of writing integers that was used by the ancient Romans, in which I denotes 1, V denotes 5, X denotes 10, L denotes 50, C denotes 100, D denotes 500, and M denotes 1000. The integers are written using the following rules: (1) the values of the letters are added if a letter is repeated or immediately followed by a letter of lesser value; (2) the value of the letter of smallest value is subtracted from the value of the letter of

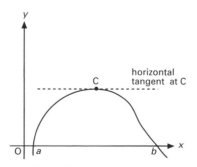

Rolle's theorem for a function f(x) that is continuous between $x = a$ and $x = b$ and for which f(a) = f(b) = 0

greater value when a letter is immediately followed by a letter of greater value.

There is no symbol for zero. The integers from 1 to 10 are written I, II, III, IV, V, VI, VII, VIII, IX, X; and, for example, 1987 is written MCMLXXXVII.

**root** In an equation, a value of the independent variable that satisfies the equation. In general, the degree of a POLYNOMIAL is equal to the number of roots. A QUADRATIC EQUATION (one of degree two) has two roots, although in some circumstances they may be equal. For a number $a$, an $n$th root of $a$ is a number that satisfies the equation

$$x^n = a$$

*See also* discriminant.

**root-mean-square** (rms) For a number of values, an average equal to the square root of (the sum of the squares of the values divided by the number of values). For example, the rms of 2, 3, 4, 5 is $\sqrt{(54/4)} = 3.674$.

**roots and coefficients** Relations between the roots of a polynomial equation and the coefficients of that equation. If the roots of the quadratic equation $ax^2 + bx + c = 0$ are denoted by $\alpha$ and $\beta$ then one has the relations $\alpha + \beta = -b/a$ and $\alpha\beta = c/a$. If the roots of the cubic equation $ax^3 + bx^2 + cx + d = 0$ are written $\alpha$, $\beta$, and $\gamma$, then $b/a = -(\alpha + \beta + \gamma)$, $c/a = \alpha\beta + \beta\gamma + \gamma\alpha$, $d/a = -\alpha\beta\gamma$. If one uses the notation $\alpha + \beta + \gamma = \Sigma\alpha$ and $\alpha\beta + \beta\gamma + \gamma\alpha = \Sigma\alpha\beta$ these results can be summarized in the form:

$$\Sigma\alpha = -b/a, \Sigma\alpha\beta = c/a.$$

If the roots of the quartic equation $ax^4 + bx^3 + cx^2 + dx + e = 0$ are denoted by $\alpha$, $\beta$, $\gamma$ and $\delta$, then using the $\Sigma$ notation, one has the relations:

$$\Sigma\alpha = -b/a, \Sigma\alpha\beta = c/a,$$
$$\Sigma\alpha\beta\gamma = -d/a, \alpha\beta\gamma\delta = e/a.$$

If the roots of the quintic equation $ax^5 + bx^4 + cx^3 + dx^2 + ex + f = 0$ are written $\alpha$, $\beta$, $\gamma$, $\delta$, $\epsilon$, then $\Sigma\alpha = -b/a$, $\Sigma\alpha\beta = c/a$, $\Sigma\alpha\beta\gamma = -d/a$, $\Sigma\alpha\beta\gamma\delta = e/a$, $\alpha\beta\gamma\delta\epsilon = -f/a$. This pattern of relations between roots and coefficients extends to higher degree.

If the coefficients are all taken to be real numbers then these results mean that the sum of the roots and the product of the roots are both real numbers, even if the roots are complex numbers. This is the case because complex roots occur in pairs that are complex conjugates of each other. The solutions of cubic and higher power equations cannot, in general, be found by using the relationships between roots and coefficients.

**rose** A curve obtained by plotting the equation

$$r = a\sin n\theta$$

in polar coordinates ($a$ is a real-number constant and $n$ is an integer constant). It has a number of petal-shaped loops, or leafs. When $n$ is even there are $2n$ loops and when $n$ is odd there are $n$ loops. For example, the graph of $r = a\sin 2\theta$ is a four-leafed rose.

**rotation** A geometrical transformation in which a figure is moved rigidly around a fixed point. If the point, the center of rotation, is labelled O, then for any point P in the figure, moving to point P$'$ after rotation, the angle POP$'$ is the same for all points in the figure. This angle is the angle of rotation. Some figures are unchanged by certain rotations. A circle is not affected by any rotation about its center. A square does not change if it is rotated through 90° about the point at which its diagonals cross. Similarly an equilateral triangle is unchanged by rotation through 120° about its centroid. These properties are known as the *rotational symmetry* of the figure. *See also* rotation of axes; transformation.

**rotational motion** Motion of a body turning about an axis. The physical quantities and laws used to describe linear motion all have rotational analogs; the equations of rotational motion are the analogs of the equations of motion (linear).

As well as the kinematic equations, the equations of rotational motion include $T = I\alpha$, the analog of $F = ma$. Here $T$ is the turning-force, or torque (the analog of force), $I$ is the moment of inertia (analogous to mass), and $\alpha$ is the angular acceleration (analogous to linear acceleration).

The kinematic equations relate the angular velocity, $\omega_1$, of the object at the start of timing to its angular velocity, $\omega_2$, at some later time, $t$, and thus to the angular displacement $\theta$. They are:

$$\omega = \omega_1 + \alpha t$$
$$\theta = (\omega_1 + \omega_2)/2t$$
$$\theta = \omega_1 t + \alpha t^2/2$$
$$\theta = \omega_2 t - \alpha t^2/2$$
$$\omega_2^2 = \omega_1^2 + 2\alpha\theta$$

**rotation group** The group consisting of the set of all possible rotations about an axis. This group is a *continuous group*, i.e. it has an infinite number of members. The rotation group in two dimensions is an ABELIAN GROUP but the rotation group in three dimensions is a non-Abelian group. In physical systems the rotation group is closely associated with the angular momentum of the system. There are many applications of the rotation group in the quantum theory of atoms, molecules, and atomic nuclei.

**rotation of axes** In coordinate geometry, the shifting of the reference axes so that they are rotated with respect to the original axes of the system by an angle ($\theta$). If the new axes are $x'$ and $y'$ and the original axes $x$ and $y$, then the coordinates $(x,y)$ of a point with the original axes are related to the new coordinates $(x',y')$ by:

$$x = x'\cos\theta - y'\sin\theta$$
$$y = x'\sin\theta - y'\cos\theta$$

**rough** In mechanics, describing a system in which frictional effects have to be taken into consideration in the calculations.

**rounding** (**rounding off**) The process of adjusting the least significant digit or digits in a number after a required number of digits has been truncated (dropped). This reduces the error arising from truncation but still leaves a *rounding error* so that the accuracy of, say, the result of a calculation will be decreased. For example, the number 2.871 329 71 could be truncated to 2.871 32 but would be rounded to 2.871 33.

**routine** A sequence of instructions used in computer programming. It may be a short program or sometimes part of a program. *See also* subroutine.

**row matrix** *See* row vector.

**row vector** (**row matrix**) A number, $(n)$, of quantities arranged in a row; i.e. a $1 \times n$ matrix. For example, the coordinates of a point in a Cartesian coordinate system with three axes is a $1 \times 3$ row vector, $(x,y,z)$.

**Runge–Kutta method** /**rûng**-ĕ kût-ă/ An iterative technique for solving ordinary DIFFERENTIAL EQUATIONS, used in computer analysis. *See also* iteration. The method is named for the German mathematicians Carl Runge (1856–1927) and Martin Kutta (1867–1944).

**Russell's paradox** /**russ**-ĕlz/ A PARADOX at the foundations of set theory which was pointed out by the British philosopher Bertrand Russell (1872–1970) in 1901.

**saddle point** A stationary point on a curved surface, representing a function of two variables, $f(x,y)$, that is not a turning point, i.e. it is neither a maximum nor a minimum value of the function. At a saddle point, the partial derivatives $\partial f/\partial x$ and $\partial f/\partial y$ are both zero, but do not change sign. The tangent plane to the surface at the saddle point is horizontal. Around the saddle point the surface is partly above and partly below this tangent plane.

**sample space** *See* probability.

**sampling** The selection of a representative subset from a whole population. Analysis of the sample gives information about the whole population. This is called *statistical inference*. For example, population parameters (such as the population mean and variance) can be estimated using sample statistics (such as the sample mean and variance). Significance (or hypothesis) tests are used to test whether observed differences between two samples are due to chance variation or are significant, as in testing a new production process against an old. The population can be finite or infinite. In sampling with replacement, each individual chosen is returned to the population before the next choice is made. In *random sampling* every member of the population has an equal chance of being chosen. In *stratified random sampling* the population is divided into strata and the random samples drawn from each are pooled. In *systematic sampling* the population is ordered, the first individual chosen at random and further individuals chosen at specified intervals, for example, every 100th person on the electoral roll. If a random sample of size $n$ is the set of numerical values $\{x_1, x_2, \ldots x_n\}$, the sample mean is:

$$\sum_1^n \bar{x} = x_i/n$$

The sample variance is:
$$\Sigma(x_i - \bar{x})^2/(n - 1)$$
$$\Sigma(x_i - \bar{x})^2/n$$
for a normal distribution. If $\mu$ is the population mean, the sample variance is:
$$\Sigma(x_i - \mu)/n$$

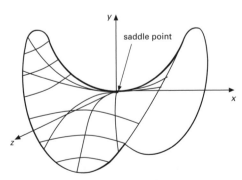

Saddle point on a surface

**sampling distribution** The distribution of a sample statistic. For example, when different samples of size $n$ are taken from the same population the means of each sample form a sampling distribution. If the population is infinite (or very large) and sampling is with replacement, the mean of the sample means is $\mu_x = \mu$ and the standard deviation of the sample means is $\sigma_x = \sigma/\sqrt{n}$, where $\mu$ and $\sigma$ are the population mean and standard deviation. When $n \geq 30$, sampling distributions are approximately normal and large-sampling theory is used. When $n < 30$, exact sample theory is used. *See also* sampling.

**satisfy** To be a solution of. For example, 3 and –3 as values of $x$ satisfy the equation $x^2 = 9$.

**scalar** /**skay**-ler/ A number or a measure in which direction is unimportant or meaningless. For instance, distance is a scalar quantity, whereas displacement is a vector. Mass, temperature, and time are scalars – they are each quoted as a pure number with a unit. *See also* vector.

**scalar product** A multiplication of two vectors to give a scalar. The scalar product of $A$ and $B$ is defined by $A.B = AB\cos\theta$, where $A$ and $B$ are the magnitudes of $A$ and $B$ and $\theta$ is the angle between the vectors. An example is a force $F$ displaced $s$. Here the scalar product is energy transferred (or work done):
$$W = F.s$$
$$W = Fs\cos\theta$$
where $\theta$ is the angle between the line of action of the force and the displacement. A scalar product is indicated by a dot between the vectors and is sometimes called a *dot product*. The scalar product is commutative
$$A.B = B.A$$
and is distributive with respect to vector addition
$$A.(B + C) = A.B + A.C$$
If $A$ is perpendicular to $B$, $A.B = 0$. In two-dimensional Cartesian coordinates with unit vectors $i$ and $j$ in the $x$- and $y$-directions respectively,
$$A.B = (a_1i + a_2j).(b_1i + b_2j) =$$

$$a_1b_1 + a_2b_2$$
*See also* vector product.

**scalar projection** The length of an orthogonal projection of one vector on another. For example, the scalar projection of $A$ on $B$ is $A\cos\theta$, or $(A.B)/b$ where $\theta$ is the smaller angle between $A$ and $B$ and $b$ is the unit vector in the direction of $B$. *Compare* vector projection.

**scale** 1. The markings on the axes of a graph, or on a measuring instrument, that correspond to values of a quantity. Each unit of length on a linear scale represents the same interval. For example, a thermometer that has markings 1 millimeter apart to represent 1°C temperature intervals has a linear scale. *See also* logarithmic scale.
2. The ratio of the length of a line between two points on a map to the distance represented. For example, a map in which two points 5 kilometers apart are shown 5 centimeters apart has a scale of 1/100 000.

**scale factor** The multiplying factor for each linear measurement of an object when it is to be enlarged about a given center of enlargement. A scale factor can be positive or negative. If the scale factor is positive, the image is on the same side of the center of enlargement as the object. If the scale factor is negative, the image will be on the opposite side of the center of enlargement and will be inverted. Fractional scale factors can be used – these give images that are smaller than the objects.

**scalene** /**skay**-leen/ Denoting a triangle with three unequal sides.

**scanner** A computer input device that produces a digital electronic version of an image (or text), so that it can be manipulated using the computer.

**scatter diagram** (Galton graph) A graphical representation of data from a bivariate distribution $(x,y)$. The variables are measured on $n$ individuals giving data $(x_2,y_1)$, ..., $(x_n,y_n)$; e.g. $x_i$ and $y_i$ are the height and weight of the $i^{th}$ individual. If $y_i$ is plotted

Scatter diagrams: the left diagram shows weak correlation; the right shows strong correlation.

against $x_i$ the resultant scatter diagram will indicate any relationship between $x$ and $y$ by showing if a smooth curve can be drawn through the points. If the points seem to lie near a straight line they are said to be linearly correlated. If they lie near another type of curve they are non-linearly correlated. Otherwise they are uncorrelated. *See also* regression line.

**Schwarz inequality** /shworts, shvarts/ An inequality that applies to both series and integrals. In the case of series the Schwarz inequality is frequently known as the CAUCHY INEQUALITY. For functions $f(x)$ and $g(x)$ the Schwarz inequality states that for definite integrals
$$\int f(x)g(x)dx \le \int f^2(x)dx \int g^2(x)dx$$
The case of equality holds if and only if $g(x) = cf(x)$, where $c$ is a constant, i.e. $g(x)$ is directly proportional to $f(x)$.

The Schwarz inequality can be deduced from the Cauchy inequality by taking the integrals as limits of series.

**scientific notation** (standard form) A number written as the product of a number between 1 and 10 with a power of 10. For example, 2342.6 in scientific notation is $2.342\ 6 \times 10^3$, and 0.0042 is written as $4.2 \times 10^{-3}$.

**screw 1.** A type of machine, related to the inclined plane, and, in practice, to the second-order lever. The efficiency of screw systems is very low because of friction.

Even so, the force ratio $(F_2/F_1)$ can be very high. The distance ratio is given by $2\pi r/p$, where $r$ is the radius and $p$ the pitch of the screw (the angle between the thread and a plane at right angles to the barrel of the screw).
**2.** A symmetry that can occur in a crystal, formed by a combination of a translation and a rotation. *See also* space group.

**scruple** A unit of mass equal to 20 grains. It is equivalent to 1.295 978 grams.

**sec** /sek/ *See* secant.

**secant** /see-kănt/ **1.** A line that intersects a given curve. The intercept is a chord of the curve.
**2.** (sec) A trigonometric function of an angle equal to the reciprocal of its cosine; i.e. $\sec\alpha = 1/\cos\alpha$. *See also* trigonometry.

**sech** /sech, shek, **sek**-aych/ A hyperbolic secant. *See* hyperbolic functions.

**second 1.** Symbol: s The SI unit of time. It is defined as the duration of 9 192 631 770 cycles of a particular wavelength of radiation corresponding to a transition between two hyperfine levels in the ground state of the cesium-133 atom.
**2.** A unit of plane angle equal to one sixtieth of a minute.

**second-order determinant** *See* determinant.

**second-order differential equation** A DIFFERENTIAL EQUATION in which the highest derivative of the dependent variable is a second derivative.

**section** *See* cross-section.

**sector** Part of a circle formed between two radii and the circumference. The area of a sector is $\frac{1}{2}r^2\theta$, where $r$ is the radius and $\theta$ is the angle, in radians, formed at the center of the circle between the two radii. A minor sector has $\theta$ less than $\pi$; a major sector has $\theta$ greater than $\pi$.

**segment** Part of a line or curve between two points, part of a plane figure cut off by a straight line, or part of a solid cut off by a plane. For example, on a graph, a line segment may show the values of a function within a certain range. The area between a chord of a circle and the corresponding arc is a segment of the circle. A cut through a cube parallel to one of its faces forms two cuboid segments.

**semantics** /si-**man**-tiks/ In computing or logic, the semantics of a language or system is the meaning of particular forms or expressions in the language or system. *Compare* syntax.

**semicircle** // Half a circle, bounded by a diameter and half of the circumference.

**semiconductor memory** *See* store.

**semigroup** /**sem**-ee-groop/ A set $G$ with a binary operation, •, that maps $G{\times}G$, the set of ordered pairs of members of $G$, into $G$, and that is associative, i.e. $a{\bullet}(b{\bullet}c) = (a{\bullet}b){\bullet}c$ for all $a$, $b$, and $c$ in $G$. A semigroup is *Abelian* or *commutative* if $a{\bullet}b = b{\bullet}a$ for all $a,b$ in $G$. Sometimes a cancellation law is assumed, i.e. $x = y$ if there is an element $z$ such that $xz = yz$ or $zx = zy$. *Compare* group.

**semilogarithmic graph** /sem-ee-lôg-ă-ri*th*-mik/ *See* log-linear graph.

**semi-regular polyhedron** A POLYHEDRON that is bounded by regular polygons but, unlike regular polyhedra (Platonic solids), there is more than one kind of polygon. Generally, there are two types of polygon. There are 13 semi-regular polyhedra. All of them can be inscribed inside a sphere. An example of a semi-regular polyhedron is the polyhedron with faces of 12 pentagons and 20 hexagons arranged like the panels on a soccer ball. Semi-regular polyhedra are also referred to as *Archimedean polyhedra* or *Archimedean solids*. *See also* tetrakaidekahedron.

**separation of variables** A method of solving ordinary differential equations. In a first-order DIFFERENTIAL EQUATION,
$$dy/dx = F(x,y)$$
if $F(x,y)$ can be written as $f(x),g(y)$, the variables in the function are separable and the equation can therefore be solved by writing it as
$$dy/g(y) = f(x)dx$$
and integrating both sides. For example
$$dy/dx = x^2y$$
can be written
$$(1/y)dy = x^2dx.$$

**sequence** (**progression**) An ordered set of numbers. Each term in a sequence can be written as an algebraic function of its position. For example, in the sequence (2, 4, 6, 8, …) the general expression for the $n$th term is $a_n = 2^n$. A *finite sequence* has a definite number of terms. An *infinite sequence* has an infinite number of terms. *Compare* series. *See also* arithmetic sequence; convergent sequence; divergent sequence; geometric sequence.

**serial access** *See* random access.

**series** The sum of an ordered set of numbers. Each term in the series can be written as an algebraic function of its position. For example, in the series $2 + 4 + 6 + 8 …$ the general expression for the $n$th term, $a_n$, is $2^n$. A *finite series* has a finite fixed number of terms. An *infinite series* has an infinite number of terms. A series with $m$ terms, or the sum of the first $m$ terms of an infinite series, can be written as $S_m$ or
$$\sum_{n=1}^{m} a_n$$

*Compare* sequence. *See also* arithmetic series; convergent series; divergent series; geometric series.

**set** Any collection of things or numbers that belong to a well-defined category. For example, 'dog' is a member, or *element*, of the set of 'types of four-legged animal'. The set of 'days in the week' has seven elements. In a set notation, this would be written as {Monday, Tuesday, ...}. This kind of set is a *finite set*. Some sets, such as the set of natural numbers, $N = \{1, 2, 3, ... \}$, have an infinite number of elements. A line segment also is an *infinite set* of points.

Another way of writing a set of numbers is by defining it algebraically. The set of all numbers between 0 and 10 could be written as $\{x: 0 < x < 10\}$. That is, all values of a variable, $x$, such that $x$ is greater than zero and less than ten. *See also* Venn diagram.

**set square** A drawing instrument, consisting of a flat right-angled triangular shape, used for drawing right angles and angles of 30°, 45°, and 60°. Some types have a moving part with a scale so that other angles can be drawn.

**sexagesimal** /sex-ă-**jess**-ă-măl/ Based on multiples of 60. The measurement of an angle in degrees, minutes, and seconds, for example, is a sexagesimal measure, because there are 60 seconds in one minute, and 60 minutes in one degree. A sexagesimal number is one that uses 60 as a base instead of 10. *See also* base.

**sextant** /sex-tănt/ A unit of plane angle equal to 60 degrees ($\pi/3$ radians).

**sextic** /sex-tik/ An equation of degree six, i.e.
$$ax^6 + bx^5 + cx^4 + dx^3 + ex^2 + fx + g = 0$$
There is no general method of solution for such an equation.

**shares** *See* stocks and shares.

**sheaf** A sheaf of planes is a set of planes that all pass through a given point, called the *center* of the sheaf. *Compare* pencil.

**shear** A TRANSFORMATION of a shape or body in which a line or plane of the shape or body is left unchanged. The unchanged line and plane are called the *invariant line* and *invariant plane* respectively. Physically, shear of a body can occur if a force is applied to the body in which the force either lies in the plane of an area of a surface of the body or a plane which is parallel to such a surface.

**s.h.m.** *See* simple harmonic motion.

**short-wavelength approximation** An approximate technique for solving differential equations or evaluating integrals involving waves in which the starting point is the solution for the problem in the limit of the wavelength going to zero and the approximate solution of the problem for nonzero wavelengths is expressed as an ASYMPTOTIC SERIES. For example, this technique can be used to study problems in wave optics, with geometrical optics, i.e. the zero-wavelength limit for light and other electromagnetic waves, being the starting point. Similarly, many problems in quantum mechanics in which the wave aspect of particles such as electrons is important can be studied using this method, with the zero wavelength limit of quantum mechanics, i.e. classical mechanics, being the starting point.

**SI** *See* SI units.

**side** One of the line segments that forms the boundary of a polygon (a many-sided plane figure). For example, a triangle has three sides, a pentagon has five.

**siemens** /see-měnz/ (mho) Symbol: S The SI unit of electrical conductance, equal to a conductance of one ohm$^{-1}$. The unit is named for the German electrical engineer Ernst Werner von Siemens (1816–92).

**sieve of Eratosthenes** *See* Eratosthenes; sieve of.

**sigma** /sig-mă/ The Greek letter Σ (upper case) or σ (lower case). In mathematics, upper-case sigma (Σ) is used to denote summation; lower-case sigma ($\sigma$) often denotes the standard deviation in statistics.

**sign** A unit of plane angle equal to 30 degrees ($\pi/6$ radian).

**signed number** A number that is denoted as positive or negative.

**significance test** *See* hypothesis test.

**significant figures** The number of digits used to denote an exact value to a specified degree of accuracy. For example, 6084.324 is a value accurate to seven significant figures. If it is written as approximately 6080, it is accurate to three significant figures The final 0 is not significant because it is used only to show the order of magnitude of the number.

**similar** Denoting two or more figures that differ in scale but not in shape. The conditions for two triangles to be similar are:
1. Three sides of one are proportional to three sides of the other.
2. The angle of one is equal to the angle of the other and the sides forming the angle in one are proportional to the same sides in the other.
3. Three angles of one are equal to three angles of the other.
*Compare* congruent.

**simple harmonic motion** (s.h.m.) Any motion that can be drawn as a sine wave. Examples are the simple oscillation (vibration) of a pendulum (with small amplitude) or a sound source and the variation involved in a simple wave motion. Simple harmonic motion is observed when the system, moved away from the central position, experiences a restoring force that is proportional to the displacement from this position.

The equation of motion for such a system can be written in the form:
$$m\,d^2x/dt^2 = \lambda x$$
$\lambda$ being a constant. During the motion there is an exchange of kinetic and potential energy, the sum of the two being constant (in the absence of damping). The period ($T$) is given by

$$T = 1/f$$
$$T = 2\pi/\omega$$
where $f$ is frequency and $\omega$ pulsatance. Other relationships are:
$$x = x_0\sin\omega t$$
$$dx/dt = \pm\omega\surd(x_0^2 - x^2)$$
$$d^2x/dt^2 = -\omega^2 x$$
Here $x_0$ is the maximum displacement; i.e. the amplitude of the vibration. In the case of angular motion, as for a pendulum, $\theta$ is used rather than $x$.

A simple harmonic motion can be represented by the motion of a point at constant speed in a circular path. The foot of the perpendicular from the point to an axis through a diameter describes a simple harmonic motion. This is used in a method of representing simple harmonic motions by rotating vectors (called *phasors*).

**simple interest** The interest earned on capital when the interest is withdrawn as it is paid, so that the capital remains fixed. If the amount of money invested (the principal) is denoted by $P$, the time in years by $T$, and the percentage rate per annum by $R$, then the simple interest is $PRT/100$. *Compare* compound interest.

**simple proposition** *See* proposition.

**simplify** To use mathematical and algebraic techniques to contract an expression. For example, simplifying the following expressions:
$$115/184 = 5/8$$
$$2x - 7x + 9x = 4x$$
$$15x^3y^2z/5x^2yz = 3xy$$

**simply connected** Describing a region for which any closed curve lying in the region can be continuously deformed into a single point without leaving the region, i.e. the region does not have any 'holes' in it. If a region is not simply connected it is said to be *multiply connected*, i.e. the region has one or more 'hole'. A *doubly-connected region* has one hole, etc. In general, an *n-tuply-connected region* has ($n$–1) holes.

The concepts of simply connected and multiply connected regions are important in topology and the theory of functions of a complex variable. The concepts of simply

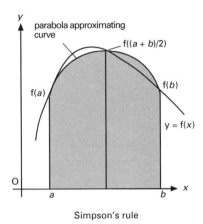

Simpson's rule

and multiply connected regions also have important applications in many branches of physics.

**Simpson's rule** A rule for finding the approximate area under a curve by dividing it into pairs of vertical columns of equal width with bases lying along the horizontal axis. Each pair of columns is bounded by the vertical lines from the $x$-axis to three corresponding points on the curve and at the top by a parabola that goes through these three points, which approximates the curve. For example, if the value of $f(x)$ is known at $x = a$, $x = b$, and at a value midway between $a$ and $b$ the integral of $f(x)dx$ between limits $a$ and $b$ is approximately equal to:
$$h\{f(a) + 4f[(a+b)/2] + f(b)\}/3$$
$h$ is half the distance between $a$ and $b$. As with the trapezium rule, which is less accurate, better approximations can be obtained by subdividing the area into 4, 6, 8, … columns until further subdivision makes no significant difference to the result. The rule is named for the British mathematician Thomas Simpson (1710–61). *Compare* trapezium rule. *See also* numerical integration.

**simultaneous equations** A set of two or more equations that together specify conditions for two or more variables. If the number of unknown variables is the same as the number of equations, then there is a unique value for each variable that satisfies all the equations. For example, the equations
$$x + 2y = 6$$
and
$$3x + 4y = 9$$
have the solution $x = -3$; $y = -4.5$. The method of solving simultaneous equations is to eliminate one of the variables by adding or subtracting the equations. For example, multiplying the first equation above by 2 and subtracting it from the second gives:
$$3x + 4y - 2x - 4y = 9 - 12$$
i.e. $x = -3$. Substituting this into either equation gives the value of $y$. Simultaneous equations can also be solved graphically. On a Cartesian coordinate graph, each equation would be shown as a straight line and the point at which the two cross is, in this case, $(-3, -4.5)$. *See also* substitution; inverse of a matrix.

**sin** /sÿn/ *See* sine.

**sine** /sÿn/ (**sin**) A trigonometric function of an angle. The sine of an angle $\alpha$ (sin$\alpha$) in a right-angled triangle is the ratio of the side opposite the angle to the hypotenuse. This definition applies only to angles between $0°$ and $90°$ (0 and $\pi/2$ radian).

More generally, in rectangular Cartesian coordinates, the $y$-coordinate of any point on the circumference of a circle of radius $r$ centred at the origin is $r$sin$\alpha$, where

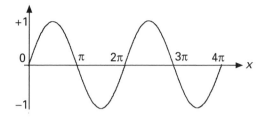

Sine curve: a graph of $y = \sin x$, with $x$ in radians.

$\alpha$ is the angle between the $x$-axis and the radius to that point. In other words, the sine function depends on the vertical component of a point on a circle. Sin$\alpha$ is zero when $\alpha$ is $0°$, rises to 1 when $\alpha = 90°$ ($\pi/2$) falls again to zero when $\alpha = 180°$ ($\pi$), becomes negative and reaches $-1$ at $\alpha = 270°$ ($3\pi/2$), and then returns to zero at $\alpha = 360°$ ($2\pi$). This cycle is repeated every complete revolution. The sine function has the following properties:

$$\sin\alpha = \sin(\alpha + 360°)$$
$$\sin\alpha = -\sin(180° + \alpha)$$
$$\sin(90° - \alpha) = \sin(90° + \alpha)$$

The sine function can also be defined as an infinite series. In the range between 1 and $-1$:

$$\sin x = x/1! - x^3/3! + x^5/5! - \ldots$$

*See also* trigonometry.

**sine rule** In any triangle, the ratio of the side length to the sine of the angle opposite that side is the same for all three sides. Thus, in a triangle with sides of lengths $a$, $b$, and $c$ and angles $\alpha$, $\beta$, and $\gamma$ ($\alpha$ opposite $a$, $\beta$ opposite $b$, and $\gamma$ opposite $c$):

$$a/\sin\alpha = b/\sin\beta = c/\sin\gamma$$

**sine wave** The waveform resulting from plotting the sine of an angle against the angle. Any motion for which distance plotted against time gives a sine wave is a simple harmonic motion.

**singular matrix** A square matrix that has a determinant equal to zero and that has no inverse matrix. *See also* determinant.

**singular point** A point on a curve $y = f(x)$ at which the derivative $dy/dx$ has the indeterminate form $0/0$. The singular points on a curve are found by writing the derivative in the form

$$dy/dx = g(x)/h(x)$$

$$\begin{pmatrix} 2 & 1 \\ 4 & 2 \end{pmatrix}$$

$$\begin{vmatrix} 2 & 1 \\ 4 & 2 \end{vmatrix} = (2 \times 2) - (4 \times 1) = 0$$

Singular 2 x 2 matrix

and then finding the values of $x$ for which $g(x)$ and $h(x)$ are both zero.

**sinh** /shÿn, sinsh, sÿn-**aych**/ A hyperbolic sine. *See* hyperbolic functions.

**sinusoidal** /sÿ-ŭ-**soid**-ăl/ Describing a quantity that has a waveform that is a sine wave.

**SI units** (Système International d'Unités) The internationally adopted system of units used for scientific purposes. It is comprised of seven base units (the meter, kilogram, second, kelvin, ampere, mole, and candela) and two supplementary units (the radian and steradian). Derived units are formed by multiplication and/or division of base units; a number have special names. Standard prefixes are used for multiples and submultiples of SI units. The SI system is a coherent rationalized system of units.

**skew lines** Lines in space that are not parallel and do not intersect. Whether a pair of lines in space are skew lines rather than the other two possibilities for lines in space, i.e. parallel lines and lines that are not parallel and intersect, can be determined from the equations for the lines. If the direction ratios of the two lines are equal then the two lines are parallel. If the equations of the two lines are: $(x-x_1)/a_1 = (y-y_1)/b_1 = (z-z_1)/c_1 = m$, and $(x-x_1)/a_2 = (y-y_2)/b_2 = (z-z_2)/c_2 = n$, where $a_1$, $b_1$, $c_1$, $m$, $a_2$, $b_2$, $c_2$ and $n$ are constants, then these two lines can only intersect if these are unique values of $m$ and $n$ such that $x = x_3$, $y = y_3$, $z = z_3$ satisfy the equations for both lines. If such unique values of $m$ and $n$ cannot be found the lines are skew lines.

**skewness** A measure of the degree of asymmetry of a distribution. If the frequency curve has a long tail to the right and a short one to the left, it is called skewed to the right or positively skewed. If the opposite is true, the curve is skewed to the left, or negatively skewed. Skewness is measured by either Pearson's first measure of skewness (mean minus mode) divided by standard deviation, or the equivalent Pearson's second measure of skewness divided by standard deviation.

**skew-symmetric matrix** A matrix $A$ which has its transpose equal to $-A$, i.e. if the entries of $A$ are denoted by $a_{ij}$ then $a_{ij} = -a_{ji}$ for all $i$ and $j$ in the matrix. It follows from the definition of a skew-symmetric matrix that the entries in the main diagonal of the matrix, i.e. all entries of the form $a_{ii}$, have to be zero.

**slant height** 1. The length of an element of a right cone; i.e. the distance from the vertex to the directrix.
2. The altitude of the faces of a right pyramid.

**slide rule** A calculating device on which sliding logarithmic scales are used to multiply and divide numbers. Most slide rules also have fixed scales showing squares, cubes, and trigonometric functions. The accuracy of the slide rule is usually to three significant figures. Slide rules have generally been replaced by electronic calculators.

**sliding friction** *See* friction.

**slope** *See* gradient.

**small circle** A circle, drawn on the surface of a sphere, whose centre is not at the centre of the sphere. *Compare* great circle.

**smooth** In mechanics, describing a system in which friction can be neglected in the calculations.

**snowflake curve** *See* fractal.

**software** The PROGRAMS that can be run on a computer, together with any associated written documentation. A *software package* is a professionally written program or group of programs that is designed to perform some commonly required task, such as statistical analysis or graph plotting. The availability of software packages means that common tasks need not be programmed over and over again. *Compare* hardware. *See also* program.

**solenoidal vector** /soh-lĕ-**noi**-dăl/ A vector $V$ for which $\text{div}V = 0$. It is possible to write a solenoidal vector in the form $V = \text{curl}A$, where $A$ is a vector called the *vector potential*. If it is known that $V$ is a solenoidal vector it is possible to construct an infinite number of $A$'s that satisfy $V = \text{curl}A$. The concept of a solenoidal vector is used extensively in the theory of electromagnetism.

**solid** A three-dimensional shape or object, such as a sphere or a cube.

**solid angle** Symbol: $\Omega$ The three-dimensional analog of angle; the region subtended at a point by a surface (rather than by a line). The unit is the steradian (sr), which is defined analogously to the radian – the solid angle subtending unit area at

## BASE AND DIMENSIONLESS SI UNITS

| Physical quantity | Name of SI unit | Symbol for SI unit |
|---|---|---|
| length | meter | m |
| mass | kilogram(me) | kg |
| time | second | s |
| electric current | ampere | A |
| thermodynamic temperature | kelvin | K |
| luminous intensity | candela | cd |
| amount of substance | mole | mol |
| *plane angle | radian | rad |
| *solid angle | steradian | sr |
| *supplementary units | | |

## DERIVED SI UNITS WITH SPECIAL NAMES

| Physical quantity | Name of SI unit | Symbol for SI unit |
|---|---|---|
| frequency | hertz | Hz |
| energy | joule | J |
| force | newton | N |
| power | watt | W |
| pressure | pascal | Pa |
| electric charge | coulomb | C |
| electric potential difference | volt | V |
| electric resistance | ohm | Ω |
| electric conductance | siemens | S |
| electric capacitance | farad | F |
| magnetic flux | weber | Wb |
| inductance | henry | H |
| magnetic flux density | tesla | T |
| luminous flux | lumen | lm |
| illuminance (illumination) | lux | lx |
| absorbed dose | gray | Gy |
| activity | becquerel | Bq |
| dose equivalent | sievert | Sv |

## DECIMAL MULTIPLES AND SUBMULTIPLES USED WITH SI UNITS

| Submultiple | Prefix | Symbol | Multiple | Prefix | Symbol |
|---|---|---|---|---|---|
| $10^{-1}$ | deci- | d | $10^{1}$ | deca- | da |
| $10^{-2}$ | centi- | c | $10^{2}$ | hecto- | h |
| $10^{-3}$ | milli- | m | $10^{3}$ | kilo- | k |
| $10^{-6}$ | micro- | μ | $10^{6}$ | mega- | M |
| $10^{-9}$ | nano- | n | $10^{9}$ | giga- | G |
| $10^{-12}$ | pico- | p | $10^{12}$ | tera- | T |
| $10^{-15}$ | femto- | f | $10^{15}$ | peta- | P |
| $10^{-18}$ | atto- | a | $10^{18}$ | exa- | E |
| $10^{-21}$ | zepto- | z | $10^{21}$ | zetta- | Z |
| $10^{-24}$ | yocto- | y | $10^{24}$ | yotta- | Y |

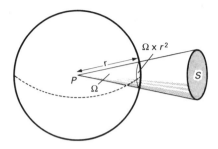

Solid angle: the surface $S$ subtends a solid angle $\Omega$ in steradians at point $P$. An area that forms part of the surface of a sphere of radius $r$, center $P$, and subtends the same solid angle $\Omega$ at $P$ is equal to $\Omega r^2$.

unit distance. As the area of a sphere is $4\pi r^2$, the solid angle at its center is $4\pi$ steradians. *See illustration overleaf.*

**solid geometry** The study of geometric figures in three dimensions.

**solid of revolution** A solid figure that can be produced by revolution of a line or curve (the *generator*) about a fixed axis. For instance, rotating a circle about a diameter generates a sphere. Rotating a circle about an axis that does not cut the circle generates a torus.

**solid-state memory** *See* store.

**soliton** /sol-ă-ton/ A type of kink solution that can occur as a solution to certain nonlinear equations for propagation. A characteristic feature of solitons is that they are very stable and can pass through each other unchanged. Since solitons are very atypical of solutions to nonlinear equations there has to be some mathematical structure in the nonlinear equation to permit their existence, and they are frequently associated with the TOPOLOGY of a system. There are many physical applications of solitons.

**solution** A value of a variable that satisfies an algebraic equation. For example, the solution of $2x + 4 = 12$ is $x = 4$. An equation may have more than one solution; for example, $x^2 = 16$ has two; $x = -4$ and $x = +4$.

**solution of triangles** Calculating the unknown sides and angles in triangles. Because the sum of angles in a triangle is always 180°, the third angle can be calculated if two are known. All the sides and angles can be calculated when two sides and the angle between them are known, but if two sides and another angle are known there are two possible solutions. Any two angles and one side are sufficient to solve a triangle. *See also* trigonometry.

**source language** (**source program**) *See* program.

**space curve** A curved line in a volume, defined in three-dimensional Cartesian coordinates by three functions:
$$x = f(t)$$
$$y = g(t)$$
$$z = h(t)$$
or by two equations of the form:
$$F(x,y,z) = 0$$
$$G(x,y,z) = 0$$

**space group** The set of all symmetry operations of a crystal, i.e. the possible rotations, reflections, and translations. There are 230 space groups in three dimensions. Space groups can also be analyzed in two dimensions (ornamental or 'wallpaper' symmetry) and in higher dimensions.

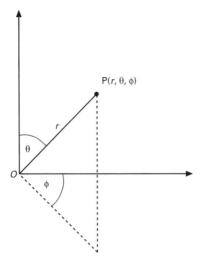

Spherical polar coordinates of point P are $(r, \theta, \phi)$.

Of the 230 space groups 73 of these groups consist only of point group operations and the translations of the crystal lattice. Such space groups are called *symmorphic*. The remaining 157 space groups consist of point-group operations and translations which are not individually in the space group. Those space groups contain GLIDE planes and SCREW axes, i.e. symmetry operations involving a translation which is not a translation of the crystal. Such space groups are said to be *non-symmorphic*.

The concept of a space group can be extended from crystals to PENROSE PATTERNS and QUASICRYSTALLINE SYMMETRY. One way in which this can be done is by projecting down to the Penrose pattern or quasicrystal from the space group of a crystal in a higher dimension.

**space–time** In Newtonian (pre-relativity) physics, space and time are separate and absolute quantities; that is they are the same for all observers in any FRAME OF REFERENCE. An event seen in one frame is also seen in the same place and at the same time by another observer in a different frame.

After Einstein had proposed his theory of relativity, Minkowski suggested that since space and time could no longer be re-garded as separate continua, they should be replaced by a single continuum of four dimensions, called space–time. In space–time the history of an object's motion in the course of time is represented by a line called the *world curve*. *See also* relativity, theory of.

**Spearman's method** /**speer**-mănz/ A way of measuring the degree of association between two rankings of $n$ objects using two different variables $x$ and $y$ which give data $(x_1, y_1), ..., (x_n, y_n)$. The objects are ranked using first the $x$'s and then the $y$'s, and the difference, $D$, between the RANKS calculated for each object.

Spearman's coefficient of rank correlation

$$\rho = 1 - (6\Sigma D^2/[n(n^2 - 1)])$$

The method is named for the British behavioral scientist Charles Spearman (1863–1945).

**special theory** *See* relativity.

**spectrum** /**spek**-trŭm/ (*pl.* **spectra**) The spectrum of an operator $A$ is the set of all complex numbers $\lambda$ such that the operator $A - \lambda I$, where $I$ is the identity operator, i.e. $I(x) \equiv x$, does not have an inverse. For example, if $A$ is a matrix, its spectrum is the set of its eigenvalues since if $\lambda$ is an eigenvalue of $A$, $\det(A - \lambda I) = 0$ and $A - \lambda I$ is therefore not invertible.

**speed** Symbol: $c$ Distance moved per unit time: $c = d/t$. Speed is a scalar quantity; the equivalent vector is velocity – a vector quantity equal to displacement per unit time. Usage can be confusing and it is common to meet the word 'velocity' where 'speed' is more correct. For instance $c_0$ is the speed of light in free space, not its velocity.

**sphere** A closed surface consisting of the locus of points in space that are at a fixed distance, the radius $r$, from a fixed point, the center. A sphere is generated when a circle is turned through a complete revolution about an axis that is one of its diameters. The plane cross-sections of a sphere are all circles. The sphere is symmetrical about any plane that passes through its center and the two mirror-image shapes on each side are called *hemispheres*. In Cartesian coordinates, the equation of a sphere of radius $r$ with its center at the origin is
$$x^2 + y^2 + z^2 = r^2$$
The volume of a sphere is $4\pi r^3/3$ and its surface area is $4\pi r^2$.

**spherical harmonic** A type of function that frequently occurs as a solution of differential equations for systems that have spherical symmetry. Spherical harmonics can be regarded as the analog for the sphere of CIRCULAR FUNCTIONS for the circle. They occur in the solutions of the equations of quantum mechanics for electrons in atoms.

**spherical polar coordinates** A method of defining the position of a point in space by its radial distance, $r$, from a fixed point, the origin O, and its angular position on the surface of a sphere centred at O. The angular position is given by two angles $\theta$ and $\phi$. $\theta$ is the angle that the radius vector makes with a vertical axis through O (from the south pole to the north pole). It is called the *co-latitude*. For points on the vertical axis above O, $\theta = 0$. For points lying in the 'equatorial' horizontal plane, $\theta = 90°$. For points on the vertical axis below O, $\theta = 180°$. $\phi$ is the angle that the radius vector makes with an axis in the equatorial plane. It is called the *azimuth*. For all points lying in the axial plane, that is, on, vertically above, or vertically below this axis, $\phi = 0$ on the positive side of O and $\phi = 180°$ on the negative side. This plane corresponds to the plane $y = 0$ in rectangular CARTESIAN COORDINATES. For points in the vertical plane at 90° to this, ($x = 0$ in rectangular Cartesian coordinates) $\phi = 90°$ in the positive half-plane and 270° in the negative half-plane. For a point P($r,\theta,\phi$), the corresponding rectangular Cartesian coordinates $(x,y,z)$ are:
$$x = r\cos\phi\sin\theta$$
$$y = r\sin\phi\sin\theta$$
$$z = r\cos\theta$$
*Compare* cylindrical polar coordinates. *See also* polar coordinates.

**spherical sector** The solid generated by rotating a sector of a circle about a diameter of the circle. The volume of a spherical sector generated by a sector of altitude $h$ (parallel to the axis of rotation) in a circle of radius $r$ is
$$(\tfrac{2}{3})\pi r^2 h$$

**spherical segment** A solid formed by cutting in one or two parallel planes through a sphere. The volume of a spherical segment bounded by circular plane cross-sections of radii $r_1$ and $r_2$ a distance $h$ apart, is:
$$\pi h(3r_1^2 + 3r_2^2 + h^2)/6$$
If the segment is bounded by one one plane of radius $r$ and the curved surface of the sphere, then the volume is:
$$\pi h(3r^2 + h^2)/6$$

**spherical triangle** A three-sided figure on the surface of a sphere, bounded by three great circles. A *right spherical triangle* has at least one right angle. A *birectangular* spherical triangle has two right angles and a *trirectangular* spherical triangle has three right angles. If one of the sides of a spherical triangle subtends an angle of 90° at the center of the sphere, then it is called a *quadrantal spherical triangle*. An *oblique spherical triangle* has no right angles.

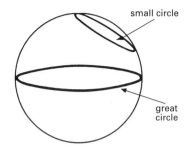

great and small circles on a sphere

spherical wedge

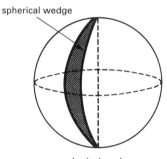

spherical triangle

spherical triangles

Spherical trigonometry

**spherical trigonometry** The study and solution of spherical triangles.

**spheroid** /**sfeer**-oid/ A body or curved surface that is similar to a sphere but is lengthened or shortened in one direction. *See* ellipsoid.

**spiral** A plane curve formed by a point winding about a fixed point at an increasing distance from it. There are many kinds of spirals, e.g. the *Archimedean spiral* is given by $r = a\theta$, the *logarithmic spiral* is given by $\log(r) = a\theta$, and the *hyperbolic spiral* is given by $r\theta = a$, where in each case $a$ is a constant, and $r$ and $\theta$ are polar coordinates.

**spline** A smooth curve passing through a fixed number of points. Splines are used in computer graphics.

**spur** *See* square matrix.

**square** 1. The second power of a number or variable. The square of $x$ is $x \times x = x^2$ ($x$ squared). The square of the square root of a number is equal to that number.
2. In geometry, a plane figure with four equal straight sides and right angles between the sides. Its area is the length of one of the sides squared. A square has four axes of symmetry – the two diagonals, which are of equal length and bisect each other perpendicularly, and the two lines joining the mid-points of opposite sides. It can be superimposed on itself after rotation through 90°.

**square matrix** A matrix that has the same number of rows and columns, that is, a square array of numbers. The diagonal from the top left to the bottom right of a square matrix is called the *leading diagonal* (or *principal diagonal*). The sum of the elements in this diagonal is called the *trace* (or *spur*) of the matrix.

$$\begin{pmatrix} 1 & 2 & 4 \\ 3 & 5 & 11 \\ 4 & 8 & 9 \end{pmatrix}$$

Square 3 x 3 matrix: the trace is 1 + 5 + 9 = 15.

**square pyramid** *See* pyramid.

**square root** For any given number, another number that when multiplied by itself equals the given number. It is denoted by the symbol $\sqrt{}$ or the index (power) ½. For example, the square root of 25, written $\sqrt{25} = 5$.

**squaring the circle** The attempt to construct a square that has the same area as a particular circle, using a ruler and compasses. An exact solution is impossible because there is no exact length for the edge, which is a multiple of the transcendental number $\sqrt{\pi}$.

**stability** A measure of how hard it is to displace an object or system from equilibrium.

Three cases are met in statics differing in the effect on the center of mass of a small displacement. They are:
1. *Stable equilibrium* – the system returns to its original state when the displacing force is removed.
2. *Unstable equilibrium* – the system moves away from the original state when displaced a small distance.
3. *Metastable or neutral equilibrium* – when displaced a small distance, the system is at equilibrium in its new position.
An object's stability is improved by: (a) lowering the center of mass; or (b) increasing the area of support; or by both.

**stable equilibrium** *See* stability.

**standard** Established as a reference.
1. Writing an equation in standard form enables comparison with other equations of the same type. For example,
$$x^2/4^2 + y^2/2^2 = 1$$
and
$$x^2/3^2 + y^2/5^2 = 1$$
are equations of hyperbolas in rectangular Cartesian coordinates, both written in standard form.

**2.** A standard measuring instrument is one against which other instruments are calibrated.

**3.** Standard form of a number. *See* scientific notation.

**standard deviation** A measure of the dispersion of a statistical sample, equal to the square root of the variance. In a sample of $n$ observations, $x_1, x_2, x_3, \ldots x_n$, the *sample standard deviation* is:

$$s = \sqrt{\left[ \sum_{1}^{n} (x_i - \bar{x})^2 / (n-1) \right]}$$

where $\bar{x}$ is the sample mean. If the mean $\mu$ of the whole population from which the sample is taken is known, then

$$s = \sqrt{\left[ \sum_{1}^{n} (x_i - \mu)^2 / n \right]}$$

**standard form** *See* scientific notation.

**standard form of a number** *See* scientific notation.

**standard pressure** An internationally agreed value; a barometric height of 760 mmHg at 0°C; 101 325 Pa (approximately 100 kPa).

This is sometimes called the *atmosphere* (used as a unit of pressure). The *bar*, used

mainly when discussing the weather, is 100 kPa exactly. *See also* STP.

**standard temperature** An internationally agreed value for which many measurements are quoted. It is the melting temperature of water, 0°C (273.15 K). *See also* STP.

**standing wave** *See* stationary wave.

**static friction** *See* friction.

**static pressure** The pressure on a surface due to a second solid surface or to a fluid that is not flowing.

**statics** A branch of MECHANICS dealing with the forces on an object or in a system in equilibrium. In such cases there is no resultant force or torque and therefore no resultant acceleration.

**stationary point** A point on a curved line at which the slope of the tangent to the curve is zero. All turning points (maximum points and minimum points) are stationary points. In this case the slope of the tangent passes through zero and changes its sign. Some stationary points are not turning points. In such cases, the curve levels out and then continues to increase or decrease

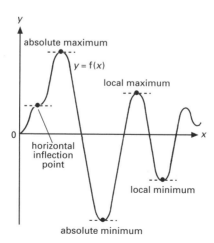

Five kinds of stationary points for a function

as before. At a stationary point the derivative $dy/dx$ of $y = f(x)$ vanishes (is zero). At a maximum, the second derivative, $d^2y/dx^2$, is negative; at a minimum it is positive. At a horizontal inflection point the second derivative is zero. Not all inflection points have $dy/dx = 0$; i.e. not all are stationary points.

At a stationary point on a curved surface, representing a function of two variables, $f(x,y)$, the partial derivatives $\partial f/\partial x$ and $\partial f/\partial y$ are both zero. This may be a maximum point, a minimum point, or a SADDLE POINT.

**stationary wave** (**standing wave**) The interference effect resulting from two waves of the same type moving with the same frequency through the same region. The effect is most often caused when a wave is reflected back along its own path. The resulting interference pattern is a stationary wave pattern. Here, some points always show maximum amplitude; others show minimum amplitude. They are called *antinodes* and *nodes* respectively. The distance between neighboring node and antinode is a quarter of a wavelength.

**statistical inference** *See* sampling.

**statistical mechanics** The evaluation of data that applies to a large number of entities by the use of statistics. It finds its main applications in chemistry and physics.

**statistics** The methods of planning experiments, obtaining data, analyzing it, and drawing conclusions from it, and making decisions on the basis of the analysis. In statistical inference, conclusions about a population are inferred from analysis of a sample. In descriptive statistics, data is summarized but no inferences are made.

**step function** A function $\theta(x)$ defined by: $\theta(x) = 1$, for $x > 0$ and $\theta(x) = 0$, for $x < 0$. In physical problems the variable $x$ is sometimes the time $t$, with the step function representing a sudden 'turning on' of some effect at $t = 0$. The step function is also sometimes known as *Heaviside unit function*. The derivative of the step func-

tion can be identified with the DIRAC DELTA FUNCTION.

**steradian** /sti-**ray**-dee-ăn/ Symbol: sr The SI unit of solid angle. The surface of a sphere, for example, subtends a solid angle of $4\pi$ at its center. The solid angle of a cone is the area intercepted by the cone on the surface of a sphere of unit radius.

**stereographic projection** /ste-ree-oh-**graf**-ik, steer-ee-/ A geometrical transformation of a sphere onto a plane. A point is taken on the surface of the sphere – the *pole* of the PROJECTION. The projection of points on the sphere onto a plane is obtained by taking straight lines from the pole through the points, and continuing them to the plane. The plane taken does not pass through the pole and is perpendicular to the diameter of the sphere through the pole.

**Stirling approximation** /**ster**-ling/ An approximate expression for the value of the factorial of a number or the GAMMA FUNCTION. The expression for $n!$ is $n! \sim n^n \exp(-n)\sqrt{(2\pi n)}$, with the expression for $\Gamma(n + 1)$ being $\Gamma(n + 1) \sim n^n \exp(-n)\sqrt{(2\pi n)}$, where $\sim$ denotes 'is asymptotic to'. This means that the ratio $n!/[n^n \exp(-n)\sqrt{(2\pi n)}]$ tends to 1 as $n\rightarrow\infty$, i.e. Stirling's approximation becomes a better approximation as $n$ increases, with the ratio of the error to the value of $n!$ tending to zero as $n\rightarrow\infty$. The approximation is named for the British mathematician James Stirling (1692–1770).

**stochastic process** /stoh-**kas**-tik/ A process that generates a series of random values of a variable and builds up a particular statistical distribution from these. For example, the POISSON DISTRIBUTION can be built up by a stochastic process that starts with values taken from tables of random numbers.

**Stokes' theorem** A result in vector calculus that states that the surface integral of the curl of a vector function is equal to the line integral of that vector function around a closed curve. If the vector function is denoted $V$, Stokes' theorem can be written in the form:

$$\int_S \text{curl}\mathbf{v}\bullet d\mathbf{S} - \oint \mathbf{v}\bullet d\mathbf{l}$$

where $S$ represents surface and $l$ represents line. Stokes' theorem has many physical applications, particularly to the theory of electricity and magnetism. There are extensions of Stokes' theorem to higher dimensions. The theorem is named for the British mathematician Sir George Gabriel Stokes (1819–1903).

**store** (**memory**) A system or device used in computing to hold information (programs and data) in such a way that any piece of information can automatically be retrieved by the computer as required. The *main store* (or *internal store*) of a computer is under the direct control of the central processor. It is the area in which programs, or parts of programs, are stored while they are being run on the computer. Data and program instructions can be extracted extremely rapidly by random access. The main store is supplemented by *backing store*, in which information can be permanently stored. The two basic forms of backing store are those in which magnetic tape is used (i.e. magnetic tape units) and those in which disks, or some other random-access device are used.

The main store is divided into a huge number of *locations*, each able to hold one unit of information, i.e. a word or a byte. The number of locations, i.e. the number of words or bytes that can be stored, gives the *capacity* of the store. Each location is identified by a serial number, known as its *address*.

There are many different ways in which memory can be classified. *Random-access memory* (RAM) and *serial-access memory* differ in the manner in which information is extracted from a store. With *volatile memory*, stored information is lost when the power supply is switched off, unlike *nonvolatile memory*. With *read-only memory* (ROM), information is stored permanently or semipermanently; it cannot be altered by programmed instructions but

can in some types be changed by special techniques.

Stores may be magnetic or electronic in character. The electronic memory now widely used in main store consists of highly complex integrated circuits. This *semiconductor memory* (or *solid-state memory*) stores an immense amount of information in a very small space; items of information can be extracted at very high speed.

**STP** (NTP) Standard temperature and pressure. Conditions used internationally when measuring quantities that vary with both pressure and temperature (such as the density of a gas). The values are 101 325 Pa (approximately 100 kPa) and 0°C (273.15 K). *See also* standard pressure; standard temperature.

**straight line** The curve that gives the shortest distance between two points in Euclidean space.

In two dimensions, i.e. a plane, a straight line can be represented using Cartesian coordinates by a linear equation of the form $ax + by + c = 0$, where $a$, $b$, and $c$ are constants and $a$ and $b$ cannot both be zero. The equation for a straight line in two dimensions can be written in several different forms. If the line has a gradient $m$ and cuts the $y$-axis at the intercept point $(0,c)$ the equation for the line can be written as $y = mx + c$. If the line has a gradient $m$ and passes through the point $(x_1, y_1)$ then its equation can be written in the form $y - y_1 = m(x - x_1)$. If the line passes through the points $(x_1, y_1)$ and $(x_2, y_2)$ then, if $x_1 \neq x_2$, the equation of the line can be written $y - y_1 = [(y_2 - y_1)/(x_2 - x_1)](x - x_1)$.

In three-dimensional space a straight line can be found by the intersection of two planes. This means that the straight line can be given in terms of two equations: $a_1x + b_1y + c_1z + d_1 = 0$ and $a_2x + b_2y + c_2z + d_2 = 0$, where these two equations are equations for two planes that intersect. *See also* vector equation of a plane.

**strain** A measure of how a solid body is deformed when a force is applied to it. The *linear strain*, sometimes called the *tensile strain*, is the ratio of the change of the body to its original length. The linear strain applies to wires. The *bulk strain*, sometimes called the *volume strain*, is the ratio of the change in the volume of the body to its original volume. When a body is subject to a *shear* force there is another type of strain called a *shear strain*, which is the angular distortion of the body measured in radians. Since linear strain and bulk strain are ratios of length and volume respectively they are dimensionless. *See also* Poisson's ratio, stress, Young modulus.

**strange attractor** A path in phase space that is not closed. Strange attractors are characteristic of chaotic behavior. *See* attractor.

**stratified random sampling** *See* sampling.

**stress** The force per unit area acting on a body and tending to deform that body. It is a term used when a solid body is subject to forces of tension, compression, or shear. The unit of stress is the pascal (Pa), or $Nm^{-2}$.

If the solid body is stretched the stress is *tensile stress*. If the body is compressed the stress is *compressive stress*. If the force tends to shear a body the stress is *shear stress*. The biggest force per unit area that a solid body can withstand without fracture is called the *breaking stress* of that body.

The *stress–strain graph* of a solid is important both in the theory of solids and in engineering. *See also* strain; Young modulus.

**Student's *t*-distribution** The distribution, written $t_n$, of a random variable
$$t = (\bar{x} - \mu)\sqrt{n}/\sigma$$
where a random sample of size $n$ is taken from a normal population $x$ with mean $\mu$ and standard deviation $\sigma$. $n$ is called the number of degrees of freedom. The mean of the distribution is 0 for $n > 1$, and the variance is $n/(n - 2)$ for $n > 2$. When $n$ is large $t$ has an approximately standard normal distribution. The probability density function, $f(t)$, has a symmetrical graph.

The values $t_n(\alpha)$ for which $P(t \leq t_n(\alpha)) = \alpha$ for various values of $n$ are available in tables. *See also* mean, standard deviation; Student's *t*-test.

**Student's *t*-test** A hypothesis test for accepting or rejecting the hypothesis that the mean of a normal distribution wih unknown variance is $\mu_0$, using a small sample. The statistic $t = (\bar{x} - \mu_0)\sqrt{n}/s$ is computed from the data $(x_1, x_2, \ldots x_n)$, where $\bar{x}$ is the sample mean, $s$ is the sample standard deviation, and $n < 30$. If the hypothesis is true $t$ has a $t_{n-1}$ distribution. If $t$ lies in the critical region $|t| > t_{n-1}(1 - \alpha/2)$ the hypothesis is rejected at significance level $\alpha$. *See also* hypothesis test; Student's *t*-distribution.

**subgroup** A subgroup $S$ of a GROUP $G$ is a subset of $G$ that is also a group under the same law of combination of elements as $G$, i.e. a group whose members are members of another group.

**subject** The main independent variable in an algebraic formula. For example, in the function
$$y = f(x) = 2x^2 + 3x,$$
$y$ is the subject of the formula.

**submatrix** /sub-**may**-triks, -**mat**-riks/ A matrix that is obtained from another matrix $M$ by deleting some rows and columns from $M$.

**subnormal** The projection on the $x$-axis of a line normal to a curve at point $P_0$ $(x_0, y_0)$ and extending from $P_0$ to the $x$-axis. The length of the subnormal is $my_0$, where $m$ is the gradient of the tangent to the curve at $P_0$.

**subroutine** /**sub**-roo-teen/ (**procedure**) A section of a computer program that performs a task that may be required several times in different parts of the program. Instead of inserting the same sequence of instructions at a number of different points, control is transferred to the subroutine and when the task is complete it is returned to the main part of the program. *See also* routine.

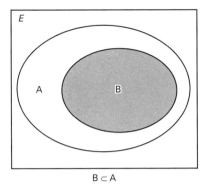

$B \subset A$

Subset: the set *B* (shaded) in the Venn diagram is a subset of *A*.

**subscript** /**sub**-skript/ A small letter or number written below, and usually to the right of, a letter for various purposes, such as to identify a particular element of a set, e.g. $x_1, x_2 \in X$, to denote a constant, e.g. $a_1, a_2$, or to distinguish between variables, e.g. $f(x_1, x_2, \ldots, x_n)$. Double subscripts are also used, for example to write a determinant with general terms; $a_{ij}$ denotes the element in the *i*th row and *j*th column. *See also* superscript.

**subset** Symbol: $\subset$ A set that forms part of another set. For example, the set of natural numbers, $N = \{1, 2, 3, 4, \ldots\}$ is a subset of the set of integers $I = \{\ldots -2, -1, 0, 1, 2, \ldots\}$, written as $N \subset I$. $\subset$ stands for the rela-tionship of *inclusion*, and $N \subset I$ can be read: *N* is included in *I*. Another symbol used is $\supset$, which means 'contains as a sub-set' as in $I \supset N$. *See also* Venn diagram. *See illustration overleaf.*

**substitution** /sub-stă-**stew**-shŏn/ A method of solving algebraic equations that involves replacing one variable by an equivalent in terms of another variable. For example, to solve the SIMULTANEOUS EQUATIONS

$$x + y = 4$$

and

$$2x + y = 9$$

we can first write *x* in terms of *y*, that is:

$$x = 4 - y$$

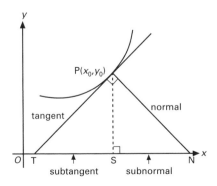

Subtangent and subnormal of a curve at a point P $(x_0, y_0)$.

The substitution of $4 - y$ for $x$ in the second equation gives:
$$2(4 - y) + y = 9$$
or $y = -1$, and therefore, from the first equation, $x = 5$. Another use of substitution of variables is in integration. *See also* integration by substitution.

**subtangent** /sub-**tan**-jĕnt/ The projection on the $x$-axis of the line tangent to a curve at a point $P_0$ $(x_0, y_0)$ and extending from $P_0$ to the $x$-axis. The length of the subtangent is $y_0/m$, where $m$ is the gradient of the tangent.

**subtend** /sŭb-**tend**/ To lie opposite and mark out the limits of a line or angle. For example, each arc of a circle subtends a particular angle at the center of the circle.

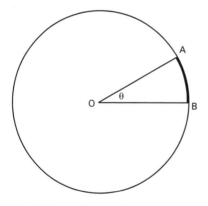

Subtend: the arc AB subtends an angle θ at the center O of the circle.

**subtraction** /sŭb-**trak**-shŏn/ Symbol: − The binary operation of finding the DIFFERENCE between two quantities. In arithmetic, unlike addition, the subtraction of two numbers is neither commutative ($4 - 5 \neq 5 - 4$) nor associative [$2 - (3 - 4) \neq (2 - 3) - 4$]. The identity element for arithmetic subtraction is zero only when it comes on the right-hand side ($5 - 0 = 5$, but $0 - 5 \neq 5$). In *vector subtraction*, two vectors are placed tail-to-tail forming two sides of a triangle. The length and direction of the third side gives the VECTOR DIFFERENCE. Just as the sign of the difference between two numbers depends on the order of subtraction, the sense of the vector difference depends on the sense of the angle between the two vectors. *Matrix subtraction*, like matrix addition, can only be carried out between matrices with the same number of rows and columns. *Compare* addition. *See also* matrix.

**subtraction of fractions** *See* fractions.

**subtrahend** /**sub**-tră-hend/ A quantity that is to be subtracted from another given quantity.

**successor** The successor of a member of a series is the next member of the series. In particular, the successor of an integer is the next integer, i.e. the successor of $n$ is $n + 1$.

**sufficient condition** *See* condition.

**sum** The result obtained by adding two or more quantities together.

**summation notation** Symbol $\Sigma$. A symbol used to indicate the sum of a series. The number of terms to be summed is indicated by putting $i$, the number of the first term to be summed, beneath the $\Sigma$ sign and the number of the last term to be summed is put above the $\Sigma$ sign. The sum to infinity of a series is indicated by putting the infinity sign $\infty$ above the $\Sigma$ sign.

**sum to infinity** In a CONVERGENT SERIES, the value that the sum of the first $n$ terms, $S_n$ approaches as $n$ becomes infinitely large.

**sup** *See* supremum.

**superelastic collision** /soo-per-i-**las**-tik/ A collision for which the restitution coefficient is greater than one. In effect the relative velocity of the colliding objects after the interaction is greater than that before. The apparent energy gain is the result of transfer from energy within the colliding objects. For example, if a collision between two trolleys causes a compressed spring in one to be released against the other, the collision may be superelastic. *See also* restitution; coefficient of.

**superscript** /soo-per-skript/ A number or letter written above and to the right or left of a letter. A superscript usually denotes a power, e.g. $x^3$, or a derivative, e.g. $f^4(x) = d^4f/dx^4$, or is sometimes used in the same sense as a SUBSCRIPT.

**supplementary angles** A pair of angles that add together to make a straight line (180° or $\pi$ radians). *Compare* complementary angles; conjugate angles.

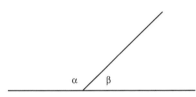

Supplementary angles: $\alpha + \beta = 180°$.

**supplementary units** The dimensionless units – the radian and the steradian – used along with base units to form derived units. *See* SI units.

**supremum** /soo-**pree**-mŭm/ (**sup**) The least upper BOUND of a set.

**surd** /serd/ *See* irrational number.

**surface** Any locus of points extending in two dimensions. It is defined as an area. A surface may be flat (a plane surface) or curved and may be finite or infinite. For example, the plane $z = 0$ in three-dimensional Cartesian coordinates is flat and infinite; the outside of a sphere is curved and finite.

**surface integral** An integral in which the integration takes place along a surface. It is possible to define a surface integral in terms of a parametric representation for the surface and then show that the value of the integral is independent of the parametric representation under general conditions. There are many mathematical and physical applications of surface integrals including the calculation of surface areas, the center of mass and moment of inertia, the flow of fluids through a surface and applications of VECTOR CALCULUS to the theory of electricity and magnetism.

**surface of revolution** A surface generated by rotating a curve about an axis. For example, rotating a parabola about its axis of symmetry produces a PARABOLOID of revolution.

**syllogism** /sil-ŏ-jiz-ăm/ In logic, a deductive argument in which a *conclusion* is derived from two propositions, the *major premiss* and the *minor premiss*, the conclusion necessarily being true if the premisses are true. For example, 'Tim wants a car or a bicycle', 'Tim does not want a car', therefore 'Tim wants a bicycle'. A *hypothetical syllogism* is a particular type of syllogism of the form '*A* implies *B*', '*B* implies *C*', therefore '*A* implies *C*'.

**symbol** A letter or character used to represent an object, operation, quantity, relation, or function. See the appendix for a list of mathematical symbols.

**symbolic logic** (**formal logic**) The branch of logic in which arguments, the terms used in them, the relationships between them, and the various operations that can be performed on them are all represented by symbols. The logical properties and IMPLICATIONS of arguments can then be more easily studied strictly and formally, using algebraic techniques, proofs, and theorems in a mathematically rigorous way. It is sometimes called mathematical logic.

The simplest system of symbolic logic is *propositional logic* (sometimes called *propositional calculus*) in which letters, e.g. *P*, *Q*, *R*, etc., stand for propositions or statements, and various special symbols stand for relationships that can hold between them. *See also* biconditional; conjunction; disjunction; negation; truth table.

**symmetrical** Denoting any figure that can be divided into two parts that are mirror images of each other. The letter A, for example, is symmetrical, and does not change when viewed in a mirror, but the letter R is not. A symmetrical plane figure has at least one line that is an axis of symmetry, which divides it into two mirror images.

**symmetric matrix** /să-**met**-rik/ A matrix $A$ that has its transpose equal to $A$, i.e. if the entries of $A$ are denoted by $a_{ij}$ then $a_{ij} = a_{ji}$ for all and $j$ in the matrix.

**symmetry** /**sim**-ĕ-tree/ The property that figures and bodies can have that these entities appear unchanged after certain transformations, called *symmetry transformations*, are performed on them. Examples of symmetry transformations include rotations about a fixed point, reflection about a mirror plane, and translation along a lattice. For example, in a square rotation about the center of the square by 90° is a symmetry transformation since the square appears unchanged after the rotation, whereas rotation about the center by 45° is not a symmetry transformation since the position of the square is different after the transformation.

Taking symmetry into account frequently simplifies the mathematical analysis of a problem considerably and sometimes enables conclusions to be drawn without performing a calculation.

Symmetry is systematically analyzed in mathematics (and its applications) using GROUP theory.

In the specific context of graphs of functions $y = f(x)$ there are symmetries associated with EVEN FUNCTIONS and ODD FUNCTIONS, with even functions having a symmetry about the $y$-axis while odd functions have a symmetry about the origin.

**syntax** /**sin**-taks/ In logic, syntax concerns the properties of formal systems which do not depend on what the symbols actually mean. It deals with the way the symbols can be connected together and what combinations of symbols are meaningful. The syntax of a formal logical language will specify precisely and rigorously which formulae are well-formed within the language, but not their intuitive meaning. *Compare* semantics.

**systematic error** *See* error.

**systematic sampling** *See* sampling.

**Système International d'Unités** /see-**stem** an-tair-nas-yo-**nal** doo-nee-**tay**/ *See* SI units.

**tan** *See* tangent.

**tangent 1.** A straight line or a plane that touches a curve or a surface without cutting through it. On a graph, the slope of the tangent to a curve is the slope of the curve at the point of contact. In Cartesian coordinates, the slope is the derivative $dy/dx$. If $\theta$ is the angle between the $x$-axis and a straight line to the point $(x,y)$ from the origin, then the trigonometric function $\tan\theta = y/x$.

**2.** (tan) A trigonometric function of an angle. The tangent of an angle $\alpha$ in a right-angled triangle is the ratio of the lengths of the side opposite to the side adjacent. This definition applies only to angles between $0°$ and $90°$ (0 and $\pi/2$ radians). More generally, in rectangular Cartesian coordinates, with origin O, the ratio of the $y$-coordinate to the $x$-coordinate of a point P $(x,y)$ is the tangent of the angle between the line OP and the $x$-axis. The tangent function, like the sine and cosine functions, is periodic, but it repeats itself every $180°$ and is not continuous. It is zero when $\alpha = 0°$, and becomes an infinitely large positive number as $\alpha$ approaches $90°$. At $+ 90°$, $\tan\alpha$ jumps from $+\infty$ to $-\infty$ and then rises to zero at $\alpha = 180°$. *See also* trigonometry.

**tangent plane** The plane through a point P on a smooth surface that is perpendicular to the normal to the surface at P. All the *tangent lines* at P lie in the tangent plane, with a tangent line at P being a tangent at P to any curve on the surface that goes through P.

It is frequently the case that all points on the surface near P are on the same side of the tangent plane at P. However, at a SADDLE POINT some of the points close to P are on one side of the tangent plane while other points close to P are on the other side of the tangent plane.

**tanh** /th'an, tansh, tan-**aych**/ A hyperbolic tangent. *See* hyperbolic functions.

**tautology** /taw-**tol**-ŏ-jee/ In LOGIC, a proposition, statement, or sentence of a form that cannot possibly be false. For ex-

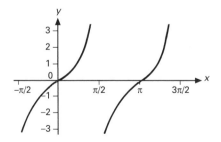

Tangent: graph of $y = \tan x$, with $x$ in radians.

ample, 'if all pigs eat mice then some pigs eat mice' and 'if I am coming then I am coming' are both true regardless of whether the component propositions 'I am coming' and 'all pigs eat mice' are true or false. More strictly, a tautology is a compound proposition that is true no matter what truth values are assigned to the simple propositions that it contains. A tautology is true purely because of the laws of logic and not because of any fact about the World (the laws of thought are tautologies). A tautology therefore contains no information. *Compare* contradiction.

**Taylor series** (**Taylor expansion**) A formula for expanding a function, f(*x*), by writing it as an infinite series of derivatives for a fixed value of the variable, *x* = *a*:

f(*x*) = f(*a*) + f'(*a*)(*x* − *a*) +
$\quad$ f''(*a*)(*x* − *a*)$^2$/2! + f'''(*a*)(*x* − *a*)$^3$/3! + ...

If *a* = 0, the formula becomes:

$\quad$ f(*x*) = f'(0) + f'(0)*x* = f''(0)*x*$^2$/2! + ...

The Taylor series is named for the British mathematician Brook Taylor (1685–1731) and the *Maclaurin series*, or *Maclaurin expansion* (a special case of the Taylor series) is named for the Scottish mathematician Colin Maclaurin (1698–1746). *See also* expansion.

*t***-distribution** *See* Student's *t*-distribution.

**tension** A force that tends to stretch a body (e.g. a string, rod, wire, etc.).

**tensor** /**ten**-ser/ A mathematical entity that is a generalization of a vector. Tensors are used to describe how all the components of a quantity in an n-dimensional system behave under certain transformations, just as a VECTOR can describe a translation from one point to another in a plane or in space.

**tera-** Symbol: T A prefix denoting 10$^{12}$. For example, 1 terawatt (TW) = 10$^{12}$ watts (W).

**terminal** A point at which a user can communicate directly with a computer both for the input and output of information. It is situated outside the computer system, often at some distance from it, and is linked to it by electric cable, telephone, or some other transmission channel. A mouse or keyboard, similar to that on a typewriter, is used to feed information (data) to the computer. The output can either be printed out or can be displayed on a screen, as with a VISUAL DISPLAY UNIT. An *interactive terminal* is one connected to a computer, which gives an almost immediate response to an enquiry from the user. An *intelligent terminal* can store information and perform simple operations on it without the assistance of the computer's central processor. *See also* input/output.

**terminating decimal** A DECIMAL that has a finite number of digits after the decimal point, also called a finite decimal.

**ternery** /**ter**-ně-ree/ Describing a number system to the base (radix) 3. It uses the numbers 0, 1, and 2 (with place values ... 243, 81, 27, 9, 3, 1). *See also* binary.

**tesla** /**tess**-lă/ Symbol: T The SI unit of magnetic flux density, equal to a flux density of one weber of magnetic flux per square meter. 1 T = 1 Wb m$^{-2}$. The unit is named for the Croatian–American physicist Nikola Tesla (1856–1943).

**tessellation** /tess-ă-**lay**-shŏn/ A regular pattern of tiles that can cover a surface without gaps. Regular polygons that will, on their own, completely tessellate a surface are equilateral triangles, squares and regular hexagons.

**tetrahedral numbers** /tet-ră-**hee**-drăl/ For a positive integer *n*, a number defined as the sum of the first *n* TRIANGULAR NUMBERS. This definition means that the *n*th tetrahedral number is given by (*n* + 1)(*n* + 2)(*n*/6). The first four triangular numbers are 1, 4, 10, and 20.

**tetrahedron** /tet-ră-**hee**-drŏn/ (**triangular pyramid**) A solid figure bounded by four triangular faces. A *regular tetrahedron* has four congruent equilateral triangles as its faces. *See also* polyhedron; pyramid.

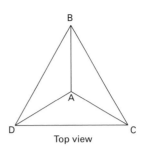

Top view

Tetrahedron

**tetrakaidekahedron** /tet-ră-kỹ-dek-ă-hee-drŏn/ A polyhedron bounded by six squares and eight hexagons. It can be formed by truncating the six corners of an octahedron in a symmetrical way. It is possible to fill the whole of space without either gaps or overlaps with tetrakaidekahedra. The tetrakaidekahedron is an example of a SEMI-REGULAR POLYHEDRON.

**theorem** /th'ee-ŏ-rĕm, th'eer-ĕm/ The conclusion which has been proved in the course of an argument upon the basis of certain given assumptions. A theorem must be a result of some general importance. *Compare* lemma.

**theorem of parallel axes** If $I_0$ is the moment of the inertia of an object about an axis, the moment of inertia $I$ about a parallel axis is given by:
$$I = I_0 md^2$$
where $m$ is the mass of the object and $d$ is the separation of the axes.

**theory of games** *See* game theory.

**therm** A unit of heat energy equal to $10^5$ British thermal units (1.055 056 joules).

**third-order determinant** *See* determinant.

**thou** /th'ow/ *See* mil.

**three-dimensional** Having length, breadth, and depth. A three-dimensional figure (solid) can be described in a coordi-nate system using three variables, for example, three-dimensional Cartesian coordinates with an $x$-axis, $y$-axis, and $z$-axis. *Compare* two-dimensional.

**thrust** A force tending to compress a body (e.g. a rod or bar) in one direction. Thrust acts in the opposite direction to tension.

**time sharing** A method of operation in computer systems in which a number of jobs are apparently executed simultaneously instead of one after another (as in BATCH PROCESSING). This is achieved by transferring each program in turn from backing store to main store and allowing it to run for a short time.

**ton** /tun/ 1. A unit of mass equal to 2240 pounds (long ton, equivalent to 1016.05 kg) or 2000 pounds (short ton, equivalent to 907.18 kg).
2. A unit used to express the explosive power of a nuclear weapon. It is equal to an explosion with an energy equivalent of one ton of TNT or approximately $5 \times 10^9$ joules.

**tonne** /tun/ (**metric ton**) Symbol: t A unit of mass equal to $10^3$ kilograms.

**topologically equivalent** *See* topology.

**topological space** /top-ŏ-**loj**-ă-kăl/ A non-empty set $A$ together with a fixed collection ($T$) of subsets of $A$ satisfying:
1. $\emptyset \in T$, $A \in T$;
2. if $U \in T$ and $V \in T$ then $U \cap V \in T$;

3. if $U_i \in T$, where $\{U_i\}$ is a finite or infinite collection of sets, then $\cap U_i \in T$.

The set of subsets $T$ is called a TOPOLOGY for $A$, and the members of $T$ are called the *open sets* of the topological space. *Compare* metric space.

**topology** /tŏ-**pol**-ŏ-jee/ A branch of geometry concerned with the general properties of shapes and space. It can be thought of as the study of properties that are not changed by continuous deformations, such as stretching or twisting. A sphere and an ellipsoid are different figures in solid (Euclidean) geometry, but in topology they are considered equivalent since one can be transformed into the other by a continuous deformation. A torus, on the other hand, is not topologically equivalent to a sphere – it would not be possible to distort a sphere into a torus without breaking or joining surfaces. A torus is thus a different type of shape to a sphere. Topology studies types of shapes and their properties. A special case of this is the investigation of networks of lines and the properties of knots.

In fact Euler's study of the KÖNIGSBERG BRIDGE PROBLEM was one of the earliest results in topology. A modern example is in the analysis of electrical circuits. A circuit diagram is not an exact reproduction of the paths of the wires, but it does show the connections between different points of the circuit (i.e. it is topologically equivalent to the circuit). In printed or integrated circuits it is important to arrange connections so that they do not cross.

Topology uses methods of higher algebra including group theory and set theory. An important notion is that of sets of points in the neighborhood of a given point (i.e. within a certain distance of the point). An *open set* is a set of points such that each point in the set has a neighborhood containing points in the set. A topological transformation occurs when there is a one-to-one correspondence between points in one figure and points in another so that open sets in one correspond to open sets in the other. If one figure can be transformed into another by such a transformation, the sets are *topologically equivalent*.

**torque** /tork/ Symbol: $T$ A turning force (or MOMENT). The torque of a force $F$ about an axis (or point) is $Fs$, where $s$ is the distance from the axis to the line of action of the force. The unit is the newton meter. Note that the unit of *work*, also the newton meter, is called the joule. Torque is *not*, however, measured in joules. The two physical quantities are not in fact the same. Work (a scalar) is the scalar product of force and displacement. Torque is the vector product $F \times s$ and is a vector at 90° to the plane of the force and displacement. *See also* couple.

**torr** A unit of pressure equal to a pressure of 101 325/760 pascals (133.322 Pa). It is equal to the mmHg. The unit is named for the Italian physicist Evangelista Torricelli (1608–47).

**torsional wave** /**tor**-shŏ-năl/ A wave motion in which the vibrations in the medium are rotatory simple harmonic motions around the direction of energy transfer.

**torus** /**tor**-ŭs, **toh**-rŭs/ (**anchor ring**) A closed curved surface with a hole in it, like a donut. It can be generated by rotating a circle about an axis that lies in the same plane as the circle but does not cut it. A cross-section of the torus in a plane perpendicular to the axis is two concentric circles. A cross-section in any plane that contains the axis is a pair of congruent circles at equal distances on both sides of the axis. The volume of the torus is $4\pi dr^2$ and its surface area is $3\pi^2 dr$, where $r$ is the radius of the generating circle and $d$ is the distance of its center from the axis.

**total derivative** A derivative that can be expressed in terms of a series of partial derivatives. For example, if the function $z = f(x,y)$ is a continuous function of $x$ and $y$, and both $x$ and $y$ are continuous functions of another variable $t$, then the total derivative of $z$ with respect to $t$ is:

$$dz/dt = (\partial z/\partial x)(dx/dt) + (\partial z/\partial y)(dy/dt)$$

*See also* chain rule; total differential.

**total differential** An infinitesimal change in a function of one or more variables. It is the sum of all the partial differentials. *See* differential.

**trace** *See* square matrix.

**track** *See* disk; drum; magnetic tape; paper tape.

**trajectory** /trǎ-**jek**-tŏ-ree/ **1.** The path of a moving object, e.g. a projectile.
**2.** A curve or surface that satisfies some given set of conditions, such as passing through a given set of points, or having a given function as its gradient.

**transcendental number** /tran-sen-**den**-tǎl/ *See* irrational number; pi.

**transfinite** /trans-**fÿ**-nÿt/ A *transfinite number* is a cardinal or ordinal number that is not an integer. *See* cardinal number; ordinal number; aleph.

*Transfinite induction* is the process by which we may reason that if some proposition is true for the first element of a well-ordered set $S$, and that if the proposition is true for a given element it is also true for the next element, then the proposition is true for every element of $S$.

**transformation** /trans-fer-**may**-shŏn/ **1.** In general, any FUNCTION or mapping that changes one quantity into another.
**2.** The changing of an algebraic expression or equation into an equivalent with a different form. For example, the equation
$$(x - 3)^2 = 4x + 2$$
can be transformed into
$$x^2 - 10x + 7 = 0$$
*See also* changing the subject of a formula.
**3.** In geometry, the changing of one shape into another by moving each point in it to a different position by a specified procedure. For example, a plane figure may be moved in relation to two rectangular axes. Another example is when a figure is enlarged. *See* translation. *See also* deformation; dilatation; enlargement; projection; rotation.

**transformation of coordinates** **1.** Changing the position of the reference axes in a cordinate system by translation, rotation, or both, usually to simplify the equation of a curve. *See* rotation of axes; translation of axes.
**2.** Changing the type of coordinate system in which a geometrical figure is described; for example, from Cartesian coordinates to POLAR COORDINATES.

**transitive relation** A relation *on a set $A$ such that if $a*b$ and $b*c$ then $a*c$ for all $a$, $b$, and $c$ in $A$. 'Greater than', 'less than', and 'equals' are examples of transitive relations. A relation that is not transitive is an *intransitive relation*.

**translation** /trans-**lay**-shŏn, tranz-/ The moving of a geometrical figure so that only its position relative to fixed axes is changed, but not its orientation, size, or shape. *See also* translation of axes. *See illustration overleaf.*

**translation of axes** In coordinate geometry, the shifting of the reference axes so that each axis is parallel to its original position and each point is given a new set of coordinates. For example, the origin O of a system of $x$- and $y$-axes, may be shifted to the point O', $(3, 2,)$ in the original system. The new axes $x'$ and $y'$ are at $x = 3$ and $y = 2$, respectively. This is sometimes done to simplify the equation of a curve. The circle $(x - 3)^2 + (y - 2)^2 = 4$ can be described by new coordinates $x' = (x - 3)$ and $y' = (y - 3)$: $(x')^2 + (y')^2 = 4$. The origin O' is then at the center of the circle. *See also* rotation of axes.

**translatory motion** /**trans**-lǎ-tor-ee, -toh-ree, **tranz**-/ (**translation**) Motion involving change of position; it compares with rotatory motion (rotation) and vibratory motion (vibration). Each is associated with kinetic energy. In an object undergoing translatory motion, all the points move in parallel paths. Translatory motion is usually described in terms of (linear) speed or velocity, and acceleration.

A transformation can be represented by a 2 x 2 matrix.
A point (x, y) is tranformed to a point (x′, y′) by multiplying the column vector of (x, y) by a matrix M

i.e. $M\begin{pmatrix} x \\ y \end{pmatrix}$

The transformation matrices are:

reflection in x-axis $\begin{pmatrix} 1 & 0 \\ 0 & -1 \end{pmatrix}$

reflection in y-axis $\begin{pmatrix} -1 & 0 \\ 0 & 1 \end{pmatrix}$

enlargement scale factor k $\begin{pmatrix} k & 0 \\ 0 & k \end{pmatrix}$

stretch in x-direction $\begin{pmatrix} k & 0 \\ 0 & 1 \end{pmatrix}$

stretch in y-direction $\begin{pmatrix} 1 & 0 \\ 0 & k \end{pmatrix}$

shear in x-direction by k $\begin{pmatrix} 1 & k \\ 0 & 1 \end{pmatrix}$

rotation by α
(anticlockwise positive) $\begin{pmatrix} \cos\alpha & -\sin\alpha \\ \sin\alpha & \cos\alpha \end{pmatrix}$

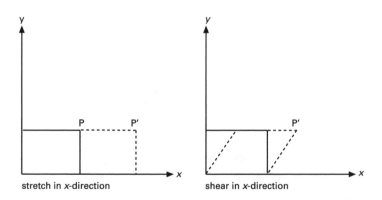

stretch in x-direction          shear in x-direction

Transformations

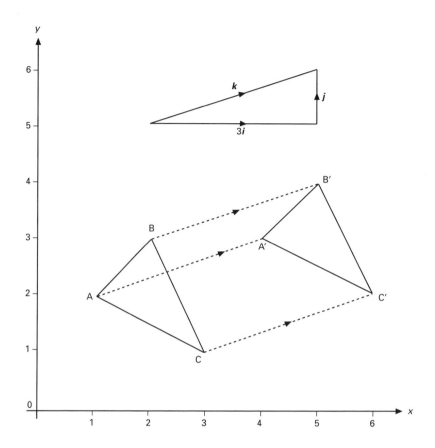

The translation of a triangle ABC
The translation vector **k** = *3i* + **j**

A translation by *a* in the *x* direction and *b* in the *y* direction
transforms a point (*x, y*) to (*x′, y′*).

In matrix terms

$$\begin{pmatrix} x' \\ y' \end{pmatrix} = \begin{pmatrix} x \\ y \end{pmatrix} + \begin{pmatrix} a \\ b \end{pmatrix}$$

Translation

**transpose of a matrix** /trans-**pohz**/ The matrix that results from interchanging the rows and columns of a given matrix. The determinant of the transpose of a square matrix is equal to that of the original matrix. The transpose of a row vector is a column vector and vise versa. If two matrices *A* and *B* are comformable (can be multiplied together), then the transpose of the matrix product *AB* = *C* is $\widetilde{C}$ = $(\widetilde{AB})$ = $\widetilde{B}\widetilde{A}$. In

$$A = \begin{pmatrix} a_1 & b_1 \\ a_2 & b_2 \\ a_3 & b_3 \end{pmatrix} \rightarrow \tilde{A} = \begin{pmatrix} a_1 & a_2 & a_3 \\ b_1 & b_2 & b_3 \end{pmatrix}$$

Transpose $\tilde{A}$ of a matrix $A$.

other words, in taking the transpose of a matrix product the order must be reversed.

**transversal** /trans-**ver**-săl, tranz-/ A line that intersects two or more other lines.

**transverse axis** /trans-**vers**, tranz-, **trans-vers, tranz-**/ *See* hyperbola.

**transverse wave** A wave motion in which the motion or change is perpendicular to the direction of energy transfer. Electromagnetic waves and water waves are examples of transverse waves. *Compare* longitudinal waves.

**trapezium** /tră-**pee**-zee-ŭm/ (*pl.* **trapeziums** or **trapezia**) A quadrilateral, none of whose sides are parallel. Note that in the UK the terms 'trapezium' and 'trapezoid' have the opposite meanings to those in the US.

**trapezoid** /**trap**-ĕ-zoid/ A quadrilateral with two parallel sides. It is sometimes part of the definition that the other sides are not parallel. The parallel sides are called the *bases* of the trapezium and the perpendicular distance between the bases is called the *altitude*. The area of a trapezium is the product of the sum of the parallel side lengths and half the perpendicular distance between them.

**trapezoid rule** A rule for finding the approximate area under a curve by dividing it into pairs of trapezium-shaped sections, forming vertical columns of equal width with bases lying on the horizontal axis. The trapezium rule is used as a method of NU-MERICAL INTEGRATION. For example, if the value of a function $f(x)$ is known at $x = a$, $x = b$, and at a value mid-way between $a$ and $b$, the integral is approximately:

$$(h/2)\{f(a) + 2f[(a + b)/2] + f(b)\}$$

$h$ is half the distance between $a$ and $b$. If this does not give a sufficiently accurate result, the area may be subdivided into 4, 6, 8, etc., columns until further subdivision makes no significant difference to the result. *See also* Simpson's rule.

**travel graph** A graph of displacement

Trapezium rule

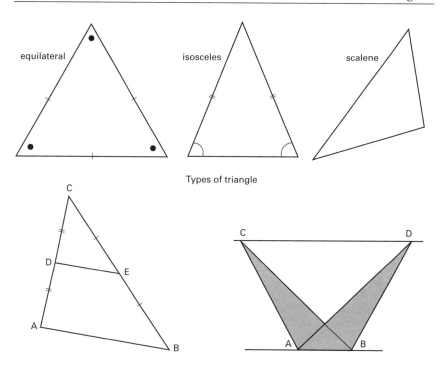

Types of triangle

The line joining the mid-points of two sides of a triangle is parallel to the third side and equal to half of it (AB = 2DE).

Triangles on the same base and between the same parallels are equal in area (area ABC = area ABD).

Triangles

(usually plotted on a $y$-axis) against time (usually plotted on the $x$-axis) which can be used to illustrate a journey of a person. The velocity of a person at any time is the gradient of the curve at that time. The concept of travel graph can be extended to graphs of velocity against time, with acceleration being the gradient of the curve and displacement being the area under the curve.

**traveling wave** *See* wave.

**traveling salesman problem** The problem of minimizing the cost or time to visit a number of towns, with the route from town $a$ to town $b$ being given a value of $c_{ab}$. This problem can, in principle, be solved by calculating the cost or time for all possible routes and seeing which route has the lowest value. However, for $n$ towns the number of routes is $(n-1)!/2$, which means that, except for small values of $n$, this number becomes too large for any computer to handle. It is possible to find algorithms that find roots that are near to the minimum, but not exactly at the minimum, which perform the calculation fairly quickly.

**traversable network** A network that can be drawn by beginning at a point on the network and not lifting the pen from the paper, without going over any line twice.

**tree diagram** A type of diagram used as an aid to solve problems in probability theory. The different ways in which an event, such as drawing balls with various colours out of a bag, can occur, are drawn as

$$\begin{pmatrix} a_{11} & a_{12} & a_{13} \\ 0 & a_{22} & a_{23} \\ 0 & 0 & a_{33} \end{pmatrix} \qquad \begin{pmatrix} a_{11} & 0 & 0 \\ a_{21} & a_{22} & 0 \\ a_{31} & a_{32} & a_{33} \end{pmatrix}$$

Triangular matrix: these are 3 x 3 matrices.

branches of a tree, with each branch showing the probability of that particular event occurring.

**trial solution** A guess at a solution to a problem such as a differential equation. The trial solution is substituted into the problem to be solved. If it does not solve the problem exactly then the trial solution is improved upon until it is a satisfactory solution to the problem.

**triangle** A plane figure with three straight sides. The area of a triangle is half the length of one side (the base) times the length of the perpendicular line (the altitude) from the base to the opposite vertex. The sum of the internal angles in a triangle is 180° (or $\pi$ radians). In an *equilateral triangle*, all three sides have the same length and all three angles are 60°. An *isosceles triangle* has two sides of equal length and two equal angles. A *scalene triangle* has no equal sides or angles. In a *right-angled triangle*, one angle is 90° (or $\pi/2$ radians), and the others therefore add up to 90°. In an *acute angled triangle*, all the angles are less than 90°. In an *obtuse-angled triangle*, one of the angles is greater than 90°. See *illustration overleaf*.

**triangle inequality** For any triangle ABC, the length of one side is always less than the sum of the other two:
AB < BC + CA

**triangle of forces** See triangle of vectors.

**triangle of vectors** A triangle describing three coplanar VECTORS acting at a point with zero resultant. When drawn to scale – shown correctly in size, direction, and sense, but not in position – they form a closed triangle. Thus three forces acting on an object at equilibrium form a *triangle of* forces. Similarly a *triangle of velocities* can be constructed.

**triangle of velocities** See triangle of vectors.

**triangular matrix** A square matrix in which either all the elements above the leading diagonal or all the elements below the leading diagonal are zero. The determinant of a triangular matrix is equal to the product of its diagonal elements.

**triangular numbers** The set of numbers {1, 3, 6, 10, ...} generated by triangular arrays of dots. Each array has one more row than the preceding one, the additional row having one more dot than the longest in the preceding array. The $n$th triangular number is $n(n + 1)/2$.

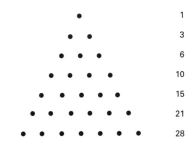

|   |
|---|
| 1 |
| 3 |
| 6 |
| 10 |
| 15 |
| 21 |
| 28 |

Triangular numbers

**triangular prism** See prism.

**triangular pyramid** See tetrahedron.

**trichotomy** /trỹ-**kot**-ŏ-mee/ The property of an ordering that exactly one of the statements $x < y$, $x = y$, $x > y$ is true for any $x$ and $y$. See linearly ordered.

**trigonometric functions** /trig-ŏ-nŏ-**met**-rik/ *See* trigonometry.

**trigonometry** /trig-ŏ-**nom**-ĕ-tree/ The study of the relationships between the sides and angles in a triangle, in terms of the trigonometric functions of angles (sine, cosine, and tangent). *Trigonometric functions*, can be defined by the ratio of sides in a right-angled triangle. In a right-angled triangle with an angle $\alpha$, $o$ is the length of the side opposite $\alpha$, $a$ the length of side adjacent to $\alpha$, and $h$ the length of the hypotenuse. For such a triangle the three trigonometric functions of $\alpha$ are defined by:

$$\sin\alpha = o/h$$
$$\cos\alpha = a/h$$
$$\tan\alpha = o/a$$

The following relationships hold for all values of the variable angle $\alpha$:

$$\cos\alpha = \sin(\alpha + 90°)$$
$$\tan\alpha = \sin\alpha/\cos\alpha$$

The trigonometric functions of an angle can also be defined in terms of a circle (hence they are sometimes called *circular functions*). A circle is taken with its center at the origin of Cartesian coordinates. If a point P is taken on this circle, the line OP makes an angle $\alpha$ with the positive direction of the $x$-axis. Then, the trigonometric functions are:

$$\tan\alpha = y/x$$
$$\sin\alpha = y/OP$$
$$\cos\alpha = x/OP$$

Here, $(x,y)$ are the coordinates of the point P and $OP = \sqrt{(x^2 + y^2)}$. The signs of $x$ and $y$ are taken into account. For example, for an angle $\beta°$ between 90° and 180°, $y$ will be positive and $x$ negative. Then:

$$\tan\beta = -\tan(180 - \beta)$$
$$\sin\beta = +\sin(180 - \beta)$$
$$\cos\beta = -\cos(180 - \beta)$$

Similar relationships can be written for the trigonometric functions of angles between 180° and 270° and between 270° and 360°.

The functions secant (sec), cosecant (cosec), and cotangent (cot), which are the reciprocals of the cosine, sine, and tangent functions respectively, obey the following rules for all values of $\alpha$:

$$\tan^2\alpha + 1 = \sec^2\alpha$$

$$1 + \cot^2\alpha = \csc^2\alpha$$

*See also* addition formulae; cosine rule; sine rule.

**trillion** In the US and Canada, a number represented by 1 followed by 12 zeros ($10^{12}$). In the UK, 1 followed by 18 zeros ($10^{18}$).

**trinomial** /trȳ-**noh**-mee-ăl/ An algebraic expression with three variables in it. For example, $2x + 2y + z$ and $3a + b = c$ are trinomials. *Compare* binomial.

**triple integral** The result of integrating the same function three times. For example, if a function $f(x,y,z)$ is integrated first with respect to $x$, holding $y$ and $z$ constant, and the result is then integrated with respect to $y$, holding $x$ and $z$ constant, and the resultant double integral is then integrated with respect to $z$ holding $x$ and $y$ constant, the triple integral is

$$\iiint f(x,y,z)\,dz\,dy\,dx$$

*See also* double integral.

**triple product** A product of three vectors. *See* triple scalar product; triple vector product.

**triple scalar product** A product of three vectors, the result of which is a scalar, defined as:

$$A.(B \times C) = ABC\sin\theta\cos\phi$$

where $\phi$ is the angle between $A$ and the vector product $(B \times C)$, $\theta$ being the angle between $B$ and $C$. The scalar triple product is equal to the volume of the parallelepiped of which $A$, $B$, and $C$ form nonparallel edges. If $A$, $B$, and $C$ are coplanar, their triple scalar product is zero.

**triple vector product** A product of three vectors, the result of which is a vector. It is a vector product of two vectors, one of which is itself a vector product. That is:

$$A \times (B \times C) = (A.C)B - (A.B)C$$

Similarly

$$(A \times B) \times C = (A.C)B - (B.C)A$$

These are equal only when $A$, $B$, and $C$ are mutually perpendicular.

**trirectangular** /trÿ-rek-**tang**-gyŭ-ler/ Having three right angles. *See* spherical triangle.

**trisection** /trÿ-**sek**-shŏn/ Division into three equal parts.

**trivial solution** A solution to an equation or set of equations that is obvious and gives no useful information about the relationships between the variables involved. For example, $x^2 + y^2 = 2x + 4y$ has a trivial solution $x = 0$; $y = 0$.

**trochoid** /**troh**-koid/ The curve described by a fixed point on the radius or extended radius of a circle as the circle rolls (in a plane) along a straight line. Let $r$ be the radius of the circle and $a$ the distance of the fixed point from the center of the circle. If $a > r$ the curve is called a *prolate cycloid*, if $a < r$ the curve is called a *curtate cycloid*, and if $a = r$ the curve is a CYCLOID.

**truncated** Describing a solid generated from a given solid by two non-parallel planes cutting the given solid.

**truncation** The dropping of digits from a number, used as an approximation when that number has more digits than it is convenient or possible to deal with. This is different to ROUNDING since in rounding the number with the required digits is the nearest number with the required digits, which is not necessarily the number obtained by truncation. For example, if truncated to one decimal place 2.791 and 2.736 both give 2.7 whereas rounding of the first number gives 2.8 and rounding of the second number gives 2.7.

**truth table** In logic, a mechanical procedure (sometimes called a matrix) that can be used to define certain logical operations, and to find the truth value of complex propositions or statements containing combinations of simpler ones. A truth table lists, in rows, all the possible combinations of truth values (T = 'true', F = 'false') of a PROPOSITION or statement, and given an initial assignment of truth or falsity to the constituent parts, mechanically

| P Q | $(P \land Q)$ | $\lor$ | $\sim P$ |
|-----|------|------|------|
| T T | T | T | F |
| T F | F | F | F |
| F T | F | T | T |
| F F | F | T | T |

Truth table

assigns a value to the whole. The truth-table definitions for conjunction, disjunction, negation, and implication are shown at those headwords.

An example of a truth table for a complex, or compound proposition is shown in the illustration. The assigning of values proceeds in this way: on the basis of the truth values of $P$ and $Q$, the simple propositions are given truth values, written under the sign $\land$ (in $P \land Q$) and under $\sim P$. Using these, truth values can then be assigned to the whole; in the example this is in effect a complex disjunction and the values are written under the sign $\lor$.

Thus in the case where $P$ is true and $Q$ is false, $P \land Q$ will be false, $\sim P$ will be false, and therefore the whole will be false. *See also* symbolic logic.

**truth value** The truth or falsity of a proposition in logic. A true statement or proposition is indicated by T and a false one by F. In computer logic, the digits 1 and 0 are often used to denote truth values. *See also* truth table.

**Turing machine** /**tewr**-ing/ An abstract model of a computer that consists of a control or processing unit and an infinitely long tape divided into single squares along its length. At any given time each square of the tape is either blank or contains a single symbol from a fixed finite list of symbols $s_1, s_2, ..., s_n$. The Turing machine moves along the tape square by square and reads, erases, and prints symbols. At any given time the Turing machine is in one of a fixed finite number of states represented by $q_1, q_2, ..., q_n$. The 'program' for the machine is made up of a finite set of instructions of the form $q_i s_j s_k X q_j$, where $X$ is either $R$ (move to the right), $L$ (move to the left), or $N$ (stay in the same position). Here, $q_i$ is the state of a machine reading $s_j$, which it changes to $s_k$, then moves left, right, or stays and completes the operation by going into state $q_j$. A Turing machine can be used to define computability. The machine is named for the British mathematician Alan Mathison Turing (1912–54). *See* computability.

**turning effect** The ability of a force to cause a body to rotate about an axis of rotation. It is described quantitatively by the *torque*, sometimes called the *moment of the force*, which is the product of the magnitude of the force, denoted $F$, and the perpendicular distance $d$ of the force from the axis of rotation. The unit of torque is the newton meter (Nm).

**turning point** A point on the graph of a function at which the slope (gradient) of the tangent to a continuous curve changes its sign. If the slope changes from positive to negative, that is, the $y$-coordinate stops increasing and starts decreasing, it is a maximum point. If the slope changes from negative to positive it is a minimum point. Turning points may be local maxima and minima or absolute maxima and minima. All turning points are STATIONARY POINTS. At a turning point the derivative, $dy/dx$, of the curve $y = f(x)$ is zero.

**two-dimensional** Having length and breadth but not depth. Flat shapes, such as circles, squares, and ellipses, can be described in a coordinate system using only two variables, for example, two-dimensional Cartesian coordinates with an $x$-axis and a $y$-axis. *See also* plane.

**unary operation** /yoo-nă-ree/ A mathematical procedure that changes one number into another. For example, taking the square root of a number is a unary operation. *Compare* binary operation.

**unbounded** An *unbounded function* is a function that is not bounded – it has no bounds or limits. Intuitively this means that for any number $N$ there is a value of the function numerically greater than $N$, i.e. there is a point $x$ such that $|f(x)| > N$. For example, the function $f(x) = x$ is unbounded on the real axis, and $f(x) = 1/x$ is unbounded on the interval $0 < x \leq 1$.

**undecidability** /un-di-sÿ-dă-**bil**-ă-tee/ In logic, if a formula or sentence can be proved within a given system the formula or sentence is said to be *decidable*. If every formula in a system can be proved then the whole system is decidable. Otherwise the system is undecidable. The process of determining which systems are decidable and which are not is an important branch of logic.

**underdamping** /un-der-**damp**-ing/ *See* damping.

**unicursal** /yoo-nă-**ker**-săl/ Describing a closed curve that, starting and finishing at the same point, can be traced in one sweep, without having any part retraced.

**uniform acceleration** Constant acceleration.

**uniform distribution** *See* distribution function.

**uniform motion** A vague phrase, usually taken to mean motion at constant velocity (constant speed in a straight line).

**uniform speed** Constant speed.

**uniform velocity** Constant velocity, describing motion in a straight line with zero acceleration.

**union** Symbol: ∪ The combined set of all the elements of two or more sets. If $A$ = {2,

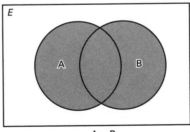

A ∪ B

Union: the shaded area in the Venn diagram is the union of sets *A* and *B*.

4, 6} and $B$ = {3, 6, 9}, then $A \cup B$ = {2, 3, 4, 6, 9}. *See also* Venn diagram.

**unique factorization theorem** A result in number theory which states that a positive integer can be expressed as a product of prime numbers in a way which is unique, except for the order in which the prime factors are written. This theorem follows from fundamental axioms for prime numbers. The factorization of a number into prime numbers (and powers of prime numbers) is called the *prime decomposition* of that number.

**uniqueness theorem** /yoo-**neek**-nis/ A theorem which shows that there can only be a single entity satisfying a given condition, for example only one solution to a given equation. The proof of such a theorem usually proceeds by assuming that there exist two distinct entities satisfying the condition and showing that this leads to a contradiction.

**unique solution** The only possible value of a variable that can satisfy an equation. For example, $x + 2 = 4$ has the unique solution $x = 2$, but $x^2 = 4$ has no unique solution because $x = +2$ and $x = -2$ both satisfy the equation.

**unit** A reference value of a quantity used to express other values of the same quantity. *See also* SI units

**unitary matrix** A matrix in which the inverse of the matrix is the complex conjugate of the transpose of the matrix. If the determinant of a unitary matrix has a value 1 then the matrix is said to be a *special unitary matrix*. The set of all unitary $N \times N$ matrices form a group called the *unitary group*, denoted U($N$), with the group of all special unitary $N \times N$ matrices forming a group called the *special unitary group*, denoted SU($N$). There are many important physical applications of unitary groups and matrices.

**unit circle** A circle in which the radius is one unit and the origin of the coordinate system is the center of the circle. The equation for a unit circle in Cartesian coordinates is $x^2 + y^2 = 1$. A unit circle in the complex plane gives the set of complex numbers $z$ for which the modulus $|z|$ has the value 1.

**unit fraction** *See* fraction.

**unit matrix** (identity matrix) Symbol: $I$ A square MATRIX in which the elements in the leading diagonal are all equal to one, and the other elements are zero. If a maxtrix $A$ with $m$ rows and $n$ columns is multiplied by an $n \times n$ unit matrix, $I$, it remains unchanged, that is $IA = A$. The unit matrix is the *identity matrix* for matrix multiplication.

**unit vector** A vector with a magnitude of one unit. Any vector $r$ can be expressed in terms of its magnitude, the scalar quantity $r$, and the unit vector $r'$ which has the same direction as $r$. The vector $r = rr'$, where $r$ is the magnitude of $r'$. In three-dimensional Cartesian coordinates with origin O unit vectors $i$, $j$, and $k$ are used in the $x$-, $y$-, and $z$-directions respectively.

**universal quantifier** In logic, a symbol meaning 'for all' and usually written $\forall$. For example $(\forall x)Fx$ means 'for all $x$, the property F is true'.

**universal set** Symbol: $E$ The set that contains all possible elements. In a particular problem, $E$ will be defined according to the scope of the problem. For example, in a calculation involving only positive numbers, the universal set, $E$, is the set of all positive numbers. *See also* Venn diagram.

**unstable equilibrium** Equilibrium such that if the system is disturbed a little, there is a tendency for it to move further from its original position rather than to return. *See* stability.

**upper bound** *See* bound.

**upthrust** An upward force on an object in a fluid. In a fluid in a gravitational field the pressure increases with depth. The pressures at different points on the object will therefore differ and the resultant is vertically upward. *See also* Archimedes' principle.

**utility programs** PROGRAMS that help in the general running of a computer system. Utility programs have a number of uses. They can be used, for example, to make copies of files (organized collections of data) and to transfer data from one storage device to another, as from a magnetic tape unit to a disk store. *See also* program.

# V

**validity** In logic, a property of arguments, inferences, or deductions. An argument is valid if it is impossible that the conclusion be false while the premises are true. That is, to assert the premises and deny the conclusion would be a contradiction.

**variable** A changing quantity, usually denoted by a letter in algebraic equations, that might have any one of a range of possible values. Calculations can be carried out on variables because certain rules apply to all the possible values. For example, to carry out the operation of squaring all the integers between 0 and 10, and equation can be written in terms of an *integer variable $n$*, : $y = n^2$, with the condition that $n$ is between 0 and 10 ($0<n<10$). $y$ is called a *dependent variable* because its value depends on the value chosen for $n$, i.e. it can only have the values 1, 4, 9, ... etc. An *independent variable* has no such relationship with another variable. For example, if one variable, $x$, denotes the number of students in a school and another, $y$ denotes the proportion of the total number of students who want to take school lunches, then $x$ and $y$ are independent variables and a change in either of them will not affect the other. However, their product, $xy$, will affect a third quantity – the number of lunches ordered. Variables may also denote quantities other than ordinary arithmetic numbers, for example, vector variables and matrix variables.

**variance** A measure of the dispersion of a statistical sample. In a sample of $n$ observations $x_1$, $x_2$, $x_3$, ... $x_n$ with a sample mean $\bar{x}$, the sample variance is
$$s^2 = [(x_1 - \bar{x})^2 + (x_2 - \bar{x})^2 + (x_3 - \bar{x})^2 + ...(x_n - \bar{x})^2]/(n-1)$$
*See also* standard deviation.

**variational principle** A mathematical principle stating that the value of some quantity either has to be a minimum or (more rarely) a maximum out of all possible values. Many physical laws can be stated as variational principles, with FERMAT'S PRINCIPLE of geometrical optics being an example of this. Variational principles are also used to calculate quantities of interest in physical science and engineering approximately.

**VDU** *See* visual display unit.

**vector** /**vek**-ter/ A measure in which direction is important and must usually be specified. For instance, displacement is a vector quantity, whereas distance is a scalar. Weight, velocity, and magnetic field strength are other examples of vectors – they are each quoted as a number with a unit and a direction. Vectors are often denoted by printing the symbol in bold italic type *F*. *Vector algebra* treats vectors symbolically in a similar way to the algebra of scalar quantities but with different rules for addition, subtraction, multiplication, etc.

Any vector can be represented in terms of component vectors. In particular, a vector in three-dimensional Cartesian coordinates can be represented in terms of three unit vector components *i, j,* and *k* directed along the $x$-, $y$-, and $z$-axes respectively. If P is a point with coordinates $(x_1, y, z_1)$, then the vector OP is equivalent to $ix_1 + jy_1 + kz_1$.

*See also* vector difference; vector sum; vector multiplication.

**vector calculus** The branch of mathematics that combines differential and integral calculus with vectors. There are three

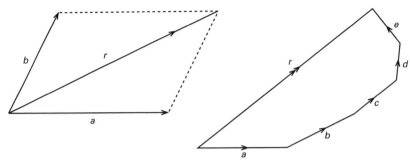

The parallelogram law: r is the resultant of a and b.

The polygon of vectors: r is the resultant.

Resolution of the vector r into different pairs of components

Vectors: addition and resolution of vectors

types of operator associated with differentiation in vector calculus: CURL, DIV, GRAD. Some of the key theorems for integrals in vector calculus are GAUSS'S THEOREM, GREEN'S THEOREM, and STOKES' THEOREM. Vector calculus was originally formulated for vectors in two and three dimensions in Euclidean space, but can be generalized to higher dimensions and curved manifolds. There are many physical applications of vector calculus, particularly in the theory of electricity and magnetism.

**vector difference** The result of subtracting two VECTORS. On a vector diagram, two vectors, *A* and *B*, are subtracted by drawing them tail-to-tail. The difference, *A* – *B* is the vector represented by the line from the head of *B* to the head of *A*. If *A* and *B* are parallel, the magnitude of the difference is the difference of the individual magnitudes. If they are antiparallel, it is the sum of the individual magnitudes.

The vector difference may also be calculated by taking the difference in the magnitudes of the corresponding components of each vector. For example, for two plane vectors in a Cartesian coordinate system.
$$A = 4i + 2j$$
$$B = 2i + j$$
where *i* and *j* are the unit vectors parallel to the *x*- and *y*-axes respectively,
$$A - B = 2i + j$$
*See also* vector sum.

**vector equation of a line** An equation for a straight line in space when it is expressed in terms of vectors. If *a* is the position vector of a point A on the line and *v* is any vector which has its direction along the line, then the line is the set of points P which have their position vector *p* given by *p* = *a* + *cv*, for some value of *c*. This type of equation is a vector equation of a line. The form of the equation means that the point P can only be on the line if *p*–*a* is propor-

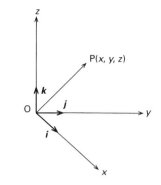

Basic vectors: the vector OP can be represented as $ix + jy + kz$.

Vector product $c = a \times b$.

Vectors: base vectors and vector product.

tional to $v$, i.e. the two vectors are in the same direction (or in opposite directions).

If a line is specified by two points A and B with position vectors $a$ and $b$ respectively then defining $v = b-a$ produces a vector equation of the line in the form $p = (1- c) a + cb$.

**vector equation of a plane** An equation for a plane when it is expressed in terms of vectors. If a point A in the plane has a position vector $a$ and $n$ is a normal vector, i.e. a vector which is perpendicular to the plane, then the plane is the set of all points P with position vector $p$ which satisfies the equation $(p-a).n = 0$. This type of equation is a vector equation of a plane. This equation can be written in the form $p.n = c$, where $c$ is a constant. The vector forms of equations of a plane can be converted into an equation for a plane in terms of the Cartesian coordinates $x$, $y$, and $z$ by taking components of the vectors.

**vector graphics** *See* computer graphics.

**vector multiplication** Multiplication of two or more vectors. This can be defined in two ways according to whether the result is a vector or a scalar. *See* scalar product; vector product; triple scalar product; triple vector product.

**vector product** A multiplication of two vectors to give a vector. The vector product of $A$ and $B$ is written $A \times B$. It is a vector of magnitude $AB\sin\theta$, where $A$ and $B$ are the magnitudes of $A$ and $B$ and $\theta$ is the angle between $A$ and $B$. The direction of the vector product is at right angles to $A$ and $B$. It points in the direction in which a right-hand screw would move turning from $A$ toward $B$. An example of a vector product is the force $F$ on a moving charge $Q$ in a field $B$ with velocity $v$ (as in the motor effect). Here

$$F = QB \times v$$

Another example is the product of a force and a distance to give a moment (turning effect), which can be represented by a vector at right angles to the plane in which the turning effect acts. The vector product is sometimes called the *cross product*. The vector product is not commutative because

$$A \times B = -(B \times A)$$

It is distributive with respect to vector addition:

$$C \times (A + B) = (C \times A) + (C \times B)$$

The magnitude of $A \times B$ is equal to the area of the parallelogram of which $A$ and $B$ form two non-parallel sides. In a three-dimensional Cartesian coordinate system with unit vectors $i$, $j$, and $k$ in the $x$, $y$, and $z$ directions respectively,

$$A \times B = (a_1i + a_2j + a_3k) \times (b_1i + b_2j + b_3k)$$

This can also be written as a determinant. *See also* scalar product.

**vector projection** The vector resulting from an orthogonal projection of one vector on another. For example, the vector projection of $A$ on $B$ is $bA\cos\theta$ where $\theta$ is the smaller angle between $A$ and $B$, and $b$ is the unit vector in the direction of $B$. *Compare* scalar projection.

**vectors, parallelogram of** *See* parallelogram of vectors.

**vectors, triangle of** *See* triangle of vectors.

**vector space** A *vector space* $V$ over a *scalar field* $F$ is a set of elements $x$, $y$, ..., called *vectors*, together with two algebraic operations called vector addition and multiplication of vectors by scalars, i.e. by elements of $F$, such that:
1. the sum of two elements is written $x + y$, and $V$ is an Abelian group with respect to addition;
2. the product of a vector, $x$, and a scalar, $a$, is written $ax$ and $a(bx) = (ab)x$, $1x = x$ for all $a$ and $b$ in $F$ and all $x$ in $V$;
3. the distributive laws hold, i.e. $a(x + y) = ax + ay$, $(a + b)x = ax + bx$ for all $a$ and $b$ in $F$ and all $x$ and $y$ in $V$.

The scalars may be real numbers, complex numbers, or elements of some other field.

**vector sum** The result of adding two VECTORS. On a vector diagram, vectors are added by drawing them head to tail. The sum is the vector represented by the straight line from the tail of the first to the head of the last. If they are parallel, the magnitude of the sum is the sum of the magnitudes of the individual vectors. If two vectors are anti-parallel, then the magnitude of the sum is the difference of the individual magnitudes.

The vector sum may also be calculated by summing the magnitudes of the corresponding components of each individual vector. For example, for two plane vectors, $A = 2i + 3j$ and $B = 6i + 4j$, in a Cartesian coordinate system with unit vectors $i$ and $j$ parallel to the $x$- and $y$-axes respectively, the vector sum, $A + B$, equals $8i + 7j$.

*See also* vector difference.

**velocities, parallelogram of** *See* parallelogram of vectors.

**velocities, triangle of** *See* triangle of vectors.

**velocity** Symbol: $v$ Displacement per unit time. The unit is the meter per second (m s$^{-1}$). Velocity is a vector quantity, speed being the scalar form. If velocity is constant, it is given by the slope of a position/time graph, and by the displacement divided by the time taken. If it is not constant, the mean value is obtained. If $x$ is the displacement, the instantaneous velocity is given by

$$v = dx/dt$$

*See also* equations of motion.

**velocity ratio** *See* distance ratio.

**Venn diagram** /ven/ A diagram used to show the relationships between SETS. The universal set, $E$, is shown as a rectangle. Inside this, other sets are shown as circles. Intersecting or overlapping circles are intersecting sets. Separate circles are sets that have no intersection. A circle inside another is a subset. A group of elements, or a subset, defined by any of these relationships may be indicated by a shaded area in the diagram. The diagram was devised by the British mathematician John Venn (1834–1923).

**vertex** (*pl.* **vertexes** or **vertices**) **1.** A point at which lines or planes meet in a figure; for example, the top point of a cone or pyramid, or a corner of a polygon or polyhedron.
**2.** One of the two points at which an axis of a conic cuts the conic. *See* ellipse; hyperbola; parabola.

**vertical** A direction that is at right angles to the horizontal (*see* horizontal); in everyday terms, upright. In geometry, measurements PERPENDICULAR to the base line are often described as being vertical.

**vertically opposite angles** One of the two pairs of equal angles formed when two straight lines cross each other.

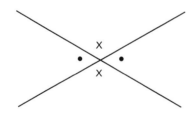

Vertically opposite angles formed at the intersection of two straight lines.

**vibration** (**oscillation**) Any regularly repeated to-and-fro motion or change. Examples are the swing of a pendulum, the vibration of a sound source, and the change with time of the electric and magnetic fields in an electromagnetic wave.

**virtual work** The work done if a system is displaced infinitesimally from its position. The virtual work is zero if the system is in equilibrium.

**visual display unit** (VDU) A computer terminal with which the user can communicate with the computer by means of a keyboard; this input and also the output from the computer appears on a cathode-ray screen. A VDU can operate both as in input and output device.

**volt** Symbol: V The SI unit of electrical potential, potential difference, and e.m.f., defined as the potential difference between two points in a circuit between which a constant current of one ampere flows when the power dissipated is one watt. One volt is one joule per coulomb ($1 \text{ V} = 1 \text{ J C}^{-1}$). The unit is named for the Italian physicist Count Alessandro Volta (1745–1827).

**volume** Symbol: V The extent of the space occupied by a solid or bounded by a closed surface, measured in units of length cubed. The volume of a box is the product of its length, its breadth, and its height. The SI unit of volume is the cubic meter ($\text{m}^3$).

**volume integral** The integral of a function over a volume. To emphasize the three-dimensional nature of the integral, a volume integral is sometimes written with three integral signs and d$x$d$y$d$z$. There are many physical applications of physical integrals, particularly to the theory of electricity and magnetism.

**volume of revolution** The volume V obtained by rotating the region bounded by $y = f(x)$, the $x$-axis and the lines $x = a$ and $x = b$ through one complete revolution, i.e. 360%, about the $x$-axis (with it being assumed that $f(x)$ is continuous in this interval). This volume is given by

$$\int_{a}^{b} \text{ab}\pi y^2 \text{d}x.$$

The expression can be derived by dividing the volume of revolution into very thin slabs. The volume of each slab is $\pi y^2 \delta x$, where $\delta x$ is the difference between $x$ values in a slab. Summing over all the slabs between $x = a$ and $x = b$ gives the desired expression for V.

**vulgar fraction** *See* fraction.

**Wallis's formula** an expression for $\pi/2$ in terms of an infinite product:
$$\frac{\pi}{2} = \frac{2.2.4.4.6.6.8.8\ldots}{1\,3\,3\,5\,5\,7\,7\,9} = \prod_{n=1}^{\infty} \frac{4n^2}{(4n^2-1)}$$
This product is sometimes known as *Wallis's product*. The formula is named for the English mathematician and theologian John Wallis (1616–1703).

**wallpaper symmetry** A name given to the types of regular repeating patterns that are possible in two dimensions, i.e. the two-dimensional analog of SPACE GROUPS. There are 17 different types of possible wallpaper symmetry.

**watt** /wot/ Symbol: W The SI unit of power, defined as a power of one joule per second. $1\text{ W} = 1\text{ J s}^{-1}$. The unit is named for the British instrument maker and inventor James Watt (1736–1819).

**wave** A method of energy transfer involving some form of vibration. For instance, waves on the surface of a liquid or along a stretched string involve regular to-and-fro motion of particles about a mean position. Sound waves carry energy by alternate compressions and rarefactions of air (or other media). In electromagnetic waves, electric and magnetic fields vary at right angles to the direction of propagation of the wave. At any particular instance, a graph of displacement against distance is a regular repeating curve – the *waveform* or wave profile of the wave. In a *traveling* (or *progressive*) wave the whole periodic displacement moves through the medium. At any point in the medium the disturbance is changing with time. Under certain conditions a STATIONARY (or standing) WAVE can be produced in which the disturbance does not change with time.

For the simple case of a plane progressive wave the displacement at a point can be represented by an equation:
$$y = a\sin 2\pi(ft - x/\lambda)$$
where $a$ is the amplitude, $f$ the frequency, $x$ the distance from the origin, and $\lambda$ the wavelength. Other relationships are:
$$y = a\sin 2\pi(vt - x)/\lambda$$
where $v$ is the speed, and
$$y = a\sin 2\pi(t/T - x/\lambda)$$
where $T$ is the period. Note that if the minus sign is replaced by a plus sign in the above equations it implies a similar wave moving in the opposite direction. For a stationary wave resulting from two waves in opposite directions, the displacement is given by:
$$Y = 2a\cos 2\pi x/\lambda$$
*See also* longitudinal wave; transverse wave; phase.

**wave equation** A second-order partial differential equation that describes wave motion. The equation
$$\partial^2 u/\partial x^2 = (1/c^2)\partial^2 u/\partial t^2$$
might represent, for example, the vertical displacement $u$ of the water surface as a plane wave of velocity $c$ passes along the surface, the horizontal position and the time being given by $x$ and $t$ respectively. The general solution of this one-dimensional wave equation is a periodic function of $x$ and $t$. There exist analogous wave equations for two and three dimensions.

**waveform** *See* wave.

**wavefront** A continuous surface associated with a wave radiation, in which all the vibrations concerned are in phase. A parallel beam has plane wavefronts; the output of a point source has spherical wavefronts.

**wavelength** Symbol: $\lambda$ The distance between the ends of one complete cycle of a wave. Wavelength relates to the wave speed ($c$) and its frequency ($\nu$) thus:
$$c = \nu\lambda$$

**wave motion** Any form of energy transfer that may be described as a wave rather than as a stream of particles. The term is also sometimes used to mean any harmonic motion.

**wave number** Symbol: $\sigma$ The reciprocal of the wavelength of a wave. It is the number of wave cycles in unit distance, and is often used in spectroscopy. The unit is the meter$^{-1}$ (m$^{-1}$). The circular wave number (Symbol: $k$) is given by
$$k = 2\pi\sigma$$

**weber** /**vay**-ber/ Symbol: Wb The SI unit of magnetic flux, equal to the magnetic flux that, linking a circuit of one turn, produces an e.m.f. of one volt when reduced to zero at a uniform rate in one second. 1 Wb = 1 V s. The unit is named for the German physicist Wilhelm Eduard Weber (1804–91).

**weight** Symbol: $W$ The force by which a mass is attracted to another, such as the Earth. It is proportional to the body's mass ($m$), the constant of proportionality being the gravitational field strength (i.e. the acceleration of free fall). Thus $W = mg$, where $g$ is the acceleration of free fall. The mass of a body is normally constant, but its weight varies with position (because it depends on $g$).

Although mass and weight are often used interchangeably in everyday language, they are different in scientific language and must not be confused.

**weighted mean** *See* mean.

**weightlessness** An apparent loss of weight experienced by an object in free fall. Thus for a person in an orbiting spacecraft, the weight in the Earth's frame of reference is the centripetal force necessary to maintain the circular orbit. In the frame of reference of the spacecraft the person feels that he or she has no weight.

**wheel and axle** A simple MACHINE consisting of a wheel on an axle that has a rope around it. An effort applied to the wheel is transmitted to a load exerted at the axle rope. The force ratio (mechanical advantage) is equal to $r_W/r_A$ where $r_W$ is the radius of the wheel and $r_A$ that of the axle.

**whole numbers** Symbol: $W$ The set of integers $\{1,2,3,...\}$, excluding zero.

**word** The basic unit in which information is stored and manipulated in a computer. Each word consists usually of a fixed number of BITS. This number, known as the *word length*, varies according to the type of computer and may be as few as eight or as many as 60. Each word is given a unique address in store. A word may represent an instruction to the computer or a piece of data. An instruction word is coded to give the operation to be performed and the address or addresses of the data on which the operation is to be performed. *See also* byte.

**word processor** A microcomputer that is programmed to help in preparing text for printing or data transmission. A general-purpose computer can be used as a word processor by means of a suitable applications program.

**work** Symbol: $W$ The work done by a force is the product of the force and the distance moved in the same direction:
$$\text{work} = \text{force} \times \text{displacement}$$
Work is in fact a process of energy transfer and, like energy, is measured in joules. If the directions of force ($F$) and motion are not the same the component of the force in the direction of the motion is used.
$$W = Fs\cos\theta$$
where $s$ is displacement and $\theta$ the angle between the directions of force and motion. Work is the scalar product of force and displacement.

**world curve** *See* space–time.

**write head** A device, part of a computer system, that records data onto a magnetic storage medium such as tape or disk. *See also* input device.

**Wronskian** /**vrons**-kee-ăn/ A determinant which can be used to examine whether the functions $f_1(x)$, $f_2(x)$, $f_n(x)$ of $x$, each of which has the non-vanishing derivatives are linearly independent. The *Wronskian* W is defined to be the determinant shown. If $W \neq 0$ then the $n$ functions are linearly in-

$$W = \begin{vmatrix} f_1(x) & f_2(x) & \cdots & f_n(x) \\ f'_1(x) & f'_2(x) & \cdots & f'_1(x) \\ \vdots & \vdots & & \vdots \\ f_1^{(n-1)}(x) & f_2^{(n-1)}(x) & & f_n^{(n-1)}(x) \end{vmatrix}$$

dependent but if $W \neq 0$ the functions are linearly dependent.

The concept of the Wronskian is used in the theory of differential equations. The Wronskian is named for the Polish mathematician J. M. H. Wronski (1778–1853).

**yard** A unit of length now defined as 0.9144 meter. *y*-coordinate *See* ordinate.

**yield** The income produced by a stock or share expressed as a percentage of its market price.

**yocto-** Symbol: y A prefix denoting $10^{-24}$. For example, 1 yoctometer (ym) = $10^{-24}$ meter (m).

**yotta-** Symbol: Y A prefix denoting $10^{24}$. For example, 1 yottameter (Ym) = $10^{24}$ meter (m).

**Young modulus** The ratio of the stress to the strain in a deformed body in the case of elongation or compression of the body. There are several other types of *elastic modulus* i.e. the ratio of stress to strain, depending on the nature of the deformation of the body. The *bulk modulus* is the ratio of the pressure applied to a body to the fraction by which the volume of the body has decreased. The *shear modulus*, also known as the *rigidy* is the ratio of the shear force per unit area divided by the deformation of the body, measured in radians. Since these elastic moduli all have ratios of stress to strain they have the dimensions of force per unit area since stresses have the dimensions of force per unit area and strains are dimensionless numbers. The modulus is named for the British physicist and physician Thomas Young (1773–1829).

**Z**

**zepto-** Symbol: z  A prefix denoting $10^{-21}$. For example, 1 zeptometer (zm) = $10^{-21}$ meter (m).

**zero** (0)  The number that when added to another number gives a sum equal to that other number. It is included in the set of integers but not in the set of whole numbers. The product of any number and zero is zero. Zero is the identity element for addition.

**zero function**  A function f($x$), for which f($x$) = 0 for all values of $x$, where $x$ belongs to the set of all real numbers.

**zero matrix**  *See* null matrix.

**zeta function**  *See* Riemann zeta function.

**zetta-** Symbol: Z  A prefix denoting $10^{21}$. For example, 1 zettameter (Zm) = $10^{21}$ meter (m).

**zone**  A part of a sphere produced by two parallel planes cutting the sphere.

**Zorn's lemma** /zornz/ (**Kuratowski–Zorn lemma**)  If a set $S$ is partially ordered and each linearly ordered subset has an upper bound in $S$, then $S$ contains at least one maximal element, i.e. an element $x$ such that there is no $y$ in $S$ for $x < y$. The lemma was first discovered in 1922 by the Polish mathematician Kazimierz Kuratowski (1896–1980) and independently in 1935 by the German-born American mathematician Max Zorn (1906–93).

# APPENDIXES

# Appendix I

## Symbols and Notation

### Arithmetic and algebra

| | |
|---|---|
| equal to | $=$ |
| not equal to | $\neq$ |
| identity | $\equiv$ |
| approximately equal to | $\simeq$ |
| approaches | $\rightarrow$ |
| proportional to | $\propto$ |
| less than | $<$ |
| greater than | $>$ |
| less than or equal to | $\leq$ |
| greater than or equal to | $\geq$ |
| much less than | $\ll$ |
| much greater than | $\gg$ |
| plus, positive | $+$ |
| minus, negative | $-$ |
| plus or minus | $\pm$ |
| multiplication | $a \times b$ |
| | $a.b$ |
| division | $a \div b$ |
| | $a/b$ |
| magnitude of $a$ | $\lvert a \rvert$ |
| factorial $a$ | $a!$ |
| logarithm (to base $b$) | $\log_b a$ |
| common logarithm | $\log_{10} a$ |
| natural logarithm | $\log_e a$ or $\ln a$ |
| summation | $\Sigma$ |
| continued product | $\Pi$ |

### Geometry and trigonometry

| | |
|---|---|
| angle | $\angle$ |
| triangle | $\triangle$ |
| square | $\square$ |
| circle | $\bigcirc$ |
| parallel to | $\parallel$ |
| perpendicular to | $\perp$ |
| congruent to | $\equiv$ |
| similar to | $\sim$ |
| sine | $\sin$ |

# Symbols and Notation

| | |
|---|---|
| cosine | cos |
| tangent | tan |
| cotangent | cot, ctn |
| secant | sec |
| cosecant | cosec, csc |
| inverse sine | $\sin^{-1}$, arc sin |
| inverse cosine | $\cos^{-1}$, arc cos |
| inverse tangent | $\tan^{-1}$, arc tan |
| Cartesian coordinates | $(x, y, z)$ |
| spherical coordinates | $(r, \theta, \phi)$ |
| cylindrical coordinates | $(r, \theta, z)$ |
| direction numbers or cosines | $(l, m, n)$ |

## Sets and logic

| | |
|---|---|
| implies that | $\Rightarrow$ |
| is implied by | $\Leftarrow$ |
| implies and is implied by (if and only if) | $\Leftrightarrow$ |
| set $a, b, c,...$ | $\{a, b, c, ...\}$ |
| is an element of | $\in$ |
| is not an element of | $\notin$ |
| such that | : |
| number of elements in set $S$ | $n(S)$ |
| universal set | $E$ or $\mathscr{E}$ |
| empty set | $\varnothing$ |
| complement of $S$ | $S'$ |
| union | $\cup$ |
| intersection | $\cap$ |
| is a subset of | $\subset$ |
| corresponds one-to-one with | $\leftrightarrow$ |
| $x$ is mapped onto $y$ | $x \rightarrow y$ |
| conjunction | $\wedge$ |
| disjunction | $\vee$ |
| negation (of $p$) | $\sim p$ or $\neg p$ |
| implication | $\rightarrow$ or $\supset$ |
| biconditional (equivalence) | $\equiv$ or $\leftrightarrow$ |
| the set of natural numbers | $\mathbb{N}$ |
| the set of integers | $\mathbb{Z}$ |
| the set of rational numbers | $\mathbb{Q}$ |
| the set of real numbers | $\mathbb{R}$ |
| the set of complex numbers | $\mathbb{C}$ |

# Symbols and Notation

## Calculus

| | |
|---|---|
| increment of $x$ | $\triangle x$ or $\delta x$ |
| limit of function of $x$ as $x$ approaches $a$ | $\displaystyle\lim_{x \to a} f(x)$ |
| derivative of f($x$) | $df(x)/dx$ or $f'(x)$ |
| second derivative of f($x$) etc. | $d^2f(x)/dx^2$ or $f''(x)$ etc. |
| indefinite integral of f($x$) with respect to $x$ | $\int f(x)dx$ |
| definite integral with limits $a$ and $b$ | $\displaystyle\int_b^a f(x)dx$ |
| partial derivative of function f($x,y$) with respect to $x$ | $\partial f(x,y)/\partial x$ |

# Appendix II

## Symbols for Physical Quantities

| | | | |
|---|---|---|---|
| acceleration | $a$ | moment of inertia | $I$ |
| angle | $\theta, \phi, \alpha,$ | momentum | $p$ |
| | $\beta$, etc. | period | $\tau$ |
| angular acceleration | $\alpha$ | potential energy | $E_p, V$ |
| angular frequency, $2\pi f$ | $\omega$ | power | $P$ |
| angular momentum | $L$ | pressure | $p$ |
| angular velocity | $\omega$ | radius | $r$ |
| area | $A$ | reduced mass | $\mu$ |
| breadth | $b$ | relative density | $d$ |
| circular wavenumber | $k$ | solid angle | $\Omega, \omega$ |
| density | $\rho$ | thickness | $d$ |
| diameter | $d$ | time | $t$ |
| distance | $s, L$ | torque | $T$ |
| energy | $W, E$ | velocity | $v$ |
| force | $F$ | viscosity | $\eta$ |
| frequency | $f, \nu$ | volume | $V$ |
| height | $h$ | wavelength | $\lambda$ |
| kinetic energy | $E_k, T$ | wavenumber | $\sigma$ |
| length | $l$ | weight | $W$ |
| mass | $m$ | work | $W, E$ |
| moment of force | $M$ | | |

# Appendix III

## Areas and Volumes

### Plane figures

| Figure | Dimensions | Perimeter | Area |
|---|---|---|---|
| triangle | sides $a$, $b$, and $c$, angle $A$ | $a + b + c$ | $\frac{1}{2}bc.\sin A$ |
| square | side $a$ | $4a$ | $a^2$ |
| rectangle | sides $a$ and $b$ | $2(a + b)$ | $a \times b$ |
| kite | diagonals $c$ and $d$ | | $\frac{1}{2}c \times d$ |
| parallelogram | sides $a$ and $b$ distances $c$ and $d$ apart | $2(a + b)$ | $a.c$ or $b.d$ |
| circle | radius $r$ | $2\pi r$ | $\pi r^2$ |
| ellipse | axes $a$ and $b$ | $2\pi\sqrt{[(a^2 + b^2)/2]}$ | $\pi ab$ |

### Solid figures

| Figure | Dimensions | Area | Volume |
|---|---|---|---|
| cylinder | radius $r$, height $h$ | $2\pi r(h + r)$ | $\pi r^2 h$ |
| cone | base radius $r$, slant height $l$, height $h$ | $\pi rl$ | $\pi r^2 h/3$ |
| sphere | radius $r$ | $4\pi r^2$ | $4\pi r^3/3$ |

# Appendix IV

## Expansions

$\sin x$      $x/1! - x^3/3! + x^5/5! - x^7/7! + x^9/9! - \ldots$

$\cos x$      $1 - x^2/2! + x^4/4! - x^6/6! + x^8/8! - \ldots$

$e^x$      $1 + x/1! + x^2/2! + x^3/3! + x^4/4! + x^5/5! + \ldots$

$\sinh x$      $x + x^3/3! + x^5/5! + x^7/7! + x^9/9! + \ldots$

$\cosh x$      $1 + x^2/2! + x^4/4! + x^6/6! + x^8/8! + \ldots$

$\log_e(1+x)$      $x - x^2/2 + x^3/3 - x^4/4 + x^5/5 - \ldots$

$\log_e(1-x)$      $-x - x^2/2 - x^3/3 - x^4/4 - x^5/5 - \ldots$

$(1 + x)^n$      $1 + nx + n(n - 1)x^2/2! + n(n - 1)(n - 2)x^3/3! + \ldots$
for $|x| < 1$

$f(a + x)$      $f(a) + xf'(a) + (x^2/2!)f''(a) + (x^3/3!)f'''(a) + (x^4/4!)f''''(a) + \ldots$
where $f'(a)$ denotes the first derivative, $f''(a)$ the second derivative, etc.

$f(x)$      $f(0) + xf'(0) + (x^2/2!)f''(0) + (x^3/3!)f'''(0) + (x^4/4!)f''''(0) + \ldots$

# Appendix V

## Derivatives

$x$ is a variable, $u$ is a function of $x$, and $a$ and $n$ are constants

| Function $f(x)$ | Derivative $df(x)/dx$ |
|---|---|
| $x$ | $1$ |
| $ax$ | $a$ |
| $ax^n$ | $nax^{n-1}$ |
| $e^{ax}$ | $ae^{ax}$ |
| $\log_e x$ | $1/x$ |
| $\log_a x$ | $(1/x)\log_e a$ |
| $\cos x$ | $-\sin x$ |
| $\sin x$ | $\cos x$ |
| $\tan x$ | $\sec^2 x$ |
| $\cot x$ | $-\mathrm{cosec}^2 x$ |
| $\sec x$ | $\tan x.\sec x$ |
| $\mathrm{cosec}\ x$ | $-\cot x.\mathrm{cosec}\ x$ |
| $\cos u$ | $-\sin u.(du/dx)$ |
| $\sin u$ | $\cos u.(du/dx)$ |
| $\tan u$ | $\sec^2 u.(du/dx)$ |
| $\log_e u$ | $(1/u)(du/dx)$ |
| $\sin^{-1}(x/a)$ | $1/\sqrt{(a^2 - x^2)}$ |
| $\cos^{-1}(x/a)$ | $-1/\sqrt{(a^2 - x^2)}$ |
| $\tan^{-1}(x/a)$ | $a/(a^2 + x^2)$ |

# Appendix VI

## Integrals

$x$ is a variable and $a$ and $n$ are constants. Note that a constant of integration $C$ should be added to each integral.

| *Function* $f(x)$ | *Integral* $\int f(x)dx$ |
|---|---|
| $a$ | $ax$ |
| $x$ | $x^2/2$ |
| $x^n$ | $x^{n+1}/(n+1)$ |
| $1/x$ | $\log_e x$ |
| $e^{ax}$ | $e^{ax}/a$ |
| $\log_e ax$ | $x\log_e ax - x$ |
| $\cos x$ | $\sin x$ |
| $\sin x$ | $-\cos x$ |
| $\tan x$ | $\log_e (\cos x)$ |
| $\cot x$ | $\log_e (\sin x)$ |
| $\sec x$ | $\log_e (\sec x + \tan x)$ |
| $\operatorname{cosec} x$ | $\log_e (\operatorname{cosec} x - \cot x)$ |
| $1/\sqrt{(a^2 - x^2)}$ | $\sin^{-1} (x/a)$ |
| $-1/\sqrt{(a^2 - x^2)}$ | $\cos^{-1} (x/a)$ |
| $a/(a^2 + x^2)$ | $\tan^{-1} (x/a)$ |

# Appendix VII

## Trigonometric Formulae

*Addition formulae*:

$$\sin(x + y) = \sin x \cos y + \cos x \sin y$$
$$\sin(x - y) = \sin x \cos y - \cos x \sin y$$
$$\cos(x + y) = \cos x \cos y - \sin x \sin y$$
$$\cos(x - y) = \cos x \cos y + \sin x \sin y$$
$$\tan(x + y) = (\tan x + \tan y)/(1 - \tan x \tan y)$$
$$\tan(x - y) = (\tan x - \tan y)/(1 + \tan x \tan y)$$

*Double-angle formulae*:

$$\sin(2x) = 2 \sin x \cos x$$
$$\cos(2x) = \cos^2 x - \sin^2 x$$
$$\tan(2x) = 2\tan x/(1 - \tan^2 x)$$

*Half-angle formulae*:

$$\sin(x/2) = \pm\sqrt{[(1 - \cos x)/2]}$$
$$\cos(x/2) = \pm\sqrt{[(1 + \cos x)/2]}$$
$$\tan(x/2) = \sin x/(1 + \cos x)$$
$$= (1 - \cos x)/\sin x$$

*Poduct formulae*:

$$\sin x \cos y = \tfrac{1}{2}[\sin(x + y) + \sin(x - y)]$$
$$\cos x \sin y = \tfrac{1}{2}[\sin(x + y) - \sin(x - y)]$$
$$\cos x \cos y = \tfrac{1}{2}[\cos(x + y) + \cos(x - y)]$$
$$\sin x \sin y = \tfrac{1}{2}[\cos(x - y) - \cos(x + y)]$$

# Appendix VIII

## Conversion Factors

### Length

| *To convert* | *into* | *multiply by* |
|---|---|---|
| inches | meters | 0.0254 |
| feet | meters | 0.3048 |
| yards | meters | 0.9144 |
| miles | kilometers | 1.60934 |
| nautical miles | kilometers | 1.85200 |
| nautical miles | miles | 1.15078 |
| kilometers | miles | 0.621371 |
| kilometers | nautical miles | 0.539957 |
| meters | inches | 39.3701 |
| meters | feet | 3.28084 |
| meters | yards | 1.09361 |

### Area

| *To convert* | *into* | *multiply by* |
|---|---|---|
| square inches | square centimeters | 6.4516 |
| square inches | square meters | $6.4516 \times 10^{-4}$ |
| square feet | square meters | $9.2903 \times 10^{-2}$ |
| square yards | square meters | 0.836127 |
| square miles | square kilometers | 2.58999 |
| square miles | acres | 640 |
| acres | square meters | 4046.86 |
| acres | square miles | $1.5625 \times 10^{-3}$ |
| square centimeters | square inches | 0.155 |
| square meters | square feet | 10.7639 |
| square meters | square yards | 1.19599 |
| square meters | acres | $2.47105 \times 10^{-4}$ |
| square meters | square miles | $3.86019 \times 10^{-7}$ |
| square kilometers | square miles | 0.386019 |

### Volume

| *To convert* | *into* | *multiply by* |
|---|---|---|
| cubic inches | liters | $1.63871 \times 10^{-2}$ |
| cubic inches | cubic meters | $1.63871 \times 10^{-5}$ |
| cubic feet | liters | 28.3168 |
| cubic feet | cubic meters | 0.0283168 |
| cubic yard | cubic meters | 0.764555 |
| gallon (US) | liters | 3.785438 |
| gallon (US) | cubic meters | $3.785438 \times 10^{-3}$ |
| gallon (US) | gallon (UK) | 0.83268 |

# Conversion Factors

## Mass

| To convert | into | multiply by |
|---|---|---|
| pounds | kilograms | 0.453592 |
| pounds | tonnes | $4.53592 \times 10^{-4}$ |
| hundredweight (short) | kilograms | 45.3592 |
| hundredweight (short) | tonnes | 0.0453592 |
| tons (short) | kilograms | 907.18 |
| tons (short) | tonnes | 0.90718 |
| kilograms | pounds | 2.204623 |
| kilograms | hundredweights (short) | 0.022046 |
| kilograms | tons (short) | $1.1023 \times 10^{-3}$ |
| tonnes | pounds | 2204.623 |
| tonnes | hundredweights (short) | 22.0462 |
| tonnes | tons (short) | 0.90718 |

The short ton is used in the USA and is equal to 2000 pounds. The short hundredweight (also known as the cental) is 100 pounds.

The long ton, which is used in the UK, is equal to 2240 pounds (1016.047 kg). The long hundredweight is 112 pounds (50.802 kg). 1 long ton equals 20 long hundredweights.

## Force

| To convert | into | multiply by |
|---|---|---|
| pounds force | newtons | 4.44822 |
| pounds force | kilograms force | 0.453592 |
| pounds force | dynes | 444822 |
| pounds force | poundals | 32.174 |
| poundals | newtons | 0.138255 |
| poundals | kilograms force | 0.031081 |
| poundals | dynes | 13825.5 |
| poundals | pounds force | 0.031081 |
| dynes | newtons | $10^{-5}$ |
| dynes | kilograms force | $1.01972 \times 10^{-6}$ |
| dynes | pounds force | $2.24809 \times 10^{-6}$ |
| dynes | poundals | $7.2330 \times 10^{-5}$ |
| kilograms force | newtons | 9.80665 |
| kilograms force | dynes | 980665 |
| kilograms force | pounds force | 2.20462 |
| kilograms force | poundals | 70.9316 |
| newtons | kilograms | 0.101972 |
| newtons | dynes | 100000 |
| newtons | pounds force | 0.224809 |
| newtons | poundals | 7.2330 |

# Conversion Factors

## Work and energy

| To convert | into | multiply by |
|---|---|---|
| British Thermal Units | joules | 1055.06 |
| British Thermal Units | calories | 251.997 |
| British Thermal Units | kilowatt-hours | $2.93071 \times 10^{-4}$ |
| kilowatt-hours | joules | 3600000 |
| kilowatt-hours | calories | 859845 |
| kilowatt-hours | British Thermal Units | 3412.14 |
| calories | joules | 4.1868 |
| calories | kilowatt-hours | $1.16300 \times 10^{-6}$ |
| calories | British Thermal Units | $3.96831 \times 10^{-3}$ |
| joules | calories | 0.238846 |
| joules | kilowatt hours | $2.7777 \times 10^{-7}$ |
| joules | British Thermal Units | $9.47813 \times 10^{-4}$ |
| joules | electron volts | $6.2418 \times 10^{18}$ |
| joules | ergs | $10^7$ |
| electronvolts | joules | $1.6021 \times 10^{-19}$ |
| ergs | joules | $10^{-7}$ |

## Pressure

| To convert | into | multiply by |
|---|---|---|
| atmospheres | pascals* | 101325 |
| bars | pascals | 100000 |
| pounds per square inch | pascals | 68894.76 |
| pounds per square inch | kilograms per square meter | 703.068 |
| pounds per square inch | atmospheres | 0.068046 |
| kilograms per square meter | pascals | 9.80661 |
| kilograms per square meter | pounds per square inch | $1.42234 \times 10^{-3}$ |
| kilograms per square meter | atmospheres | $9.67841 \times 10^{-5}$ |
| pascals | kilograms per square meter | 0.101972 |
| pascals | pounds per square inch | $1.45038 \times 10^{-4}$ |
| pascals | atmospheres | $9.86923 \times 10^{-6}$ |

*1 pascal = 1 newton per square meter

# Appendix IX

## Powers and Roots

| $n$ | $n^2$ | $n^3$ | $\sqrt{n}$ | $\sqrt[3]{n}$ |
|---|---|---|---|---|
| 1 | 1 | 1 | 1.000 | 1.000 |
| 2 | 4 | 8 | 1.414 | 1.260 |
| 3 | 9 | 27 | 1.732 | 1.442 |
| 4 | 16 | 64 | 2.000 | 1.587 |
| 5 | 25 | 125 | 2.236 | 1.710 |
| 6 | 36 | 216 | 2.449 | 1.817 |
| 7 | 49 | 343 | 2.646 | 1.913 |
| 8 | 64 | 512 | 2.828 | 2.000 |
| 9 | 81 | 729 | 3.000 | 2.080 |
| 10 | 100 | 1 000 | 3.162 | 2.154 |
| 11 | 121 | 1 33` | 3.317 | 2.224 |
| 12 | 144 | 1 728 | 3.464 | 2.289 |
| 13 | 169 | 2 197 | 3.606 | 2.351 |
| 14 | 196 | 2 744 | 3.742 | 2.410 |
| 15 | 225 | 3 375 | 3.873 | 2.466 |
| 16 | 256 | 4 096 | 4.000 | 2.520 |
| 17 | 289 | 4 913 | 4.123 | 2.571 |
| 18 | 324 | 5 832 | 4.243 | 2.621 |
| 19 | 361 | 6 859 | 4.359 | 2.668 |
| 20 | 400 | 8 000 | 4.472 | 2.714 |
| 21 | 441 | 9 261 | 4.583 | 2.759 |
| 22 | 484 | 10 648 | 4.690 | 2.802 |
| 23 | 529 | 12 167 | 4.796 | 2.844 |
| 24 | 576 | 13 824 | 4.899 | 2.844 |
| 25 | 625 | 15 625 | 5.000 | 2.924 |
| 26 | 676 | 17 576 | 5.099 | 2.962 |
| 27 | 729 | 19 683 | 5.196 | 3.000 |
| 28 | 784 | 21 952 | 5.292 | 3.037 |
| 29 | 841 | 24 389 | 5.385 | 3.072 |
| 30 | 900 | 27 000 | 5.477 | 3.107 |
| 31 | 961 | 29 791 | 5.568 | 3.141 |
| 32 | 1 024 | 32 768 | 5.657 | 3.175 |

# Powers and Roots

| $n$ | $n^2$ | $n^3$ | $\sqrt{n}$ | $\sqrt[3]{n}$ |
|---|---|---|---|---|
| 33 | 1 089 | 35 937 | 5.745 | 3.208 |
| 34 | 1 156 | 39 304 | 5.831 | 3.240 |
| 35 | 1 225 | 42 875 | 5.916 | 3.271 |
| 36 | 1 296 | 46 656 | 6.000 | 3.302 |
| 37 | 1 369 | 50 653 | 6.083 | 3.332 |
| 38 | 1 444 | 54 872 | 6.164 | 3.362 |
| 39 | 1 521 | 59 319 | 6.245 | 3.391 |
| 40 | 1 600 | 64 000 | 6.325 | 3.420 |
| 41 | 1 681 | 68 921 | 6.403 | 3.448 |
| 42 | 1 764 | 74 088 | 6.481 | 3.476 |
| 43 | 1 849 | 79 507 | 6.557 | 3.503 |
| 44 | 1 936 | 85 184 | 6.633 | 3.530 |
| 45 | 2 025 | 91 125 | 6.708 | 3.557 |
| 46 | 2 116 | 97 336 | 6.782 | 3.583 |
| 47 | 2 209 | 103 823 | 6.856 | 3.609 |
| 48 | 2 304 | 110 592 | 6.928 | 3.634 |
| 49 | 2 401 | 117 649 | 7.000 | 3.659 |
| 50 | 2 500 | 125 000 | 7.071 | 3.684 |
| 51 | 2 601 | 132 651 | 7.141 | 3.708 |
| 52 | 2 704 | 140 608 | 7.211 | 3.733 |
| 53 | 2 809 | 148 877 | 7.280 | 3.756 |
| 54 | 2 916 | 157 464 | 7.348 | 3.780 |
| 55 | 3 025 | 166 375 | 7.416 | 3.803 |
| 56 | 3 136 | 175 616 | 7.483 | 3.826 |
| 57 | 3 249 | 185 193 | 7.550 | 3.849 |
| 58 | 3 364 | 195 112 | 7.616 | 3.871 |
| 59 | 3 481 | 205 379 | 7.681 | 3.893 |
| 60 | 3 600 | 216 000 | 7.746 | 3.915 |
| 61 | 3 721 | 226 981 | 7.810 | 3.936 |
| 62 | 3 844 | 238 328 | 7.874 | 3.958 |
| 63 | 3 969 | 250 047 | 7.937 | 3.979 |
| 64 | 4 096 | 262 144 | 8.000 | 4.000 |
| 65 | 4 225 | 274 625 | 8.062 | 4.021 |

# Powers and Roots

| $n$ | $n^2$ | $n^3$ | $\sqrt{n}$ | $\sqrt[3]{n}$ |
|---|---|---|---|---|
| 66 | 4 356 | 287 496 | 8.124 | 4.041 |
| 67 | 4 489 | 300 763 | 8.185 | 4.062 |
| 68 | 4 624 | 314 432 | 8.246 | 4.082 |
| 69 | 4 761 | 328 509 | 8.307 | 4.102 |
| 70 | 4 900 | 343 000 | 8.367 | 4.121 |
| 71 | 5 041 | 357 911 | 8.426 | 4.141 |
| 72 | 5 184 | 373 248 | 8.485 | 4.160 |
| 73 | 5 329 | 389 017 | 8.544 | 4.179 |
| 74 | 5 476 | 405 224 | 8.602 | 4.198 |
| 75 | 5.625 | 421 875 | 8.660 | 4.217 |
| 76 | 5 776 | 438 976 | 8.718 | 4.236 |
| 77 | 5 929 | 456 533 | 8.775 | 4.254 |
| 78 | 6 084 | 474 552 | 8.832 | 4.273 |
| 79 | 6 241 | 493 039 | 8.888 | 4.291 |
| 80 | 6 400 | 512 000 | 8.944 | 4.309 |
| 81 | 6 561 | 531 441 | 9.000 | 4.327 |
| 82 | 6 724 | 551 368 | 9.055 | 4.344 |
| 83 | 6 889 | 571 787 | 9.110 | 4.362 |
| 84 | 7 056 | 592 704 | 9.165 | 4.380 |
| 85 | 7 225 | 614 125 | 9.220 | 4.397 |
| 86 | 7 396 | 636 056 | 9.274 | 4.414 |
| 87 | 7 569 | 658 503 | 9.327 | 4.431 |
| 88 | 7 744 | 681 472 | 9.381 | 4.448 |
| 89 | 7 921 | 704 969 | 9.434 | 4.465 |
| 90 | 8 100 | 729 000 | 9.487 | 4.481 |
| 91 | 8 281 | 753 571 | 9.539 | 4.498 |
| 92 | 8 464 | 778 688 | 9.592 | 4.514 |
| 93 | 8 649 | 804 357 | 9 644 | 4.531 |
| 94 | 8 836 | 830 584 | 9.695 | 4.547 |
| 95 | 9 025 | 857 375 | 9.747 | 4.563 |
| 96 | 9 216 | 884 736 | 9.798 | 4.579 |
| 97 | 9 409 | 912 673 | 9.849 | 4.595 |
| 98 | 9 604 | 941 192 | 9.899 | 4.610 |
| 99 | 9 801 | 970 299 | 9.950 | 4.626 |
| 100 | 10 000 | 1 000 000 | 10.000 | 4.642 |

# Appendix X

## The Greek Alphabet

| | | | | | |
|---|---|---|---|---|---|
| A | α | alpha | N | ν | nu |
| B | β | beta | Ξ | ξ | xi |
| Γ | γ | gamma | O | o | omikron |
| Δ | δ | delta | Π | π | pi |
| E | ε | epsilon | P | ρ | rho |
| Z | ζ | zeta | Σ | σ | sigma |
| H | η | eta | T | τ | tau |
| Θ | θ | theta | Y | υ | upsilon |
| I | ι | iota | Φ | φ | phi |
| K | κ | kappa | X | χ | chi |
| Λ | λ | lambda | Ψ | ψ | psi |
| M | μ | mu | Ω | ω | omega |

# Appendix XI

## Web Sites

**Organizations:**

| | |
|---|---|
| American Mathematical Society | www.ams.org |
| Mathematical Association of America | www.maa.org |
| International Mathematical Union | www.mathunion.org |
| London Mathematical Society | www.lms.ac.uk |
| Mathematics Foundation of America | www.mfoa.org |

**General resources:**

| | |
|---|---|
| Mathematics on the Web | www.ams.org/mathweb |
| Mathematics WWW Virtual Library | euclid.math.fsu.edu/Science/math.html |
| MathGate Homepage | www.mathgate.ac.uk |
| Mathematical Resources | www.ama.caltech.edu/resources.html |
| Math-Net Links | www.math-net.de/links/ show?collection=math |
| Mathematics Resources on the WWW | mthwww.uwc.edu/wwwmahes/ homepage.htm |

**History and biography:**

| | |
|---|---|
| History of Mathematics Archive | www-history.mcs.st-and.ac.uk/history |

# Bibliography

Bittinger, Marvin L. *Basic Mathematics*. 9th ed. New York: Addison Wesley, 2002.

Boyer, Carl B. and Isaac Asimov. *A History of Mathematics*. 2nd ed. New York: Wiley, 1991.

Courant, Richard, Herbert Robbins, and Ian Stewart. *What Is Mathematics?: An Elementary Approach to Ideas and Methods*. 2nd ed. Oxford, U.K.: Oxford University Press, 1996.

Devlin, Keith. *The Language of Mathematics : Making the Invisible Visible*. New York: Owl Books, 2000.

Dunham, William. *Journey Through Genius: The Great Theorems of Mathematics*. London: Penguin Books, 1991.

Hall, James W. and Brian A. Mercer. *Beginning & Intermediate Algebra*. New York: McGraw-Hill, 2002.

Johnson, David B. and Thomas A. Mowry. *Mathematics : A Practical Odyssey*. 5th ed. Pacific Grove, CA: Brooks Cole, 2003.

Jurgensen, Ray and Richard G. Brown. *Geometry*. New York: Houghton Mifflin, 2000.

Martin-Gay, K. Elayn. *Prealgebra*. 4th ed. New York: Prentice Hall, 2003.

Martin-Gay, K. Elayn. *Intermediate Algebra*. 4th ed. New York: Prentice, 2004.

Penrose, Sir Roger. *The Road to Reality: The Mathematics and Physics of the Universe*. London: Vintage, 2002.

Stewart, Ian. *Concepts of Modern Mathematics*. New York: Dover Publications, 1995.

Stewart, Ian. *Does God Play Dice? The New Mathematics of Chaos*. 2nd ed. London: Penguin, 1997.

Stewart, James. *Calculus*. 5th ed. Pacific Grove, CA: Brooks Cole, 2002.

Tannenbaum, Peter. *Excursions in Modern Mathematics*. 5th ed. New York: Prentice Hall, 2003.